Partial-Update Adaptive Filters and Adaptive Signal Processing

Design, Analysis and Implementation

Kutluyıl Doğançay

ELSEVIER

AMSTERDAM • BOSTON • HEIDELBERG • LONDON • NEW YORK • OXFORD
PARIS • SAN DIEGO • SAN FRANCISCO • SINGAPORE • SYDNEY • TOKYO
Academic Press is an imprint of Elsevier

Academic Press is an imprint of Elsevier
Linacre House, Jordan Hill, Oxford OX2 8DP, UK
30 Corporate Drive, Suite 400, Burlington, MA 01803, USA

British Library Cataloguing in Publication Data
A catalogue record for this book is available from the British Library

Library of Congress Cataloguing in Publication Data
A catalogue record for this book is available from the Library of Congress

ISBN: 978-0-1237-4196-7

For information on all Academic Press publications
visit our web site at http://books.elsevier.com

Printed and bound in Great Britain by
CPI Antony Rowe, Chippenham and Eastbourne

Transferred to Digital Printing, 2010

To Hien and Deniz

Acknowledgements

I am indebted to a large number of people for educating me and being a great source of inspiration in signal processing research. Oğuz Tanrıkulu introduced me to partial-update adaptive filters almost a decade ago and was my co-inventor on two related US patents (one issued and one pending at the time of this writing). I am grateful to my commissioning editor Tim Pitts for his encouragement and support throughout the preparation of the manuscript. My sincere thanks go to Jackie Holding, Melanie Benson and Kate Dennis at Elsevier for their excellent help with the production of the book.

Contents

Preface

Adaptive signal processing is at the heart of many electronic systems and devices in telecommunications, multimedia, radar, sonar and biomedical engineering, to name but a few. It is a key enabler for the wireless and mobile telecommunications revolution we are currently experiencing. The important role played by adaptive signal processing as a fundamental design and analysis tool will continue to grow unabated into next generation telecommunications, industrial and entertainment systems thanks to advances in microelectronics and very large-scale integration (VLSI).

A main objective for adaptive signal processing is to identify and track changes in an unknown time-varying system with no, or limited, knowledge of signal statistics and system properties. This objective characterizes vast majority of signal processing problems encountered in practice. An important consideration in practical applications is the interplay between computational complexity and performance of adaptive signal processing systems. A promising approach to controlling computational complexity is to employ partial-coefficient-update techniques whereby, rather than update the whole coefficient vector, a subset of the adaptive filter coefficients is updated at each update iteration. In certain applications, such as acoustic and network echo cancellation, the complexity reduction achievable by partial coefficient updates is significant.

This book aims to present a state-of-the-art approach to the design, analysis and implementation of partial-update adaptive signal processing algorithms. A promising feature of partial-update adaptive signal processing is its potential to improve the adaptive filter performance in certain telecommunications applications while also permitting some complexity reduction in adaptive filter implementation. The book provides a unifying treatment of key partial-coefficient-update techniques. The description of concepts and algorithms is accompanied by detailed convergence and complexity analysis. Throughout the book several simulation examples are presented to illustrate the algorithms and their application to echo cancellation, channel equalization and multiuser detection.

The book will be of particular interest to practising signal processing engineers as complexity-to-performance ratio is an important design consideration in practical systems. Advanced analysis and design tools included in the book will provide an essential background for postgraduate students and researchers exploring complex adaptive systems comprising several smaller and possibly distributed adaptive subsystems with complexity constraints.

Chapter 1 presents an overview of adaptive signal processing and motivates partial-update adaptive signal processing as a low-complexity implementation option in the face of resource constraints. In the context of adaptive system identification, partial coefficient updating is introduced as an attractive approach to complexity reduction. Simple adaptive system identification examples are presented to illustrate the potential performance improvement achievable for partial-update adaptive signal processing.

Chapter 2 provides a comprehensive overview of partial-coefficient-update techniques. The least-mean-square (LMS) and normalized LMS (NLMS) algorithms are utilized to demonstrate the application of partial-update techniques, and to provide a platform for convergence and stability analysis under the assumption of sufficiently small step-size parameters.

Chapter 3 investigates steady-state and convergence performance of several partial-update LMS and NLMS algorithms. In the absence of exact performance analysis the concept of energy conservation is utilized to derive approximate closed-form expressions. The accuracy of the analytical results is verified by several simulation examples.

Chapter 4 studies several stochastic gradient algorithms with partial coefficient updates. Both time-domain and transform-domain adaptive filters are included. The computational complexity of each algorithm is examined thoroughly. Finally, the convergence performance of the partial-update algorithms is compared in a channel equalization example.

Chapter 5 presents selected applications of partial-update adaptive signal processing. The applications include acoustic/network echo cancellation, blind fractionally spaced channel equalization, and blind adaptive linear multiuser detection. The performance improvement achieved by selective-partial-update algorithms compared with full-update algorithms is demonstrated in blind channel equalization and linear multiuser detection.

Kutluyıl Doğançay
Adelaide, Australia
2008

Chapter | one

Introduction

1.1 ADAPTIVE SIGNAL PROCESSING

In practice most systems are inherently time-varying and/or nonlinear. The signals associated with these systems often have time-varying characteristics. Adaptive signal processing is a branch of statistical signal processing that deals with the challenging problem of estimation and tracking of time-varying systems. By virtue of its applicability to time-varying and/or nonlinear systems, adaptive signal processing finds application in a broad range of practical fields such as telecommunications, radar and sonar signal processing, biomedical engineering and entertainment systems. In order to make the estimation and tracking task tractable, the unknown system is usually modelled as a time-varying *linear* system or in some cases as a finitely parameterized nonlinear system such as the Volterra filter. This simplified system modelling is guided by prior knowledge of system characteristics. An important objective of adaptive signal processing is to learn the unknown and possibly time-varying signal statistics in conjunction with system estimation.

This chapter presents a brief overview of fundamental principles of adaptive signal processing and motivates partial-update adaptive signal processing as a low-complexity implementation option in the face of resource constraints. Adaptive system identification is shown to be the central theme of adaptive signal processing. In the context of adaptive system identification partial coefficient updating is proposed as an attractive approach to complexity reduction. The chapter sets the scene for the remainder of the book by presenting simple adaptive system identification examples that illustrate the potential benefits of partial-update adaptive signal processing in addition to allowing compliance with the existing resource constraints.

1.2 EXAMPLES OF ADAPTIVE FILTERING

A fundamental building block for an adaptive signal processing system is the adaptive filter. The objective of an adaptive filter is to learn an unknown system from

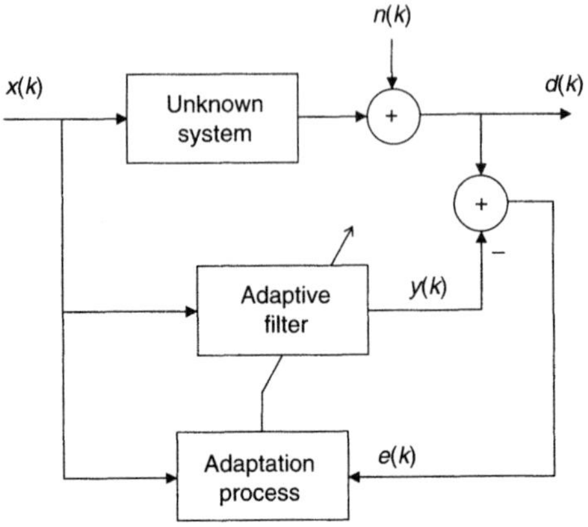

Figure 1.1 Adaptive system identification.

observations of the system input and/or output signals utilizing any *a priori* knowledge of the system and signal characteristics. The task of learning an unknown system is fundamental to many signal processing problems and comes in many disguises in applications of adaptive filters. In this section we take a brief look at two broad examples of adaptive filters. These examples serve the purpose of elucidating the basic components of adaptive filters.

1.2.1 Adaptive system identification

Most adaptive filtering problems are either (1) a special case of adaptive system identification or (2) utilize adaptive system identification as a means of solving another signal processing problem. In this sense, adaptive system identification provides the basis for a wide range of adaptive signal processing applications. It is, therefore, essential that we have a good understanding of the underlying principles and assumptions relating to adaptive system identification.

As depicted in Figure 1.1, in adaptive system identification, the objective is to estimate an unknown system from its input and output observations given by $x(k)$ and $d(k)$, respectively. Throughout this book we restrict our attention to discrete-time signals and systems, so the independent time index k is an integer. A model for the adaptive filter is chosen based on prior knowledge of the unknown system characteristics, as well as complexity considerations. In its simplest and most preferred form, the adaptive filter is a finite impulse response (FIR) filter of length N with adjustable impulse response coefficients (adaptive filter coefficients):

$$\boldsymbol{w}(k) = [w_1(k), w_2(k), \ldots, w_N(k)]^T \tag{1.1}$$

Here T denotes the transpose operator. Equation (1.1) is perhaps the most widely used adaptive filter model mainly because of its applicability to a wide range of practical problems. Throughout this book the adaptive filter coefficients $w_i(k)$ are assumed to be real-valued unless otherwise specified. It is often straightforward to extend the analysis to adaptive filters with complex coefficients.

In a system identification context, the adaptive filter attempts to learn the unknown system by using a model of the unknown system represented by $\boldsymbol{w}(k)$. The difference between the noisy response of the unknown system (the desired response $d(k)$) and the response of the adaptive filter $y(k)$ is called the error signal $e(k)$:

$$e(k) = d(k) - y(k) \tag{1.2}$$

At each iteration k the adaptive filter updates its coefficients in order to minimize an appropriate norm of the error signal $e(k)$. When the error norm is minimized in a statistical sense, the corresponding $\boldsymbol{w}(k)$ gives an estimate of the unknown system parameters. If the unknown system is time-varying, i.e. its parameters change with time, the adaptive filter can track these changes by updating its coefficients in accordance with the error signal. It can take several iterations for the adaptation process to converge (i.e. to learn unknown system parameters). The time taken by the adaptation process to converge provides an indication of the convergence rate.

There are two main tasks performed by the adaptive filter; viz. adaptation process and filtering process. In Figure 1.1 these processes are identified by the adaptation process and adaptive filter blocks. For a linear adaptive filter as given by (1.1), the filtering process involves convolution. If the number of adaptive filter coefficients is large, the convolution operation may prove to be computationally expensive. Reduced complexity convolution techniques based on fast Fourier transform (FFT), such as overlap-add and overlap-save, may be used to ease computational demand. The adaptation process can also become computationally expensive for long adaptive filters due to the arithmetic operations required to update the adaptive filter coefficients. The computational complexity of the adaptation process depends on the adaptation algorithm employed.

Prediction of random signals and noise cancellation are two special cases of adaptive system identification. Figure 1.2(a) shows a one-step predictor which estimates the present value of the random signal $x(k)$ based on past values $x(k-1), \ldots, x(k-N)$. If $x(k)$ is a stable autoregressive (AR) process of order N:

$$x(k) = a_1 x(k-1) + a_2 x(k-2) + \cdots + a_N x(k-N) + v(k) \tag{1.3}$$

where $v(k)$ is white noise, the adaptive filter $\boldsymbol{w}(k)$ in Figure 1.2(a) estimates the AR coefficients $[a_1, a_2, \ldots, a_N]^T$. After convergence the prediction error $e(k)$ is equal to $v(k)$, which implies whitening of the coloured noise signal $x(k)$. Referring to Figure 1.1, we observe that the adaptive system identification setup can be converted

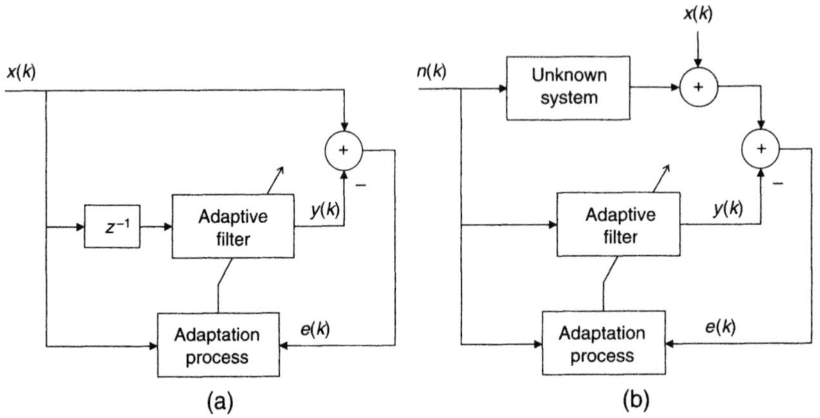

Figure 1.2 Special cases of system identification. (a) One-step prediction of a random signal by an adaptive filter which identifies the autoregressive model of the random signal with $y(k)$ giving the prediction output and $e(k)$ the prediction error; (b) Adaptive noise cancellation where the adaptive filter estimates the unknown system filtering the input noise signal $n(k)$ with $x(k)$ denoting the signal of interest and $e(k)$ the 'cleaned' signal.

to a one-step predictor by replacing the unknown system with a direct connection (short-circuiting the unknown system), setting $n(k) = 0$ and inserting a one-sample delay z^{-1} at the input of the adaptive filter. Swapping $x(k)$ and $n(k)$ in Figure 1.1 and referring to $x(k)$ as the signal of interest and $n(k)$ as the interfering noise changes the system identification problem to a noise cancellation problem with $e(k)$ giving the 'cleaned' signal (see Figure 1.2(b)). The noise signal $n(k)$ is the reference signal and the unknown system represents any filtering that $n(k)$ may undergo before interfering with the signal of interest $x(k)$. The sum of $x(k)$ and filtered $n(k)$ is the primary signal. The unknown system is identified by an adaptive filter. Subtracting the adaptive filter output from the reference signal gives the error signal. Minimization of the error norm implies minimization of the difference between the adaptive filter output and the filtered reference signal. If the adaptive filter provides a perfect estimate of the unknown system, then the error signal becomes identical to the signal of interest. This provides perfect noise removal.

1.2.2 Adaptive inverse system identification

Figure 1.3 illustrates the adaptive inverse system identification problem. Comparison of Figures 1.2 and 1.3 reveals that adaptive inverse system identification requires an adaptive filter to be connected to the input and noisy output of the unknown system in the reverse direction. The use of the D-sample delay z^{-D} for the desired adaptive filter response ensures that the adaptive filter will be able to approximate the inverse

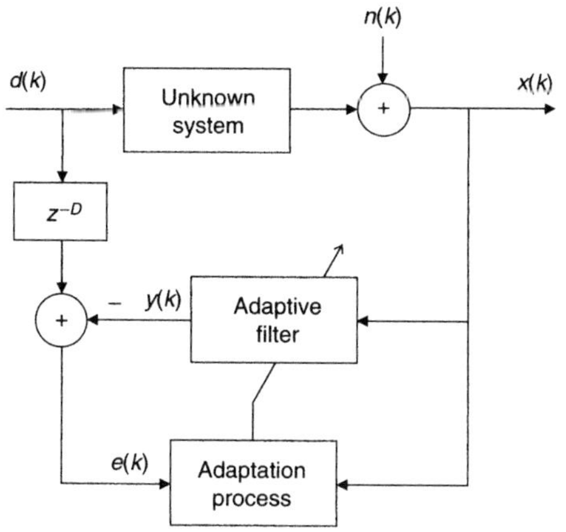

Figure 1.3 Adaptive inverse system identification.

system for non-minimum-phase or maximum-phase linear systems. For such systems the stable inverse system is non-causal and has infinite impulse response (IIR). A (causal) FIR adaptive filter can approximate the stable inverse only if it is sufficiently long (i.e. N is sufficiently large) and D is non-zero. If the unknown system is a minimum-phase linear system, then no delay for the desired response is required (i.e. one can set $D = 0$).

The adaptive filter uses an appropriate norm of the error signal $e(k) = d(k - D) - y(k)$ as a measure of accuracy for inverse system identification. The adaptive filter coefficients are adjusted iteratively in order to minimize the error norm in a statistical sense. At the end of the minimization process the adaptive filter converges to an estimate of the inverse of the unknown system. Due to the presence of noise at the unknown system output, the inverse system estimate is not identical to the zero-forcing solution that ignores the output noise.

An important application of inverse system identification is channel equalization. Fast data communications over bandlimited channels causes intersymbol interference (ISI). The ISI can be eliminated by an adaptive channel equalizer connected in cascade with the communication channel (unknown system). The adaptive channel equalizer is an adaptive filter performing inverse channel identification for the unknown communication channel. The delayed channel input signal $d(k - D)$ is made available to the channel equalizer during intermittent training intervals so that the adaptation process can learn any variations in the inverse system model due to channel fading. Outside training intervals the equalizer usually switches to a decision directed mode whereby the error signal $e(k)$ is obtained from the difference between the equalizer output $y(k)$ and its nearest neighbour in the channel input signal constellation.

1.3 RAISON D'ÊTRE FOR PARTIAL COEFFICIENT UPDATES

1.3.1 Resource constraints

Cost, performance, portability and physical size considerations impose stringent resource constraints on adaptive signal processing systems. Real-time hardware multipliers, power, memory and bandwidth are scarce system resources. The number of hardware multipliers utilized and power consumption are closely coupled (Liu *et al.*, 1998). In general, more hardware multipliers and more memory implies more power consumption. For portable battery powered electronic devices, reducing the power consumption enables longer battery life and leads to more compact and light-weight physical devices as required by mobile communications and computing systems. The relation between power consumption and hardware multipliers suggests that one way to reduce power consumption at the software level would be to limit the use of hardware multipliers. As a parenthetical remark, we note that the number of multiply operations is usually considered to be the key complexity indicator in digital signal processing systems (Eyre and Bier, 2000). Bandwidth is another scarce resource which plays a crucial role, especially in distributed adaptive systems such as sensor networks.

One focus of this book is on dealing with resource constraints imposed by limited hardware multipliers when designing an adaptive signal processing system. When dealing with the limitation on hardware multipliers we also indirectly address some of the important design considerations such as cost, low power consumption and physical size. The following scenarios illustrate the linkage between these interdependent design considerations:

- The required number of hardware multipliers entails upgrading of the current digital signal processing (DSP) chip to a more expensive one exceeding the anticipated system cost, so there is a need to cut the system cost possibly by doing away with some of the hardware multipliers.
- If all of the required hardware multipliers are included in the adaptive system design, a larger and heavier battery will be needed along with a cooling fan, thereby increasing the physical size of the device.
- There is not sufficient space on the circuit board to include all of the required multipliers.

The complexity constraint posed by scarce resources necessitates the development of resource allocation schemes that assign available resources to adaptation processes on the basis of a well-defined allocation or selection criterion. In general, the selection criterion must reflect the merit of a given resource assignment in terms of performance measures. The most commonly employed performance measure is the convergence rate for a given steady-state error level. Figure 1.4 depicts an adaptive filter operating

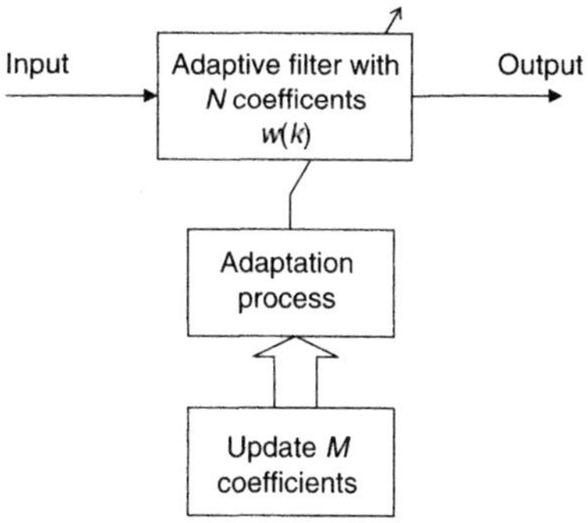

Figure 1.4 Resource constrained adaptive filtering with scarce hardware multipliers. If the number of available multipliers is not sufficient for full update, a decision needs to be made as to which M coefficients out of N should be updated by the adaptation process at every update iteration.

under a resource constraint arising from the availability of a limited number of hardware multipliers for the adaptation process. The filtering process is assumed to have all the computational resources required for its operation. Thus, only the adaptation process is affected by the limited hardware multipliers. The adaptive filter has N coefficients while the adaptation process can update M coefficients due to a limited number of hardware multipliers for coefficient updates. If $M < N$ (i.e. the number of hardware multipliers is not sufficient to allow all adaptive filter coefficients to be updated) then a decision needs to be made as to which M coefficients of the adaptive filter should be selected for update. This selection process implies that M coefficients are updated while the remaining $N - M$ coefficients stay unchanged at each coefficient update. Intuitively, the selected M coefficients should change at each iteration so as to allow all adaptive filter coefficients to be updated over time.

We will call the process of updating a subset of the adaptive filter coefficients *partial coefficient updates* or *partial updates*. The main reason for partial coefficient updates is the limited availability of hardware multipliers often driven by cost, space and power consumption considerations. This problem is particularly acute in DSP and field-programmable gate array (FPGA) implementation of adaptive signal processing algorithms. Even though the notion of partial coefficient updates is very effective in terms of addressing scarce hardware resources, it can lead to degradation in convergence performance. After all, only a subset of all coefficients is updated, so one would expect some performance degradation as a consequence. Another focus of this book is on the exploration of coefficient selection techniques and the impact of partial updates on convergence performance.

In a more formal setup, partial updates are characterized by a time-varying diagonal matrix premultiplying the update vector in an adaptive filter. The diagonal entries take on the values zero or one. To illustrate the use of partial updates, we consider the following stochastic gradient algorithm which is better known as the least-mean-square (LMS) algorithm:

$$\underbrace{\begin{bmatrix} w_1(k+1) \\ \vdots \\ w_N(k+1) \end{bmatrix}}_{\boldsymbol{w}(k+1)} = \underbrace{\begin{bmatrix} w_1(k) \\ \vdots \\ w_N(k) \end{bmatrix}}_{\boldsymbol{w}(k)} + \mu e(k) \underbrace{\begin{bmatrix} x(k) \\ \vdots \\ x(k-N+1) \end{bmatrix}}_{\boldsymbol{x}(k)}, \quad k = 0, 1, 2, \ldots \tag{1.4}$$

where $\boldsymbol{w}(k)$ is the $N \times 1$ adaptive filter coefficients vector at time k, $\mu > 0$ is a constant step-size parameter, $e(k)$ is the error signal given by the difference between the desired response and the adaptive filter output, and $\boldsymbol{x}(k)$ is the $N \times 1$ adaptive filter input regressor vector. If the hardware constraints allow only M out of N coefficients to be updated at each iteration k, this can be accommodated by modifying the adaptation algorithm in (1.4) to

$$\boldsymbol{w}(k+1) = \boldsymbol{w}(k) + \mu e(k) \boldsymbol{I}_M(k) \boldsymbol{x}(k), \quad k = 0, 1, 2, \ldots \tag{1.5}$$

where $\boldsymbol{I}_M(k)$ is a diagonal matrix with M ones and $N - M$ zeros on its diagonal indicating which M coefficients are to be updated at iteration k:

$$\boldsymbol{I}_M(k) = \begin{bmatrix} i_1(k) & & \boldsymbol{0} \\ & \ddots & \\ \boldsymbol{0} & & i_N(k) \end{bmatrix}, \quad \sum_{j=1}^{N} i_j(k) = M, \quad i_j(k) \in \{0, 1\} \tag{1.6}$$

If $i_j(k) = 1$, $j = 1, \ldots, N$, then the coefficient $w_j(k)$ gets an update at iteration k, otherwise it remains unchanged. Thus the unity diagonal entries indicate the M adaptive filter coefficients selected for update. The chief advantage of partial updates in (1.5) is reduced complexity in line with hardware complexity constraints. A potential drawback of partial updates is the reduction in convergence speed. In this context, partial updates can be considered to provide a means of trading computational complexity for convergence performance. The extent of degradation in convergence performance directly depends on how the updated coefficients are selected (i.e. how the diagonal entries of $\boldsymbol{I}_M(k)$ are determined at every k).

To achieve the best convergence performance under the given hardware constraints, the coefficient selection process in Figure 1.4 can be cast as an optimization problem where the objective is to determine the subset of M coefficients at each update k that maximizes the convergence speed. This involves quantification of convergence speed and an exhaustive search over all possible $\boldsymbol{I}_M(k)$. The number of possible $\boldsymbol{I}_M(k)$

matrices for given N and M is

$$\binom{N}{M} = \frac{N!}{M!(N-M)!} \tag{1.7}$$

Searching for the optimum coefficients over these subsets could be computationally expensive and time consuming, thereby defeating the purpose of partial updates. However, the optimization problem can be significantly simplified by exploiting the relationship between convergence speed and update vector norm, resulting in replacement of the exhaustive search for coefficient subsets with a simple sorting operation. We will have more to say about this when discussing M-max updates and selective partial updates in Chapter 2.

1.3.2 Convergence performance

To illustrate the effects of partial coefficient updates on the convergence of an adaptive filter, we consider average cost function surfaces for conventional full-update LMS and a number of partial-update variants of LMS. Our main objective here is to observe the changes to the cost function surface caused by partial updates without going into an in-depth analysis of particular partial-update techniques used. Based on the observed changes to cost functions we will argue that certain partial-update techniques have the potential to maintain a convergence rate close to, and sometimes better than, that of a conventional adaptive filter employing no partial updates. Even though this may sound counterintuitive, some particular partial-update techniques are capable of outperforming the corresponding full-update adaptive filter at a fraction of the total computational complexity.

For a sufficiently small step-size parameter μ a stochastic gradient algorithm can be approximated by an averaged system where the update term of the stochastic gradient algorithm is replaced with its average. This fundamental result from averaging theory (Anderson *et al.*, 1986; Benveniste *et al.*, 1990; Mareels and Polderman, 1996; Solo and Kong, 1995) allows us to approximate LMS and partial-update LMS algorithms with appropriate steepest descent algorithms. The analysis of steepest descent algorithms is considerably easier than stochastic gradient algorithms. We will explore the impact of partial updates on the convergence performance of the approximate steepest descent algorithms in order to gain insight into the convergence performance of the partial-update LMS algorithms.

Supposing μ is sufficiently small, the partial-update LMS algorithm in (1.5) can be approximated by:

$$\begin{aligned} \boldsymbol{w}^a(k+1) &= \boldsymbol{w}^a(k) + \mu \text{avg}\{e(k)\boldsymbol{I}_M(k)\boldsymbol{x}(k)\} \\ \boldsymbol{w}^a(k+1) &= \boldsymbol{w}^a(k) + \mu E\{e(k)\boldsymbol{I}_M(k)\boldsymbol{x}(k)\} \end{aligned} \tag{1.8}$$

where avg$\{\cdot\}$ is the time average and $E\{\cdot\}$ is the expectation. The input signal $x(k)$ is assumed to be correlation-ergodic so that expectation and averaging can be used interchangeably. Using $e(k) = d(k) - \boldsymbol{x}^T(k)\boldsymbol{w}^a(k)$ where $d(k)$ is the desired filter response, (1.8) can be rewritten as:

$$\begin{aligned}\boldsymbol{w}^a(k+1) &= \boldsymbol{w}^a(k) + \mu E\{\boldsymbol{I}_M(k)\boldsymbol{x}(k)(d(k) - \boldsymbol{x}^T(k)\boldsymbol{w}^a(k))\} \\ &= \boldsymbol{w}^a(k) + \mu(\boldsymbol{p}_M - \boldsymbol{R}_M\boldsymbol{w}^a(k)) \end{aligned} \quad (1.9)$$

where

$$\boldsymbol{p}_M = E\{\boldsymbol{I}_M(k)\boldsymbol{x}(k)d(k)\} \quad (1.10a)$$

$$\boldsymbol{R}_M = E\{\boldsymbol{I}_M(k)\boldsymbol{x}(k)\boldsymbol{x}^T(k)\} \quad (1.10b)$$

are the cross-correlation vector between the partial-update regressor vector and the desired filter response, and the correlation matrix of the partial-update regressor, respectively. If $M = N$ (i.e. all coefficients are updated) (1.9) becomes identical to the steepest descent algorithm. The cost function minimized by the averaged system (1.9) is:

$$J_M(k) = (\boldsymbol{w}^a(k))^T\boldsymbol{R}_M\boldsymbol{w}^a(k) - 2(\boldsymbol{w}^a(k))^T\boldsymbol{p}_M + c \quad (1.11)$$

where c is a constant independent of $\boldsymbol{w}^a(k)$ and $\nabla J_M(k) = -2(\boldsymbol{p}_M - \boldsymbol{R}_M\boldsymbol{w}^a(k))$ as desired (see (1.9)). It is of particular interest to see how different partial-update techniques for selecting $\boldsymbol{I}_M(k)$ affect the correlation matrix $\boldsymbol{R}_M$ as it plays a crucial role in determining the convergence behaviour of the adaptive filter.

1.3.3 System identification with white input signal

We will illustrate the impact of partial updates on the cost function $J_M(k)$ with two numerical examples. The adaptive system identification problem considered in the examples is shown in Figure 1.5. The unknown system is a two-tap filter with impulse response $\boldsymbol{h} = [1, -0.7]^T$. The additive noise corrupting the system output $n(k)$ is white Gaussian with zero mean and variance $\sigma_n^2 = 0.01$. The input signal $x(k)$ and system noise $n(k)$ are statistically independent.

We analyse three cases of coefficient updates; viz., full updates, M-max updates and sequential partial updates. The coefficient selection matrix $\boldsymbol{I}_M(k)$ takes different forms for these three cases. In full updates all coefficients of the adaptive filter are updated, i.e. $M = N = 2$ and

$$\boldsymbol{I}_M(k) = \begin{bmatrix} 1 & 0 \\ 0 & 1 \end{bmatrix} \quad (1.12)$$

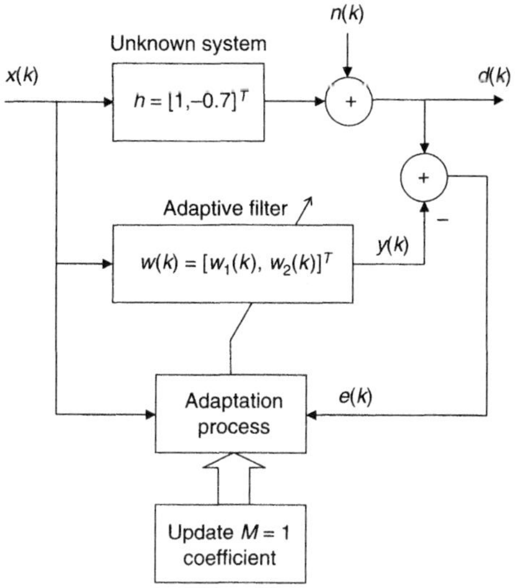

Figure 1.5 System identification example using an adaptive filter with $N = 2$ coefficients. The adaptation process can update only $M = 1$ coefficient due to hardware constraints.

M-max partial updates is a data-dependent technique that selects the adaptive filter coefficients with the M largest updates in absolute value. For $M = 1$ and $N = 2$ it takes the following form:

$$\boldsymbol{I}_M(k) = \begin{bmatrix} i_1(k) & 0 \\ 0 & i_2(k) \end{bmatrix}, \quad \{i_1(k), i_2(k)\} = \begin{cases} \{1, 0\} & \text{if } |x(k)| > |x(k-1)| \\ \{0, 1\} & \text{otherwise} \end{cases} \tag{1.13}$$

This coefficient selection matrix sets the diagonal element corresponding to the larger of the regressor elements in absolute value to be one, and the other diagonal element to be zero. In sequential partial updates predetermined subsets of adaptive filter coefficients are selected in a round-robin fashion. For $M = 1$ and $N = 2$ a sequential-partial-update coefficient selection matrix can de defined as:

$$\boldsymbol{I}_M(k) = \begin{bmatrix} i_1(k) & 0 \\ 0 & i_2(k) \end{bmatrix}, \quad \{i_1(k), i_2(k)\} = \begin{cases} \{1, 0\} & k \text{ is even} \\ \{0, 1\} & k \text{ is odd} \end{cases} \tag{1.14}$$

The sequential-partial-update coefficient selection matrix given above selects the first adaptive filter coefficient for update at even iterations and the second coefficient at odd iterations.

The input signal $x(k)$ is a white Gaussian signal with zero mean and unit variance. Figure 1.6 shows the time-averaged mean-square error (MSE) curves for LMS, M-max LMS, sequential-partial-update LMS and their averaged approximations. The

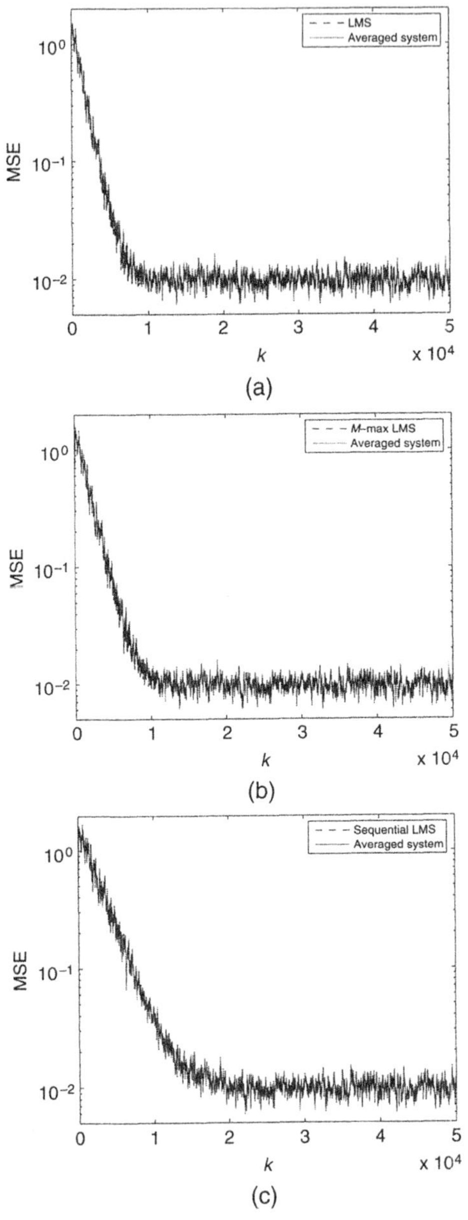

Figure 1.6 Time-averaged MSE curves for stochastic gradient algorithms and their averaged approximations. (a) LMS; (b) M-max LMS; and (c) sequential-partial-update LMS.

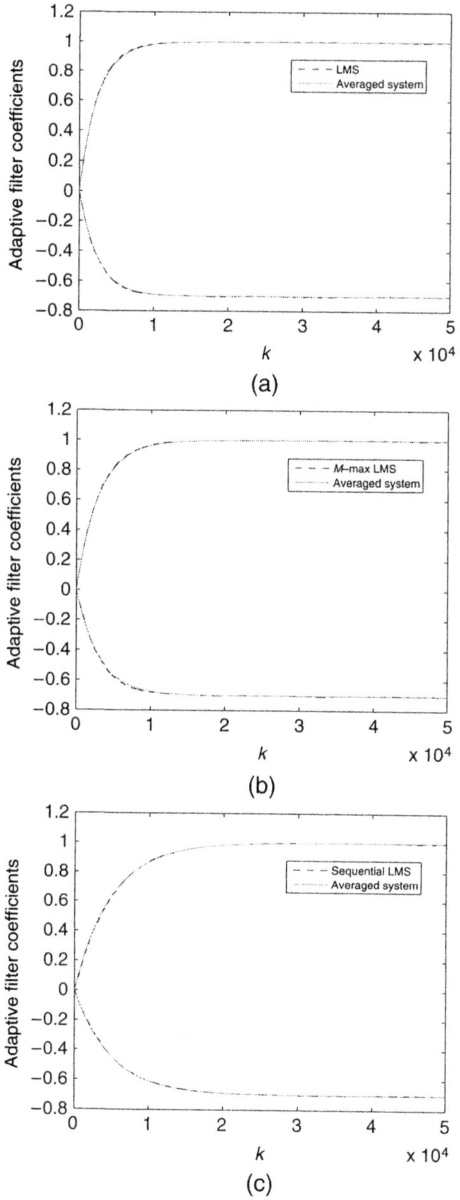

Figure 1.7 Evolution of adaptive filter coefficients for stochastic gradient algorithms and their averaged approximations. (a) LMS; (b) M-max LMS; and (c) sequential-partial-update LMS.

time-averaged MSE is obtained by passing $e^2(k)$ through a lowpass comb filter :

$$H_C(z) = \frac{1}{L}\frac{1 - z^{-L}}{1 - z^{-1}} \tag{1.15}$$

where L is the length of the filter impulse response which is a rectangular window. The larger L the smoother the time-averaged MSE becomes. We use $L = 100$ in the simulations.

All three adaptive filtering algorithms and averaged systems are initialized to $\boldsymbol{w}(0) = \boldsymbol{w}^a(0) = [0, 0]^T$, which is a common initialization strategy for LMS, and use the same step-size parameter $\mu = 4 \times 10^{-4}$. For this step-size the stochastic gradient algorithms and averaged systems have almost identical MSE. The evolution of adaptive filter coefficients for the stochastic gradient algorithms and averaged systems are shown in Figure 1.7. Again the difference between the stochastic gradient algorithms and the averaged systems is negligible, thereby justifying the approximation.

Figure 1.8 compares the MSE curves of the full-update LMS, M-max LMS and sequential-partial-update LMS algorithms. The full-update LMS algorithm has the fastest converge rate closely followed by the M-max LMS algorithm. The sequential-partial-update LMS algorithm takes twice as long as the full-update LMS to converge to the steady-state MSE. What is remarkable here is the closeness of the MSE curves for the full-update LMS and M-max LMS.

The cost functions of the averaged systems have been computed to shed some light on the observed differences in convergence rates. The converge of adaptive filter coefficients over cost function contours is depicted in Figure 1.9 for the three stochastic gradient algorithms. We make the following observations:

- The contours of cost functions for all three adaptive filters are circular.
- The spacing between contours evaluated at the same levels is increased slightly for M-max and significantly for sequential partial updates with respect to the full update LMS algorithm. This indicates that the cost function gets gradually flatter for M-max and is the flattest for sequential partial updates.
- The increased spacing between contours appears to be proportional to the reduction in convergence rates.
- All three adaptive filters exhibit linear coefficient trajectories. Coefficient trajectories are perpendicular to the cost function contours.

The eigenvalue spread of the correlation matrix $\boldsymbol{R}_M$ determines the shape of the contours of the cost function $J_M(k)$ and is defined by:

$$\chi(\boldsymbol{R}_M) = \frac{\lambda_{\max}}{\lambda_{\min}} \tag{1.16}$$

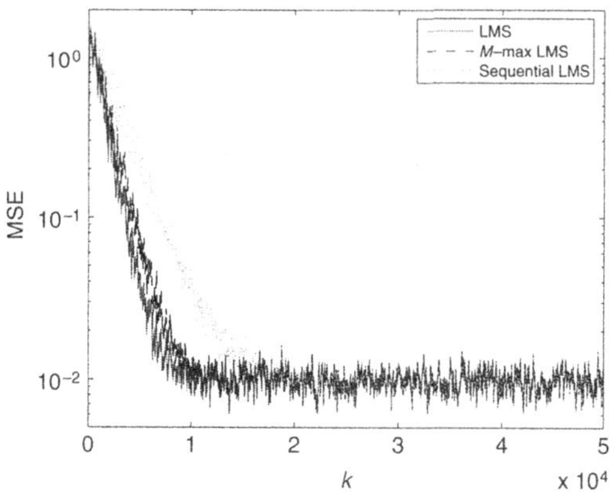

Figure 1.8 Time-averaged MSE curves for LMS, M-max LMS and sequential-partial-update LMS.

where $\lambda_{\max}$ is the maximum eigenvalue and $\lambda_{\min}$ is the minimum eigenvalue of $\boldsymbol{R}_M$. For $N = 2$ the cost function contours are, in general, ellipses centred at the minimum of the cost function surface. The minor and major axes of the ellipses are given by the eigenvectors of $\boldsymbol{R}_M$ and their lengths are proportional to the eigenvalues of $\boldsymbol{R}_M$. Circular contours, therefore, imply an eigenvalue spread of one. We will next check this for each adaptive algorithm.

For white Gaussian input, the full-update LMS algorithm has the input correlation matrix:

$$\boldsymbol{R}_M = \boldsymbol{R} = \sigma_x^2 \begin{bmatrix} 1 & 0 \\ 0 & 1 \end{bmatrix} \tag{1.17}$$

where σ_x^2 is the variance of the input signal. The eigenvalue spread for the full-update LMS is therefore $\chi(\boldsymbol{R}_M) = 1$.

The input correlation matrix of the M-max LMS algorithm is given by:

$$\boldsymbol{R}_M = \begin{bmatrix} E\{x^2(k) \mid |x(k)| > |x(k-1)|\} & E\{x(k)x(k-1) \mid |x(k)| > |x(k-1)|\} \\ E\{x(k)x(k-1) \mid |x(k)| < |x(k-1)|\} & E\{x^2(k-1) \mid |x(k)| < |x(k-1)|\} \end{bmatrix}. \tag{1.18}$$

The off-diagonal entries of $\boldsymbol{R}_M$ are zero because $x(k)$ is uncorrelated. The conditional expectations on the diagonal are identical. Since $x(k)$ and $x(k-1)$ are uncorrelated zero-mean Gaussian random variables with variance σ_x^2, the probability that $|x(k-1)| < |x(k)|$ for $x(k) = \alpha$ is given by:

$$\Pr\{|x(k-1)| < |\alpha| \mid x(k) = \alpha\} = \Pr\{|x(k-1)| < |\alpha|\} \tag{1.19a}$$

$$= \frac{2}{\sqrt{2\pi\sigma_x^2}} \int_0^{|\alpha|} \exp\left(-\frac{x^2}{2\sigma_x^2}\right) \mathrm{d}x \tag{1.19b}$$

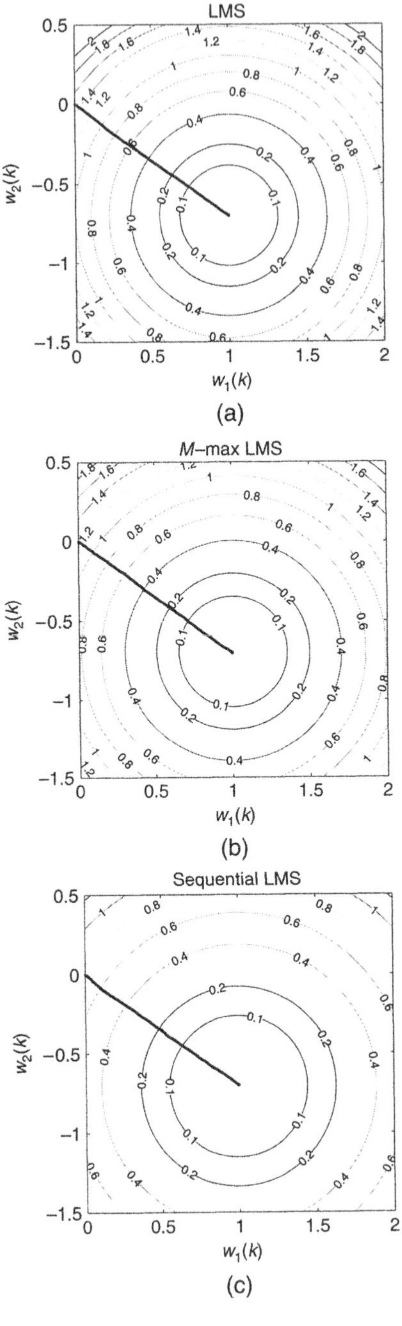

Figure 1.9 Convergence of adaptive filter coefficients over cost function contours. (a) LMS; (b) M-max LMS; and (c) sequential-partial-update LMS.

The conditional expectation then becomes:

$$E\{x^2(k) \mid |x(k)| > |x(k-1)|\}$$
$$= \frac{1}{\sqrt{2\pi\sigma_x^2}} \int_{-\infty}^{\infty} x^2 \left(\frac{2}{\sqrt{2\pi\sigma_x^2}} \int_0^{|x|} \exp\left(-\frac{y^2}{2\sigma_x^2}\right) \mathrm{d}y \right)$$
$$\times \exp\left(-\frac{x^2}{2\sigma_x^2}\right) \mathrm{d}x \tag{1.20a}$$
$$= \frac{2}{\pi\sigma_x^2} \int_{x=0}^{\infty} \int_{y=0}^{x} x^2 \exp\left(-\frac{x^2+y^2}{2\sigma_x^2}\right) \mathrm{d}x\mathrm{d}y. \tag{1.20b}$$

Numerical computation of the above expression gives $E\{x^2(k) \mid |x(k)| > |x(k-1)|\} = 0.8183\sigma_x^2$, yielding:

$$\boldsymbol{R}_M = \sigma_x^2 \begin{bmatrix} 0.8183 & 0 \\ 0 & 0.8183 \end{bmatrix} \tag{1.21}$$

For this correlation matrix the eigenvalue spread is $\chi(\boldsymbol{R}_M) = 1$.

The input correlation matrix for the sequential-partial-update LMS algorithm is simply half of the input correlation matrix for the full-update LMS by virtue of the fact that each adaptive filter coefficient is updated half of the time independent of instantaneous input signal values:

$$\boldsymbol{R}_M = \sigma_x^2 \begin{bmatrix} 1/2 & 0 \\ 0 & 1/2 \end{bmatrix} \tag{1.22}$$

The eigenvalue spread for the sequential-partial-update LMS is also $\chi(\boldsymbol{R}_M) = 1$.

A comparison of the input correlation matrices $\boldsymbol{R}_M$ for the three adaptive filters reveals that they all have unity eigenvalue spread and consequently their cost functions exhibit circular contours. The cost function of the M-max LMS is flatter than the full-update LMS as a result of having an input correlation matrix 0.8381 times that of the full-update LMS. The sequential-partial-update LMS has the flattest cost function since its input correlation matrix is half that of the full-update LMS. This also explains the observed differences in convergence rates. Note that the convergence rate cannot be increased simply by using a larger step-size parameter as this would also increase the steady-state (converged) MSE, as we shall see in Chapter 3.

1.3.4 System identification with correlated input signal

The second set of simulations uses a correlated Gaussian input signal $x(k)$ derived from a white Gaussian signal through linear filtering as shown in Figure 1.10. The

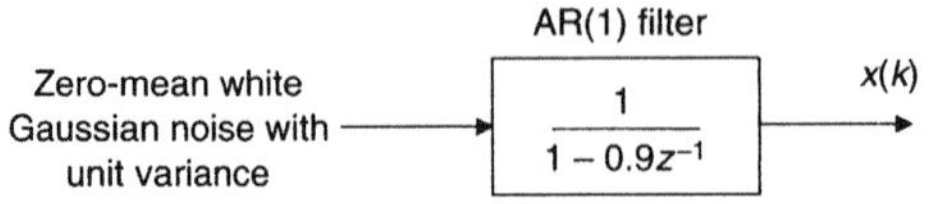

Figure 1.10 Correlated input signal is an AR(1) Gaussian process.

resulting input signal is an AR process of order one. Figure 1.11 shows the time-averaged MSE curves for the LMS, M-max LMS, sequential-partial-update LMS and their averaged approximations. The convergence of adaptive filter coefficients is depicted in Figure 1.12. The close match between the stochastic gradient and averaged algorithms is clearly visible. When it comes to the convergence rates, we make an unexpected observation that the M-max LMS outperforms the full-update LMS despite updating one out of two coefficients at every k (see Figure 1.13). The sequential-partial-update LMS algorithm is still twice as slow as the full-update LMS algorithm. We will now justify these observations analytically.

In Figure 1.14 we see that the cost function contours are now elliptical as a result of input signal correlation. For the full-update LMS, the input correlation matrix is simply given by the covariance matrix of the input signal:

$$\boldsymbol{R}_M = \boldsymbol{R} = \begin{bmatrix} r_{11} & r_{12} \\ r_{21} & r_{22} \end{bmatrix} \tag{1.23}$$

where:

$$r_{11} = r_{22} = \sigma_x^2 \sum_{n=0}^{\infty} (0.9^2)^n = 5.2632\sigma_x^2 \tag{1.24a}$$

$$r_{12} = r_{21} = \sigma_x^2 0.9 \sum_{n=0}^{\infty} (0.9^2)^n = 4.7368\sigma_x^2 \tag{1.24b}$$

Thus we have:

$$\boldsymbol{R}_M = \sigma_x^2 \begin{bmatrix} 5.2632 & 4.7368 \\ 4.7368 & 5.2632 \end{bmatrix} \tag{1.25}$$

The eigenvalue spread for $\boldsymbol{R}_M$ is $\chi(\boldsymbol{R}_M) = 19$. As the eigenvalue spread gets larger the eccentricity of elliptical contours increases in line with the strong correlation of the input signal.

For the M-max LMS algorithm the input correlation matrix is defined by (1.18). Using Bayes' rule, the probability that $|x(k-1)| < |x(k)|$ when $x(k) = \alpha$ can be

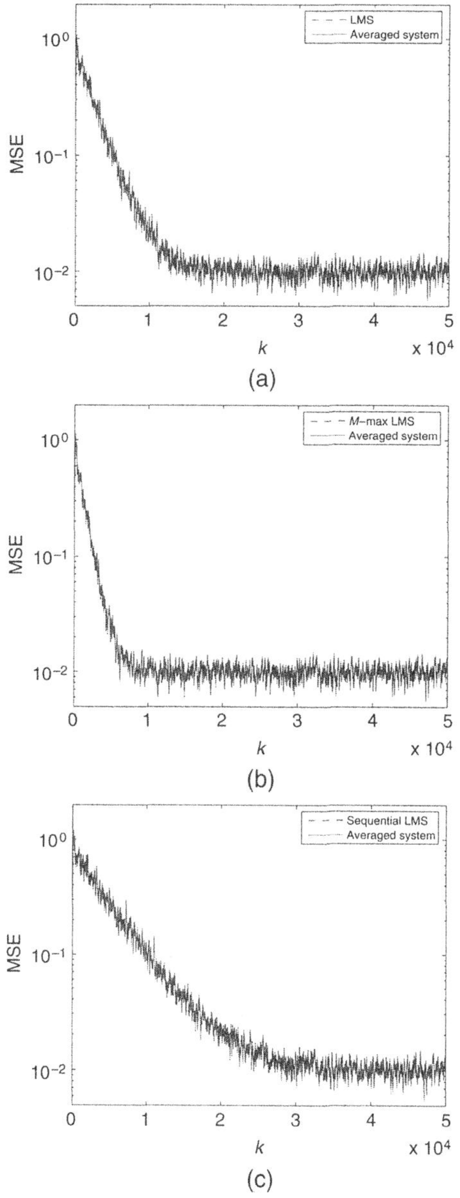

Figure 1.11 Time-averaged MSE curves for correlated input. (a) LMS; (b) M-max LMS; and (c) sequential-partial-update LMS.

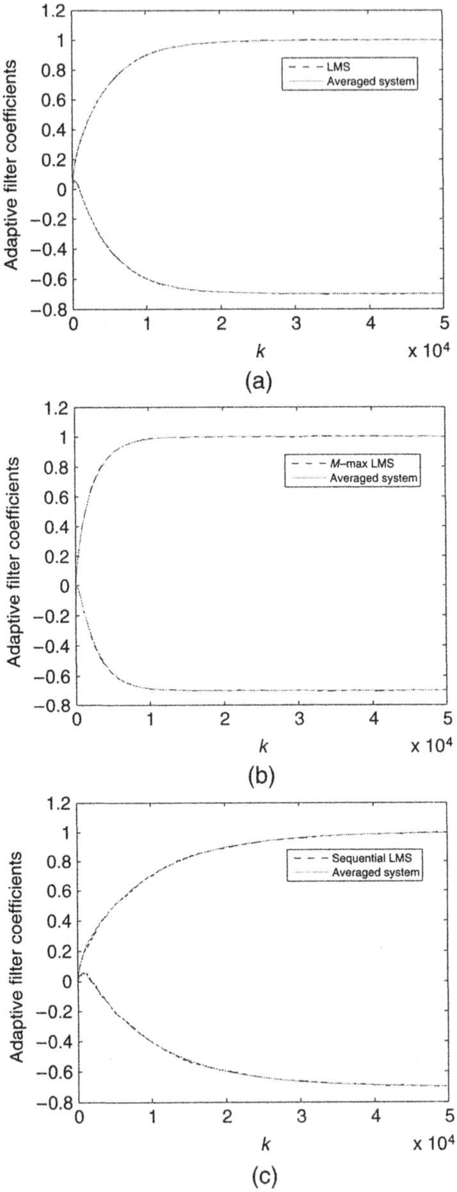

Figure 1.12 Evolution of adaptive filter coefficients for correlated input. (a) LMS; (b) M-max LMS; and (c) sequential-partial-update LMS.

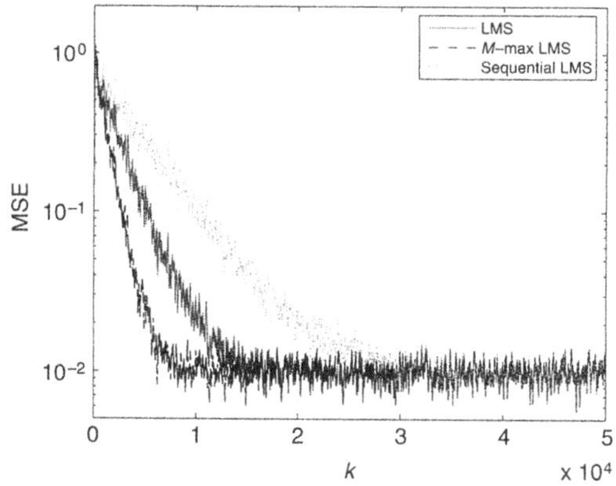

Figure 1.13 Time-averaged MSE curves for correlated input.

written as

$$\begin{aligned}\Pr\{|x(k-1)| < |\alpha| \mid x(k) = \alpha\} \\ = \frac{\Pr\{|x(k-1)| < |\alpha|, x(k) = \alpha\}}{\Pr\{x(k) = \alpha\}}\end{aligned} \quad (1.26a)$$

$$= \sqrt{\frac{r_{11}}{2\pi|\boldsymbol{R}|}} \frac{\int_{-|\alpha|}^{|\alpha|} \exp\left(-\frac{1}{2}[\alpha, x]\boldsymbol{R}^{-1}[\alpha, x]^T\right) \mathrm{d}x}{\exp\left(-\frac{\alpha^2}{2r_{11}}\right)} \quad (1.26b)$$

It follows that:

$$\begin{aligned}E\{x^2(k) \mid |x(k)| > |x(k-1)|\} \\ = \frac{2}{\sqrt{2\pi r_{11}}} \int_{x=-\infty}^{\infty} x^2 \exp\left(-\frac{x^2}{2r_{11}}\right) \\ \times \left(\sqrt{\frac{r_{11}}{2\pi|\boldsymbol{R}|}} \exp\left(\frac{x^2}{2r_{11}}\right) \int_{y=0}^{|x|} \exp\left(-\frac{1}{2}[x, y]\boldsymbol{R}^{-1}[x, y]^T\right) \mathrm{d}y\right) \mathrm{d}x\end{aligned} \quad (1.27a)$$

$$= \frac{1}{2\pi|\boldsymbol{R}|^{1/2}} \int_{x=-\infty}^{\infty} \int_{y=-|x|}^{|x|} x^2 \exp\left(-\frac{1}{2}[x, y]\boldsymbol{R}^{-1}[x, y]^T\right) \mathrm{d}x\mathrm{d}y \quad (1.27b)$$

$$= \frac{1}{\pi|\boldsymbol{R}|^{1/2}} \int_{x=-\infty}^{\infty} x^2 \int_{y=0}^{|x|} \exp\left(-\frac{1}{2}[x, y]\boldsymbol{R}^{-1}[x, y]^T\right) \mathrm{d}x\mathrm{d}y \quad (1.27c)$$

Numerical computation of the integrals for the given input correlation matrix $\boldsymbol{R}$ yields $E\{x^2(k) \mid |x(k)| > |x(k-1)|\} = 3.3619\sigma_x^2$. The off-diagonal elements

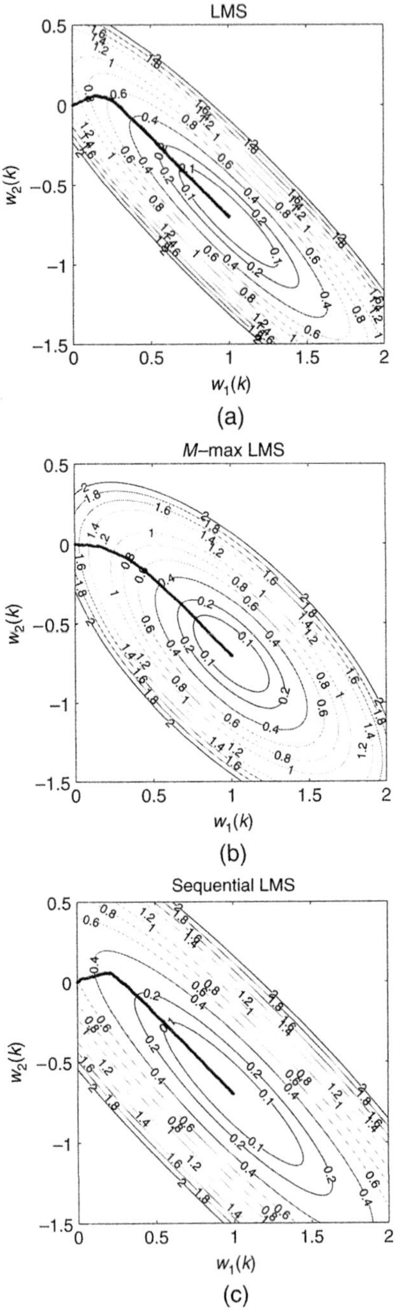

Figure 1.14 Convergence of adaptive filter coefficients over cost function contours for correlated input. (a) LMS; (b) M-max LMS; and (c) sequential-partial-update LMS.

$E\{x(k)x(k-1) \mid |x(k)| > |x(k-1)|\}$ and $E\{x(k)x(k-1) \mid |x(k)| < |x(k-1)|\}$ are simply one-half of $E\{x(k)x(k-1)\}$. Noting that the diagonal elements are identical, we obtain:

$$\boldsymbol{R}_M = \sigma_x^2 \begin{bmatrix} 3.3619 & 2.3684 \\ 2.3684 & 3.3619 \end{bmatrix} \tag{1.28}$$

For the M-max LMS algorithm the eigenvalue spread of the input correlation matrix is $\chi(\boldsymbol{R}_M) = 5.7678$.

The coefficient selection criterion (1.14) used by the sequential-partial-update LMS algorithm simply updates each adaptive filter coefficient half of the time, resulting in (cf. (1.22))

$$\boldsymbol{R}_M = \frac{1}{2}\boldsymbol{R} = \sigma_x^2 \begin{bmatrix} 2.6316 & 2.3684 \\ 2.3684 & 2.6316 \end{bmatrix} \tag{1.29}$$

The eigenvalue spread for the sequential-partial-update LMS algorithm is $\chi(\boldsymbol{R}_M) = 19$ which is the same as for the full-update LMS. Note that multiplying $\boldsymbol{R}_M$ by a positive scalar does not change its eigenvalue spread.

Referring to Figure 1.14 we observe that the full-update LMS and sequential-partial-update LMS algorithms have contours of the same shape and orientation since they have the same eigenvalue spread and their input correlation matrices only differ by a scaling factor. The spacing between the contours of the sequential-partial-update LMS is larger due to the scaling factor of $1/2$ which leads to flattening of the cost function surface. As evidenced by the MSE curves in Figure 1.13, the flattened cost function exhibits itself as slower converge. In fact, for stationary input signals the sequential-partial-update LMS algorithm will always converge slower than the full-update LMS no matter where it is initialized.

The M-max LMS algorithm has rounder cost function contours than the full-update and sequential-partial-update LMS algorithms because of the reduced eigenvalue spread for the input correlation matrix of M-max updates. Unlike sequential partial updates, the method of M-max updates has the effect of making the input correlation matrix more diagonally dominant, which in turn leads to a reduction in the input signal correlation as 'seen' by the adaptation algorithm. The reduced input correlation is a desirable property as it often leads to faster convergence (see Figure 1.13). The reduced eigenvalue spread also explains the capability of the M-max LMS algorithm to outperform the full-update LMS. We therefore conclude that it is possible to turn resource constraints into an advantage by an appropriate use of partial coefficient updates.

Chapter | two

Approaches to partial coefficient updates

2.1 INTRODUCTION

With the advent of digital signal processing systems, several schemes for controlling the computational complexity of adaptive filters by means of partial coefficient updates have emerged. Early approaches were based on the intuitive notion of round-robin updating of coefficient subsets (sequential partial updates) and updating all the coefficients at periodic intervals (periodic partial updates). As we shall see in this chapter, these so-called data-independent approaches suffer from convergence rate reduction, often proportional to the size of coefficient subsets in the case of sequential partial updates and the update frequency for periodic partial updates. More recently, data-dependent partial update techniques have been proposed with improved convergence performance. These data-dependent techniques require sorting of data leading to some complexity overheads.

In this chapter, we provide a comprehensive overview of both data-independent and data-dependent approaches to partial coefficient updates; viz., periodic partial updates, sequential partial updates, stochastic partial updates, M-max updates, selective partial updates, and set membership partial updates. The LMS and normalized LMS (NLMS) algorithms are utilized to demonstrate the application of partial-update techniques. Several simulation examples are presented to illustrate the convergence and stability properties of partial-update LMS and NLMS algorithms. In particular, we observe that partial-update adaptive filters may become susceptible to stability problems when the input signal is cyclostationary or periodic. The method of stochastic partial updates offers a solution to such stability problems. The convergence properties of partial coefficient updates are studied under the assumption of sufficiently small step-size parameter. Resorting to averaging theory, this assumption allows us to replace a partial-update adaptive filter with its equivalent averaged system, which is often much easier to analyse. Furthermore, the small step-size assumption is well-suited to the stability analysis since our objective is to determine input signals for which the partial-update adaptive filters become unstable no matter how small the step-size parameter is. The convergence analysis for large step-sizes is investigated in Chapter 3. In addition to LMS and NLMS, we study the max-NLMS and sign-NLMS algorithms in the context of selective partial updates. The method of set membership partial updates combines the data-dependent method of selective partial updates with set membership filtering, resulting in sparse time updates with improved convergence performance. At the end of the chapter, we briefly review block partial updates as a means of reducing memory requirements for data-dependent partial-update techniques and comment on other methods of complexity reduction.

2.2 PERIODIC PARTIAL UPDATES

The method of periodic partial updates (Douglas, 1997) allows the update complexity to be spread over a number of iterations in order to reduce the average update complexity per iteration. In the early days of digital signal processing it was successfully employed in adaptive network cancellation applications (Messerschmitt *et al.*, 1989).

Consider the following generic adaptive filter:

$$\begin{aligned} \boldsymbol{w}(k+1) &= \boldsymbol{w}(k) + \boldsymbol{f}(k), \quad k = 0, 1, 2, \ldots \\ y(k) &= h(\boldsymbol{w}(k), \boldsymbol{x}(k)) \end{aligned} \tag{2.1}$$

where

$$\boldsymbol{w}(k) = [w_1(k), w_2(k), \ldots, w_N(k)]^T \tag{2.2}$$

is the $N \times 1$ adaptive filter coefficient vector,

$$\boldsymbol{f}(k) = [f_1(k), f_2(k), \ldots, f_N(k)]^T \tag{2.3}$$

is the $N \times 1$ update vector,

$$\boldsymbol{x}(k) = [x(k), x(k-1), \ldots, x(k-N+1)]^T \tag{2.4}$$

is the $N \times 1$ input regressor vector and $y(k)$ is the adaptive filter output at iteration k. Using periodic partial updates for the adaptive filter coefficients in (2.1) results in:

$$\begin{aligned} \boldsymbol{w}((k+1)S) &= \boldsymbol{w}(kS) + \boldsymbol{f}(kS), \quad k = 0, 1, 2, \ldots \\ \boldsymbol{w}(kS+i) &= \boldsymbol{w}(kS), \quad i = 0, 1, \ldots, S-1 \\ y(k) &= h(\boldsymbol{w}(\lfloor k/S \rfloor S), \boldsymbol{x}(k)) \end{aligned} \tag{2.5}$$

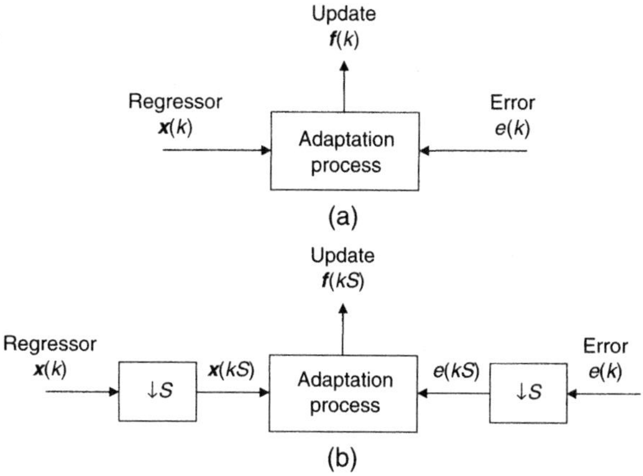

Figure 2.1 (a) Full coefficient updates. Time available for updates is one iteration; (b) Periodic partial updates. Time available for updates is S iterations. The adaptive filter coefficients are frozen between successive updates.

where S is the *period* of coefficient updates and $\lfloor\cdot\rfloor$ denotes truncation towards minus infinity (e.g. $\lfloor 4.23\rfloor = 4$). The method of periodic partial updates 'decimates' the coefficient updates $\boldsymbol{w}(k)$ by a factor of S (see Figure 2.1). The adaptive filter updates its coefficients every Sth iteration at $k = 0, S, 2S, 3S, \ldots$, while the adaptive filter coefficients are frozen between successive updates, i.e. $\boldsymbol{w}(kS) = \boldsymbol{w}(kS+1) = \cdots = \boldsymbol{w}(kS+S-1)$ (see (2.5)). The decimation of updates by S allows the adaptation process to have a duration of S iterations to complete the calculation of the update vector. This reduces the average computational requirements per iteration by a factor of S. Setting $S = 1$ gives the full-update adaptive filter with no complexity reduction.

The application of periodic partial updates to the LMS adaptation process gives:

$$\begin{aligned} \boldsymbol{w}((k+1)S) &= \boldsymbol{w}(kS) + \mu e(kS)\boldsymbol{x}(kS), \quad k = 0, 1, 2, \ldots \\ \boldsymbol{w}(kS+i) &= \boldsymbol{w}(kS), \quad i = 0, 1, \ldots, S-1 \end{aligned} \tag{2.6}$$

If the resource constraints permit M out of N coefficients to be updated at each time index k, the required complexity reduction can be accomplished by setting:

$$S = \left\lceil \frac{N}{M} \right\rceil \tag{2.7}$$

where $\lceil\cdot\rceil$ denotes truncation towards infinity (e.g. $\lceil 4.23\rceil = 5$). The LMS filtering equation is given by:

$$y(k) = \boldsymbol{w}^T(\lfloor k/S\rfloor S)\boldsymbol{x}(k) \tag{2.8}$$

Equation (2.6) implies that the update vector is zero at iterations that are not an integer multiple of S, i.e. $k \bmod S \neq 0$ where mod is the modulo operator. This is equivalent to having $\boldsymbol{x}(k) = \mathbf{0}$ if $k \bmod S \neq 0$.

Under the assumption of sufficiently small step-size parameter, the periodic-partial-update LMS can be replaced by the following averaged system:

$$\boldsymbol{w}^a((k+1)S) = \boldsymbol{w}^a(kS) + \mu(\tilde{\boldsymbol{p}} - \tilde{\boldsymbol{R}}\boldsymbol{w}^a(kS)), \quad k = 0, 1, 2, \ldots \tag{2.9}$$

where, for a stationary input signal,

$$\tilde{\boldsymbol{p}} = E\{\boldsymbol{x}(kS)d(kS)\} \tag{2.10a}$$

$$\tilde{\boldsymbol{R}} = E\{\boldsymbol{x}(kS)\boldsymbol{x}^T(kS)\} \tag{2.10b}$$

are the cross-correlation vector between the periodic-partial-update regressor vector and the corresponding desired filter response, and the autocorrelation matrix of the periodic-partial-update regressor, respectively. The autocorrelation matrix of the periodic-partial-update regressor $\tilde{\boldsymbol{R}}$ is identical to the autocorrelation matrix of the input signal $\boldsymbol{R} = E\{\boldsymbol{x}(k)\boldsymbol{x}^T(k)\}$. However, because the adaptive filter coefficients are updated every Sth iteration, the periodic-partial-update LMS will take S times as long as the full-update LMS to converge.

We digress briefly to analyse the stability of the equivalent averaged system. The averaged system with periodic partial updates is a steepest descent algorithm which produces the optimum (minimum mean-square error (MSE)) solution:

$$\boldsymbol{w}_o = \tilde{\boldsymbol{R}}^{-1}\tilde{\boldsymbol{p}}. \tag{2.11}$$

This solution is sometimes referred to as the Wiener solution. Denoting the difference between $\boldsymbol{w}_o$ and $\boldsymbol{w}^a(kS)$ (coefficient error) by:

$$\Delta\boldsymbol{w}^a(kS) = \boldsymbol{w}_o - \boldsymbol{w}^a(kS) \tag{2.12}$$

and subtracting both sides of (2.9) from $\boldsymbol{w}_o$, we obtain:

$$\Delta\boldsymbol{w}^a((k+1)S) = \Delta\boldsymbol{w}^a(kS) - \mu(\tilde{\boldsymbol{p}} - \tilde{\boldsymbol{R}}\boldsymbol{w}^a(kS)) \tag{2.13a}$$

$$= (\boldsymbol{I} - \mu\tilde{\boldsymbol{R}})\Delta\boldsymbol{w}^a(kS) \tag{2.13b}$$

$$= (\boldsymbol{I} - \mu\tilde{\boldsymbol{R}})^{k+1}\Delta\boldsymbol{w}^a(0) \tag{2.13c}$$

Applying similarity transformation to $\tilde{\boldsymbol{R}}$ to diagonalize it results in:

$$\boldsymbol{Q}^T\tilde{\boldsymbol{R}}\boldsymbol{Q} = \underbrace{\begin{bmatrix} \lambda_1 & & & \boldsymbol{0} \\ & \lambda_2 & & \\ & & \ddots & \\ \boldsymbol{0} & & & \lambda_N \end{bmatrix}}_{\boldsymbol{\Lambda}} \tag{2.14}$$

where $\boldsymbol{Q}$ is a unitary matrix (i.e., $\boldsymbol{Q}\boldsymbol{Q}^T = \boldsymbol{I}$) and the λ_i are the eigenvalues of $\tilde{\boldsymbol{R}}$. Premultiplying (2.13b) by $\boldsymbol{Q}^T$ yields

$$\boldsymbol{Q}^T\Delta\boldsymbol{w}^a((k+1)S) = (\boldsymbol{I} - \mu\boldsymbol{Q}^T\tilde{\boldsymbol{R}}\boldsymbol{Q})\boldsymbol{Q}^T\Delta\boldsymbol{w}^a(kS) \tag{2.15}$$

$$\boldsymbol{v}^a((k+1)S) = (\boldsymbol{I} - \mu\boldsymbol{\Lambda})\boldsymbol{v}^a(kS) \tag{2.16}$$

whence we obtain:

$$\boldsymbol{v}^a((k+1)S) = (\boldsymbol{I} - \mu\boldsymbol{\Lambda})^{k+1}\boldsymbol{v}^a(0) \tag{2.17a}$$

$$= \begin{bmatrix} (1-\mu\lambda_1)^{k+1} & & & \boldsymbol{0} \\ & (1-\mu\lambda_2)^{k+1} & & \\ & & \ddots & \\ \boldsymbol{0} & & & (1-\mu\lambda_N)^{k+1} \end{bmatrix} \boldsymbol{v}^a(0) \tag{2.17b}$$

This expression describes the evolution of the coefficient errors rotated by $\boldsymbol{Q}^T$ for the averaged system. The coefficients are guaranteed to converge, i.e., $\boldsymbol{v}^a(k) \to \boldsymbol{0}$ as $k \to \infty$, if

$$|1 - \mu\lambda_i| < 1 \quad \forall i \in \{1, \ldots, N\}. \tag{2.18}$$

For the full-update LMS, which is obtained by setting $S = 1$, the condition for converge is also given by $|1 - \mu\lambda_i| < 1$ where the λ_i are now the eigenvalues of $\boldsymbol{R}$. The convergence properties of the full-update and partial-update LMS algorithms are determined from the respective correlation matrices of the coefficient update vector. Assuming that all λ_i are positive, the stability bound on the step-size parameter μ is

$$0 < \mu < \frac{2}{\lambda_{\max}} \tag{2.19}$$

where $\lambda_{\max}$ is the maximum eigenvalue. In particular, if any of the eigenvalues of the coefficient update correlation matrix is negative, then the condition for convergence is not satisfied and the adaptive filter diverges. The convergence speed of individual coefficients becomes extremely slow if the eigenvalues associated with them is almost zero. This is also reflected by a very large eigenvalue spread for the coefficient update correlation matrix.

The MSE of the steepest descent algorithm is:

$$E\{\epsilon^2(kS)\} = E\{(d(kS) - (\boldsymbol{w}^a(kS))^T \boldsymbol{x}(kS))^2\} \tag{2.20a}$$

$$= \sigma_d^2 - 2(\boldsymbol{w}^a(kS))^T \tilde{\boldsymbol{p}} + (\boldsymbol{w}^a(kS))^T \tilde{\boldsymbol{R}} \boldsymbol{w}^a(kS) \tag{2.20b}$$

The minimum MSE is:

$$\sigma_d^2 - \tilde{\boldsymbol{p}}^T \boldsymbol{w}_o \tag{2.21}$$

which is obtained when $\boldsymbol{w}^a(kS) = \boldsymbol{w}_o$ or $\boldsymbol{v}^a(k) \to \boldsymbol{0}$, i.e., the steepest descent algorithm has converged to the optimum solution in (2.11). The minimum MSE or the steady-state MSE is invariant to partial update techniques as long as converge to the optimum solution $\boldsymbol{w}_o$ is achieved.

The periodic-partial-update LMS algorithm is known to suffer from excruciatingly slow convergence for certain non-stationary input signals despite no such drastic slowdown in the convergence rate of the full-update LMS. Referring to the definition of the periodic-partial-update correlation matrix $\tilde{\boldsymbol{R}}$ in (2.10), it is easy to see that one can construct cyclostationary input signals with cyclic period S such that $\chi(\tilde{\boldsymbol{R}}) \to \infty$ as a result of some of the eigenvalues of $\tilde{\boldsymbol{R}}$ becoming almost zero while $\boldsymbol{R}$ remains perfectly well-conditioned. For non-stationary input signals, the ensemble average is replaced with time average, yielding:

$$\boldsymbol{R} = \lim_{K\to\infty} \frac{1}{K} \sum_{k=0}^{K-1} \boldsymbol{x}(k)\boldsymbol{x}^T(k), \quad \tilde{\boldsymbol{R}} = \lim_{K\to\infty} \frac{1}{K} \sum_{k=0}^{K-1} \boldsymbol{x}(kS)\boldsymbol{x}^T(kS) \tag{2.22}$$

We observe that the eigenvalues of $\tilde{\boldsymbol{R}}$ are always non-negative since (i) $\tilde{\boldsymbol{R}}$ is non-negative definite, i.e., $\boldsymbol{w}^T \tilde{\boldsymbol{R}} \boldsymbol{w} \geq 0$, and (ii) $\tilde{\boldsymbol{R}} = \tilde{\boldsymbol{R}}^T$ (Diniz, 2002). Thus the periodic-partial-update LMS cannot become divergent as long as the step-size parameter is within the stability bound.

2.2.1 Example 1: Convergence performance

We will illustrate the convergence implications of periodic partial updates with simulation examples. Consider a system identification problem where the unknown system to be identified is given by:

$$H(z) = 1 - 0.7z^{-1} + 0.4z^{-2} - 0.3z^{-3} + 0.2z^{-4} \tag{2.23}$$

The input signal $x(k)$ is white Gaussian with zero mean and unit variance. The output signal $y(k)$ is corrupted by additive zero-mean white Gaussian noise $n(k)$ with variance 0.001. As usual, the signals $x(k)$ and $n(k)$ are assumed to be statistically independent. Figure 2.2(a) shows time-averaged learning curves of the LMS and periodic-partial-update LMS algorithms with parameters $\mu = 5 \times 10^{-4}$, $N = 5$ and $S = 5$. Both algorithms are initialized to a zero vector. It is clear from Figure 2.2(a) that the convergence rate of the periodic-partial-update LMS is S times slower than the full-update LMS. Figure 2.2(b) shows the learning curves of the two algorithms for a coloured Gaussian input signal obtained from:

$$x(k) = 0.5x(k-1) + \nu(k) \tag{2.24}$$

where $\nu(k)$ is white Gaussian noise with zero mean and unit variance. The input signal $x(k)$ is an AR(1) Gaussian process. Due to the increase in the eigenvalue spread of the input correlation matrix, both algorithms exhibit a slower converge overall. Again the periodic-partial-update LMS is S times slower than the LMS algorithm.

2.2.2 Example 2: Convergence difficulties

We next consider a cyclostationary input signal defined by:

$$x(k) = a_i u(k), \quad i = \begin{cases} 1 & \text{if } k \bmod 5 = 1 \\ 2 & \text{if } k \bmod 5 = 2 \\ 3 & \text{if } k \bmod 5 = 3 \\ 4 & \text{if } k \bmod 5 = 4 \\ 5 & \text{if } k \bmod 5 = 0 \end{cases} \tag{2.25}$$

where $u(k)$ is an uncorrelated binary-phase-shift-keying (BPSK) signal taking on values ± 1 with equal probability, and the a_i are constant weights that multiply $u(k)$ periodically, creating a cyclostationary signal with cyclic period 5. For $S = 5$ the periodic-partial-update regressor vector becomes

$$\boldsymbol{x}(5k) = \begin{bmatrix} a_5 u(5k) \\ a_4 u(5k-1) \\ \vdots \\ a_1 u(5k-4) \end{bmatrix} \tag{2.26}$$

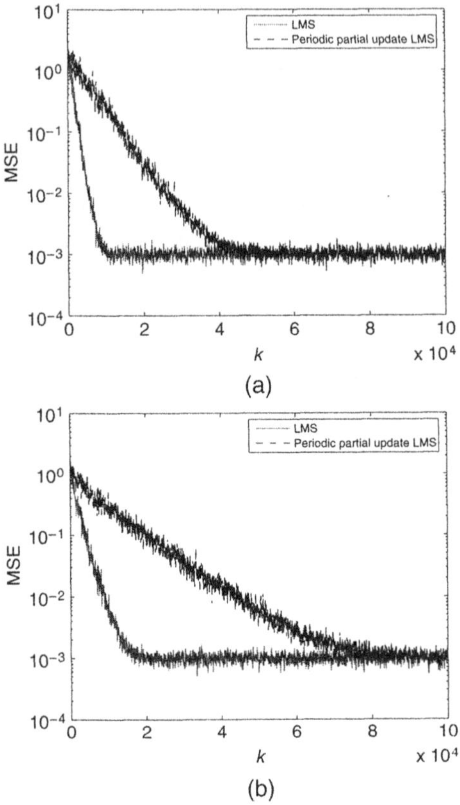

Figure 2.2 Time-averaged MSE curves for periodic-partial-update LMS and full-update LMS for: (a) white Gaussian input; and (b) coloured Gaussian input.

The coefficient update correlation matrix $\tilde{\boldsymbol{R}}$ is given by:

$$\tilde{\boldsymbol{R}} = \begin{bmatrix} a_5^2 & & & & \mathbf{0} \\ & a_4^2 & & & \\ & & a_3^2 & & \\ & & & a_2^2 & \\ \mathbf{0} & & & & a_1^2 \end{bmatrix}. \tag{2.27}$$

The eigenvalues of $\tilde{\boldsymbol{R}}$ are simply equal to its diagonal entries, i.e. $\lambda_1 = a_5^2$, $\lambda_2 = a_4^2, \ldots, \lambda_5 = a_1^2$. The eigenvalue spread of $\tilde{\boldsymbol{R}}$ is:

$$\chi(\tilde{\boldsymbol{R}}) = \frac{\max_i a_i^2}{\min_i a_i^2} \tag{2.28}$$

which depends on how the a_i^2 are distributed.

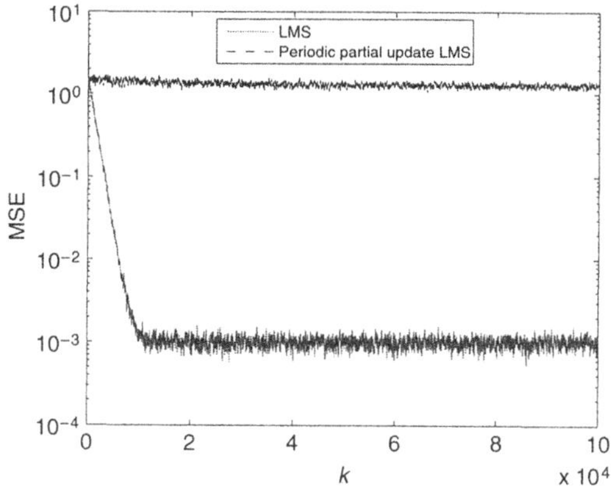

Figure 2.3 Time-averaged MSE curves for periodic-partial-update LMS and full-update LMS for a cyclostationary input.

On the other hand, for the cyclostationary input signal in (2.25) the input autocorrelation matrix of the full-update LMS algorithm is:

$$\boldsymbol{R} = \frac{1}{5}\sum_{i=1}^{5} a_i^2 \boldsymbol{I} \tag{2.29}$$

for which we have $\chi(\boldsymbol{R}) = 1$ irrespective of the a_i. As a result, the full-update LMS does not incur any convergence penalty due to an uneven distribution of the a_i^2.

The learning curves for the full-update LMS and the periodic-partial-update LMS have been simulated using the cyclostationary input signal in (2.25). In the simulations we use $a_1 = 0.03$, $a_2 = -2$, $a_3 = -0.5$, $a_4 = -0.05$ and $a_5 = 0.01$, which yields an eigenvalue spread of $\chi(\tilde{\boldsymbol{R}}) = 4 \times 10^4$ for the periodic-partial-update LMS. This represents an ill-conditioned $\tilde{\boldsymbol{R}}$. The small eigenvalues a_1^2, a_4^2 and a_5^2 are responsible for the periodic-partial-update LMS to stall its convergence. Figure 2.3 depicts the time-averaged learning curves for the two algorithms. Note that the periodic-partial-update LMS fails to converge altogether whereas the full-update LMS appears not to have been affected by the cyclostationary input signal as expected.

2.3 SEQUENTIAL PARTIAL UPDATES

The sequential partial updates method (Douglas, 1997) aims to reduce the computational complexity associated with the adaptation process by updating a subset of the adaptive filter coefficients at each iteration. Unlike periodic partial updates, in which the complete update of coefficients is delayed until the entire update vector is computed, the sequential partial updates method adapts a partition of the coefficient vector within the complexity constraints at each iteration. In this sense the use of sequential partial updates results in 'decimation' of the adaptive filter coefficient vector. What distinguished sequential partial updates from other data-dependent methods is that the coefficient subsets to be updated are selected deterministically in a round-robin fashion. Therefore, the updates follow a periodic pattern regardless of the input signal.

Applying sequential partial updates to the generic adaptive filter in (2.1) yields:

$$\begin{aligned} \boldsymbol{w}(k+1) &= \boldsymbol{w}(k) + \boldsymbol{I}_M(k)\boldsymbol{f}(k), \quad k = 0, 1, 2, \ldots \\ y(k) &= h(\boldsymbol{w}(k), \boldsymbol{x}(k)) \end{aligned} \tag{2.30}$$

where

$$\boldsymbol{I}_M(k) = \begin{bmatrix} i_1(k) & 0 & \cdots & 0 \\ 0 & i_2(k) & \ddots & \vdots \\ \vdots & \ddots & \ddots & 0 \\ 0 & \cdots & 0 & i_N(k) \end{bmatrix}, \quad \sum_{j=1}^{N} i_j(k) = M, \quad i_j(k) \in \{0, 1\} \tag{2.31}$$

is the coefficient selection matrix. For given N and M, $\boldsymbol{I}_M(k)$ is not uniquely specified. Consider the distinct M-subsets (i.e. subsets with M members) of the coefficient index set $\mathcal{S} = \{1, 2, \ldots, N\}$, denoted by $\mathcal{I}_1, \mathcal{I}_2, \ldots, \mathcal{I}_C$ where $C = \binom{N}{M}$. Suppose that $B = N/M$ is an integer. Then the method of sequential partial updates can be implemented by using any B M-subsets of $\mathcal{S}$ as long as the following two conditions are satisfied:

- the union of B M-subsets is $\mathcal{S}$ so that no adaptive filter coefficient is left out, and
- no M-subset pairs share a common member, which ensures that each coefficient gets an equal chance of update.

In terms of M-subsets, $\mathcal{J}_1, \mathcal{J}_2, \ldots, \mathcal{J}_B$, satisfying the above conditions, the coefficient selection matrix for sequential partial updates can be expressed as

$$\boldsymbol{I}_M(k) = \begin{bmatrix} i_1(k) & 0 & \cdots & 0 \\ 0 & i_2(k) & \ddots & \vdots \\ \vdots & \ddots & \ddots & 0 \\ 0 & \cdots & 0 & i_N(k) \end{bmatrix}, \quad i_j(k) = \begin{cases} 1 & \text{if } j \in \mathcal{J}_{(k \bmod B)+1} \\ 0 & \text{otherwise.} \end{cases} \tag{2.32}$$

As an example, for $N = 4$ and $M = 2$, we have $C = 6$ and

$$\mathcal{I}_1 = \{1, 2\}, \mathcal{I}_2 = \{1, 3\}, \mathcal{I}_3 = \{1, 4\}, \mathcal{I}_4 = \{2, 3\}, \mathcal{I}_5 = \{2, 4\}, \\ \mathcal{I}_6 = \{3, 4\}. \tag{2.33}$$

The B coefficient subsets to be updated where $B = 2$ can be arranged into periodic sequences with corresponding coefficient selection matrices $\boldsymbol{I}_M(k)$ given by (2.32). There are three possible ways to assign $\mathcal{J}_1$ and $\mathcal{J}_2$: (i) $\mathcal{J}_1 = \mathcal{I}_1$, $\mathcal{J}_2 = \mathcal{I}_6$, (ii) $\mathcal{J}_1 = \mathcal{I}_2$, $\mathcal{J}_2 = \mathcal{I}_5$, (iii) $\mathcal{J}_1 = \mathcal{I}_3$, $\mathcal{J}_2 = \mathcal{I}_4$. Note that in general there is no unique way to design periodic coefficient subsets as long as the M-subsets used in sequential partial updates satisfy the two conditions stated above.

Updating M coefficients in an adaptive filter of length N at every iteration leads to a complexity reduction in the adaptation process roughly proportional to B. The resource constraints placed on the number of coefficients that can be updated is therefore easily accommodated by sequential partial updates. In practical implementations the choice of M-subsets is guided by software considerations. The usual choice is to partition the adaptive filter coefficient vector into B vectors of length M:

$$\boldsymbol{w}(k) = \begin{bmatrix} \boldsymbol{w}_1(k) \\ \boldsymbol{w}_2(k) \\ \vdots \\ \boldsymbol{w}_B(k) \end{bmatrix} \tag{2.34}$$

with corresponding update partitions:

$$\boldsymbol{f}(k) = \begin{bmatrix} \boldsymbol{f}_1(k) \\ \boldsymbol{f}_2(k) \\ \vdots \\ \boldsymbol{f}_B(k) \end{bmatrix} \tag{2.35}$$

and iterate over them in a round-robin fashion as illustrated in Figure 2.4. This corresponds to having:

$$\begin{aligned} \mathcal{J}_1 &= \{1, 2, \ldots, M\} \\ \mathcal{J}_2 &= \{M+1, M+2, \ldots, 2M\} \\ &\vdots \\ \mathcal{J}_B &= \{(B-1)M+1, (B-1)M+2, \ldots, N\} \end{aligned} \tag{2.36}$$

In some applications N/M may turn out to be non-integer. In such cases we set $B = \lceil N/M \rceil$ resulting in $B - 1$ coefficient partitions of length M, $\mathcal{J}_1, \ldots, \mathcal{J}_{B-1}$, and one partition of length $N \bmod (B-1) = N - (B-1)M$ given by $\mathcal{J}_B$. This arrangement of subsets always creates a partition with a smaller length than the rest of partitions.

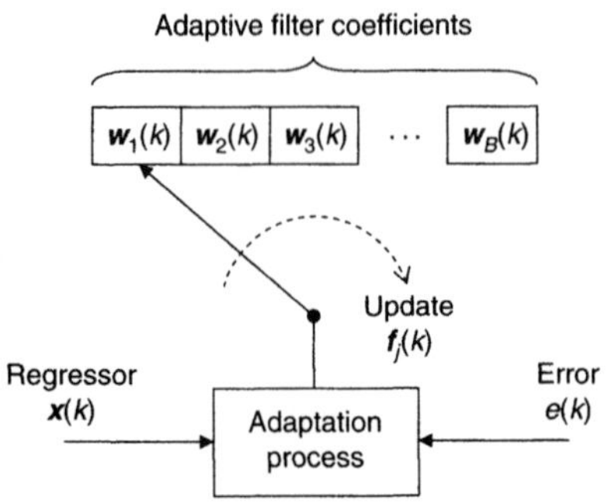

Figure 2.4 Sequential partial updates using partitions of the adaptive filter coefficient vector. The partition to be updated at iteration k is $j = (k \bmod B) + 1$, which produces the periodic sequence $1, 2, \ldots, B, 1, 2, \ldots, B, \ldots$ as desired.

The sequential-partial-update LMS algorithm is given by:

$$\boldsymbol{w}(k+1) = \boldsymbol{w}(k) + e(k)\boldsymbol{I}_M(k)\boldsymbol{x}(k), \quad k = 0, 1, 2, \ldots \tag{2.37}$$

$$y(k) = \boldsymbol{w}^T(k)\boldsymbol{x}(k) \tag{2.38}$$

For a sufficiently small step-size μ, the equivalent averaged system is:

$$\boldsymbol{w}^a(k+1) = \boldsymbol{w}^a(k) + \mu(\boldsymbol{p}_M - \boldsymbol{R}_M\boldsymbol{w}^a(k)), \quad k = 0, 1, 2, \ldots \tag{2.39}$$

where

$$\boldsymbol{p}_M = E\{\boldsymbol{I}_M(k)\boldsymbol{x}(k)d(k)\} \tag{2.40a}$$

$$\boldsymbol{R}_M = E\{\boldsymbol{I}_M(k)\boldsymbol{x}(k)\boldsymbol{x}^T(k)\} \tag{2.40b}$$

If the input signal is stationary, it is easy to show that the correlation matrix $\boldsymbol{R}_M$ can be expressed in terms of the autocorrelation matrix of the input signal $\boldsymbol{R} = E\{\boldsymbol{x}(k)\boldsymbol{x}^T(k)\}$ as:

$$\boldsymbol{R}_M = \frac{1}{B}\boldsymbol{R} \tag{2.41}$$

This implies that, similar to periodic partial updates, the method of sequential partial updates suffers from reduced convergence rate proportional to B when it is applied to the LMS algorithm.

We will now demonstrate the convergence properties of the sequential-partial-update LMS using stationary and non-stationary input signals. For a non-stationary

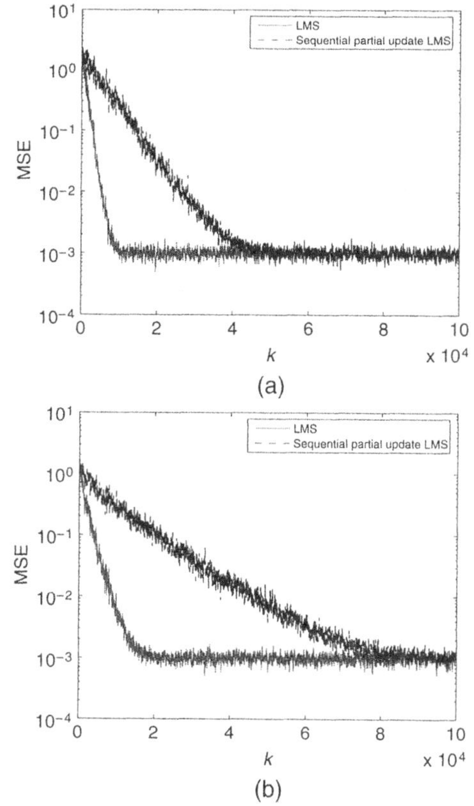

Figure 2.5 Time-averaged MSE curves for sequential-partial-update LMS and full-update LMS for: (a) white Gaussian input; and (b) coloured Gaussian input.

input signal $x(k)$, the ensemble average needs to be replaced with time average:

$$\boldsymbol{R}_M = \lim_{K\to\infty} \frac{1}{K} \sum_{k=0}^{K-1} \boldsymbol{I}_M(k)\boldsymbol{x}(k)\boldsymbol{x}^T(k) \tag{2.42}$$

Unlike $\tilde{\boldsymbol{R}}$, the eigenvalues of $\boldsymbol{R}_M$ cannot be guaranteed to be non-negative. Indeed, in what follows we will show how to design a cyclostationary input signal that makes the sequential-partial-update LMS unstable.

2.3.1 Example 1: Convergence performance

In this and the following simulation examples we consider the same system identification problem that was investigated in periodic partial updates. The unknown system to be identified is given by (2.23). Figure 2.5(a) and (b) show the time-averaged learning curves of the full-update LMS and the sequential-partial-update LMS algorithms for the same white and coloured Gaussian input signals used in periodic-partial-update LMS simulations. The adaptive algorithms have the parameters $\mu = 5 \times 10^{-4}$, $N = 5$ and $M = 1$ (i.e. one out of 5 coefficients is updated at each iteration). Sequential partial updates are selected using (2.32) with $\mathcal{J}_i = \{i\}, i = 1, \ldots, 5$. Both algorithms are initialized to a zero vector. As is evident from Figure 2.5(a) and (b), for the stationary input signals considered the convergence rate of the sequential-partial-update LMS is $B = 5$ times slower than the full-update LMS.

2.3.2 Example 2: Cyclostationary inputs

Consider a shifted version of the cyclostationary input signal in (2.25) given by:

$$x(k) = a_i u(k), \quad i = (k \bmod 5) + 1 \tag{2.43}$$

The new signal has the same cyclic period as the previous one and it is obtained by advancing the signal in (2.25) by one sample. This ensures that the same input sequence is seen by both periodic and sequential partial updates for the complete update of the coefficients so that a fair comparison can be made. For the input signal in (2.43) the coefficient update correlation matrix $\boldsymbol{R}_M$ is:

$$\boldsymbol{R}_M = \frac{a_1^2}{5} \boldsymbol{I} \tag{2.44}$$

with eigenvalue spread $\chi(\boldsymbol{R}_M) = 1$. While no undesirable increase in the eigenvalue spread is observed, a_1 still determines by how much $\boldsymbol{R}_M$ is scaled down.

If $a_1^2 \gg (a_2^2 + \cdots + a_5^2)$, then $\boldsymbol{R}_M \approx \boldsymbol{R}$ where $\boldsymbol{R}$ is given in (2.29) (i.e. the full-update LMS and the sequential-partial-update LMS will perform similarly). Otherwise, the converge rate of the sequential-partial-update LMS is penalized roughly in proportion to the ratio:

$$\frac{\sum_{i=1}^{5} a_i^2}{a_1^2} \tag{2.45}$$

Using the cyclostationary input signal in (2.43) with $a_1 = 0.03$, $a_2 = -2$, $a_3 = -0.5$, $a_4 = -0.05$ and $a_5 = 0.01$, the learning curves for the full-update LMS and the sequential-partial-update LMS have been simulated. The ratio in (2.45) is equal to 4.7261×10^3, which implies an extremely slow convergence. Figure 2.6 depicts the time-averaged learning curves. The sequential-partial-update LMS fails to converge after 100 000 iterations. To show the effect of a_1 we have increased a_1 to 10 while keeping the rest of the a_i unchanged. This has resulted in the ratio given in (2.45) being equal to 1.0425, which suggests a very similar performance for both

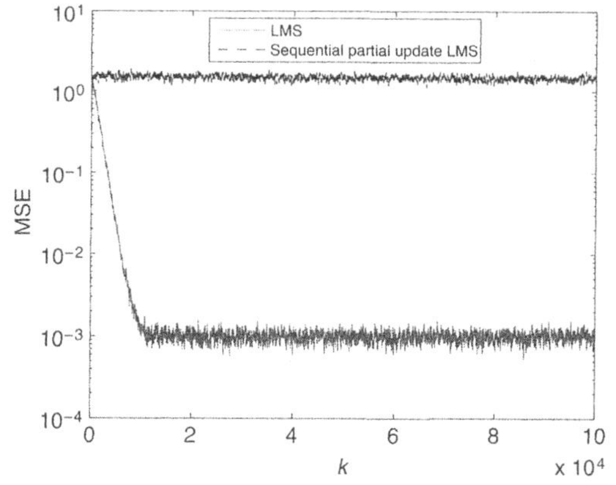

Figure 2.6 Time-averaged MSE curves for sequential-partial-update LMS and full-update LMS for a cyclostationary input.

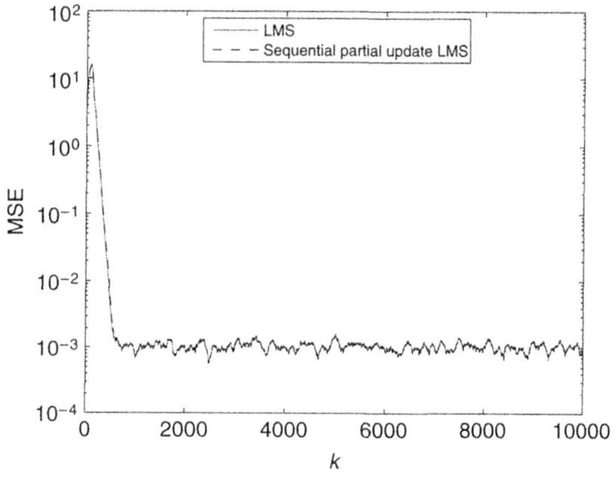

Figure 2.7 Time-averaged MSE curves for sequential-partial-update LMS and full-update LMS for a cyclostationary input with $a_1 = 10$.

the full-update LMS and the sequential-partial-update LMS algorithms as verified by Figure 2.7.

2.3.3 Example 3: Instability

It is possible to find non-stationary input signals for which the method of sequential partial updates becomes unstable no matter how small the step-size parameter is. The key to constructing such signals is to ensure that the resulting coefficient update correlation matrix $\boldsymbol{R}_M$ has some negative eigenvalues, which violates the stability condition in (2.18). In this context let the input signal have the following cyclostationary structure:

$$\begin{aligned} x(5k) &= a_1 \nu_1(k) \\ x(5k+1) &= a_2 \nu_1(k) \\ x(5k+2) &= a_3 \nu_2(k) \\ x(5k+3) &= a_4 \nu_3(k) \\ x(5k+4) &= a_5 \nu_4(k) \end{aligned} \tag{2.46}$$

for $k = 0, 1, 2, \ldots$. Here $\nu_1(k)$ is a stationary coloured Gaussian signal with zero mean and autocorrelation function $r_{\nu_1}(l) = E\{\nu_1(k)\nu_1(k-l)\}$, and $\nu_2(k)$, $\nu_3(k)$ and $\nu_4(k)$ are mutually independent zero-mean white Gaussian signals with unit variance. For this input signal we have:

$$\boldsymbol{R}_M = \frac{1}{5}\begin{bmatrix} a_1^2 r_{\nu_1}(0) & 0 & 0 & 0 & a_1 a_2 r_{\nu_1}(1) \\ a_1 a_2 r_{\nu_1}(0) & a_1^2 r_{\nu_1}(0) & 0 & 0 & 0 \\ 0 & a_1 a_2 r_{\nu_1}(0) & a_1^2 r_{\nu_1}(0) & 0 & 0 \\ 0 & 0 & a_1 a_2 r_{\nu_1}(0) & a_1^2 r_{\nu_1}(0) & 0 \\ 0 & 0 & 0 & a_1 a_2 r_{\nu_1}(0) & a_1^2 r_{\nu_1}(0) \end{bmatrix} \tag{2.47}$$

which is in general not symmetric.

Set the constants in (2.46) to $a_1 = 0.5$ and $a_2 = a_3 = a_4 = a_5 = 1$, and let $\nu_1(k)$ be an AR(1) process given by filtering of a zero-mean white Gaussian noise with unit variance by the following first-order AR system:

$$\frac{1}{1 + 0.8z^{-1}} \tag{2.48}$$

In general, the autocorrelation function of an AR(1) process:

$$x(k) = -ax(k-1) + \nu(k) \tag{2.49}$$

can be expressed as (Diniz, 2002)

$$r_x(l) = \frac{(-a)^{|l|}}{1 - a^2}\sigma_\nu^2 \tag{2.50}$$

where σ_ν^2 is the variance of the white noise $\nu(k)$. Thus for $a = 0.8$, we have $r_{\nu_1}(0) = 25/9$ and $r_{\nu_1}(1) = -20/9$. Substituting these values into (2.47) yields:

$$R_M = \begin{bmatrix} 5/36 & 0 & 0 & 0 & -2/9 \\ 5/18 & 5/36 & 0 & 0 & 0 \\ 0 & 5/18 & 5/36 & 0 & 0 \\ 0 & 0 & 5/18 & 5/36 & 0 \\ 0 & 0 & 0 & 5/18 & 5/36 \end{bmatrix} \tag{2.51}$$

with eigenvalues:

$$\begin{aligned} \lambda_1 &= -0.1268 \\ \lambda_2 &= 0.0568 + \jmath 0.2527 \\ \lambda_3 &= 0.0568 - \jmath 0.2527 \\ \lambda_4 &= 0.3538 + \jmath 0.1561 \\ \lambda_5 &= 0.3538 - \jmath 0.1561 \end{aligned} \tag{2.52}$$

where $\jmath \triangleq \sqrt{-1}$. Note that $\lambda_1 < 0$ and thus for this input signal the sequential-partial-update LMS algorithm does not satisfy the stability condition in (2.18). The learning curves for the full-update LMS and the sequential-partial-update LMS algorithms are shown in Figure 2.8. The divergence of the sequential-partial-update LMS is clearly visible. The full-update LMS has no convergence problems for this signal since its coefficient update correlation matrix is:

$$R = \frac{1}{5} \begin{bmatrix} r & a_1 a_2 r_{\nu_1}(0) & 0 & 0 & a_1 a_2 r_{\nu_1}(1) \\ a_1 a_2 r_{\nu_1}(0) & r & a_1 a_2 r_{\nu_1}(0) & 0 & 0 \\ 0 & a_1 a_2 r_{\nu_1}(0) & r & a_1 a_2 r_{\nu_1}(0) & 0 \\ 0 & 0 & a_1 a_2 r_{\nu_1}(0) & r & a_1 a_2 r_{\nu_1}(0) \\ a_1 a_2 r_{\nu_1}(1) & 0 & 0 & a_1 a_2 r_{\nu_1}(0) & r \end{bmatrix},$$

$$r = \sum_{i=1}^{5} a_i^2 E\{\nu_i^2\} \tag{2.53}$$

which is well-conditioned with $\chi(R) = 2.3025$.

In the case of non-stationary inputs what makes sequential partial updates vulnerable to instability is periodic selection of the coefficient subsets. This periodicity can be avoided by changing the B M-subsets on a regular basis or by adopting a random coefficient selection process, which is discussed in the next section. Another way to avoid divergence at the expense of increased steady-state MSE and significantly slow convergence rate is to use the sequential-partial-update version of the so-called leaky LMS algorithm as suggested in (Douglas, 1997):

$$w(k+1) = (I - \mu\epsilon I_M(k))w(k) + \mu e(k) I_M(k) x(k), \quad k = 0, 1, 2, \ldots \tag{2.54}$$

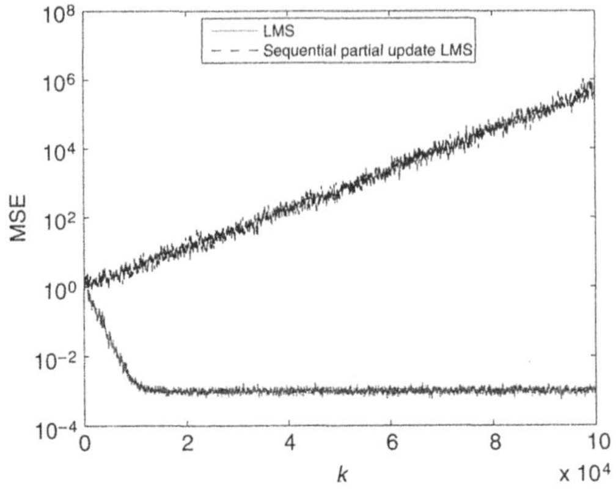

Figure 2.8 Time-averaged MSE curves for sequential-partial-update LMS and full-update LMS for a cyclostationary input that has $\boldsymbol{R}_M$ with a negative eigenvalue.

where $0 \leq \epsilon < 1/\mu$. The input correlation matrix for the sequential-partial-update leaky LMS algorithm is:

$$\boldsymbol{R}_M + \epsilon E\{\boldsymbol{I}_M(k)\} = \boldsymbol{R}_M + \frac{\epsilon}{B}\boldsymbol{I}. \tag{2.55}$$

If the regularization parameter ϵ is chosen appropriately, the regularized input correlation matrix in (2.55) can be guaranteed to have eigenvalues obeying the stability condition (2.18). For example, if we set $\epsilon = 0.14B$, then the input correlation matrix in (2.51) becomes:

$$\boldsymbol{R}_M + \epsilon E\{\boldsymbol{I}_M(k)\} = \begin{bmatrix} 5/36 & 0 & 0 & 0 & -2/9 \\ 5/18 & 5/36 & 0 & 0 & 0 \\ 0 & 5/18 & 5/36 & 0 & 0 \\ 0 & 0 & 5/18 & 5/36 & 0 \\ 0 & 0 & 0 & 5/18 & 5/36 \end{bmatrix} + 0.14\boldsymbol{I} \tag{2.56}$$

with eigenvalues:

$$\begin{aligned} \lambda_1 &= 0.14 - 0.1268 = 0.0132 \\ \lambda_2 &= 0.14 + 0.0568 + \jmath 0.2527 = 0.1968 + \jmath 0.2527 \\ \lambda_3 &= 0.14 + 0.0568 - \jmath 0.2527 = 0.1968 - \jmath 0.2527 \\ \lambda_4 &= 0.14 + 0.3538 + \jmath 0.1561 = 0.4938 + \jmath 0.1561 \\ \lambda_5 &= 0.14 + 0.3538 - \jmath 0.1561 = 0.4938 - \jmath 0.1561 \end{aligned} \tag{2.57}$$

The regularized input correlation matrix satisfies (2.18). However, the minimum regularization parameter ϵ/B required to make λ_1 of $\boldsymbol{R}_M$ positive is rather large compared to the original eigenvalues of $\boldsymbol{R}_M$. An undesirable consequence of this is that the adaptive filter will converge to a biased estimate, leading to a significant increase in the steady-state MSE. Figure 2.9 confirms these predictions. Even though instability is avoided, the steady-state MSE has increased intolerably. If we were to use the minimum regularization to make λ_1 positive, this would lead to a very large eigenvalue spread, which means a very slow convergence rate. Because of these major shortcomings, the sequential-partial-update leaky LMS algorithm does not usually provide a satisfactory solution to the instability problem.

2.4 STOCHASTIC PARTIAL UPDATES

The method of stochastic partial updates (Godavarti and Hero, 2005) is a randomized version of sequential partial updates in that the adaptive filter coefficient subsets are selected randomly rather than in a deterministic fashion. The main motivation for stochastic partial updates is to eliminate instability problems experienced by sequential partial updates for certain non-stationary inputs. Compared with the sequential-partial-update coefficient selection matrix in (2.32), the method of stochastic partial updates uses the following coefficient selection matrix:

$$\boldsymbol{I}_M(k) = \begin{bmatrix} i_1(k) & 0 & \cdots & 0 \\ 0 & i_2(k) & \ddots & \vdots \\ \vdots & \ddots & \ddots & 0 \\ 0 & \cdots & 0 & i_N(k) \end{bmatrix}, \quad i_j(k) = \begin{cases} 1 & \text{if } j \in \mathcal{J}_{m(k)} \\ 0 & \text{otherwise} \end{cases} \tag{2.58}$$

where $m(k)$ is an independent random process with probability mass function:

$$\Pr\{m(k) = i\} = \pi_i, \quad i = 1, \ldots, B, \quad \sum_{i=1}^{B} \pi_i = 1 \tag{2.59}$$

Assume that the coefficient subsets are given by the partitions of the coefficient vector as in (2.36) with no loss of generality. For a stationary input signal $x(k)$, the coefficient update correlation matrix for the stochastic-partial-update LMS algorithm can be expressed as:

$$\boldsymbol{R}_M = E\{\boldsymbol{I}_M(k)\boldsymbol{x}(k)\boldsymbol{x}^T(k)\} \tag{2.60a}$$

$$= E\left\{\begin{bmatrix} \pi_1 \boldsymbol{x}_1(k) \\ \pi_2 \boldsymbol{x}_2(k) \\ \vdots \\ \pi_B \boldsymbol{x}_B(k) \end{bmatrix} \boldsymbol{x}^T(k)\right\} \tag{2.60b}$$

where the $\boldsymbol{x}_i(k)$, $i = 1, \ldots, B$, are partitions of the regressor vector $\boldsymbol{x}(k)$ corresponding to the subsets $\mathcal{J}_i$:

$$\boldsymbol{x}(k) = \begin{bmatrix} x(k) \\ \vdots \\ x(k-M+1) \\ \cdots\cdots \\ x(k-M) \\ \vdots \\ x(k-2M+1) \\ \cdots\cdots \\ \vdots \\ \cdots\cdots \\ x(k-(B-1)M) \\ \vdots \\ x(k-N+1) \end{bmatrix} = \begin{bmatrix} \boldsymbol{x}_1(k) \\ \cdots \\ \boldsymbol{x}_2(k) \\ \cdots \\ \vdots \\ \cdots \\ \boldsymbol{x}_B(k) \end{bmatrix} \tag{2.61}$$

For non-stationary inputs the coefficient update correlation matrix takes the form:

$$\boldsymbol{R}_M = \lim_{K\to\infty} \frac{1}{K} \sum_{k=0}^{K-1} \boldsymbol{I}_M(k)\boldsymbol{x}(k)\boldsymbol{x}^T(k) \tag{2.62a}$$

$$= \lim_{K\to\infty} \frac{1}{K} \begin{bmatrix} \pi_1 \boldsymbol{x}_1(k) \\ \pi_2 \boldsymbol{x}_2(k) \\ \vdots \\ \pi_B \boldsymbol{x}_B(k) \end{bmatrix} \boldsymbol{x}^T(k) \tag{2.62b}$$

Non-uniform probability masses π_i are often not desirable since they tend to increase the eigenvalue spread of $\boldsymbol{R}_M$. The preferred choice is, therefore, a uniform probability mass function giving $\pi_i = 1/B$, $i = 1, \ldots, B$. In this case the coefficient update correlation matrix becomes:

$$\boldsymbol{R}_M = \frac{1}{B}\boldsymbol{R} \tag{2.63}$$

for both stationary and non-stationary inputs. As we have seen previously, periodic and sequential-partial-update algorithms do not have this property since their coefficient update correlation matrix is not always related to $\boldsymbol{R}$ by (2.63) for non-stationary inputs. Based on (2.63) one can conclude that if the full-update LMS algorithm is stable for a given input signal, so is the stochastic-partial-update LMS. This

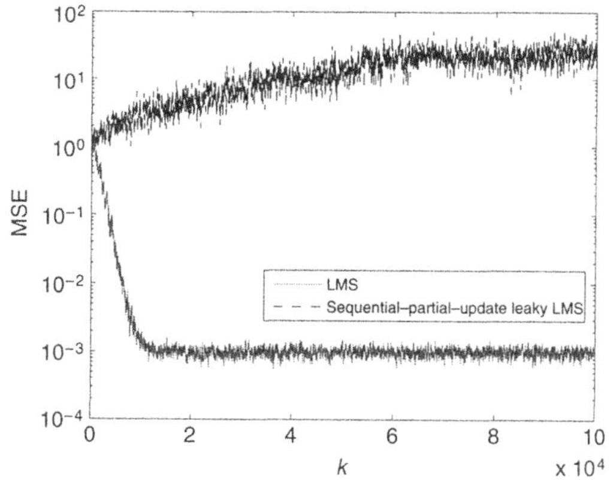

Figure 2.9 Time-averaged MSE curves for sequential-partial-update leaky LMS and full-update LMS for the cyclostationary input in Figure 2.8. The regularization required to make the eigenvalues positive has resulted in a significant estimation bias, which is seen to have led to a very large steady-state MSE.

feature of stochastic partial updates is highly desirable mainly because of its stability implications. In terms of convergence rate, the stochastic-partial-update LMS is still slower than the full-update LMS by a factor of B as a consequence of scaling down of the autocorrelation matrix $\boldsymbol{R}$. The complexity reduction achieved by stochastic partial updates is the same as sequential partial updates if one ignores the overheads for generation of the random signal $m(k)$.

2.4.1 System identification example

We present simulation examples to verify the properties of stochastic partial updates. The simulations have been carried out for the same system identification problem that was described in Section 2.3.1 with identical input signals and parameters. The step-size used is sufficiently small to justify approximation of the stochastic-partial-update LMS by the corresponding averaged system.

For stationary white and coloured Gaussian input signals, the time-averaged learning curves of the full-update LMS and the stochastic-partial-update LMS are shown in Figure 2.10. As for periodic and sequential partial updates, the method of stochastic partial updates exhibits a slower converge rate by a factor of $B = 5$. Figure 2.11 shows the learning curves for the cyclostationary input signal in (2.43) that makes the sequential-partial-update LMS fail to converge due to increased eigenvalue spread (see Figure 2.6). This input signal causes no undue convergence degradation for the stochastic-partial-update LMS. Increasing the weight coefficient a_1 to 10 results in the sequential-partial-update and full-update LMS to perform similarly. However, because of the relation between the correlation matrices of the full-update and stochastic-

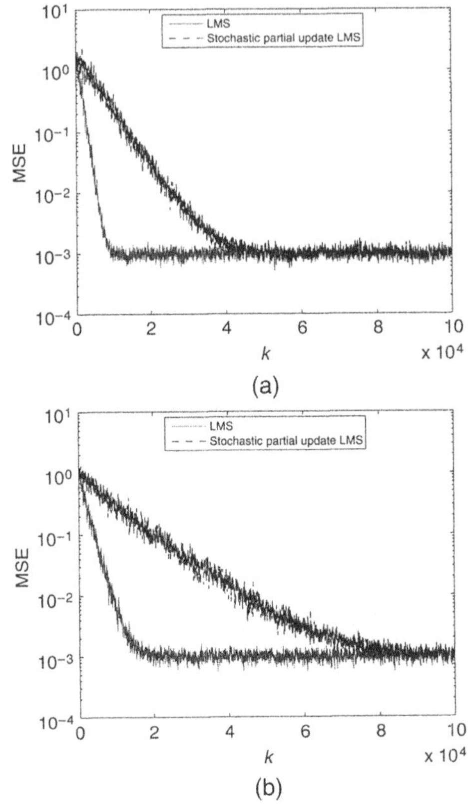

Figure 2.10 Time-averaged MSE curves for stochastic-partial-update LMS and full-update LMS for: (a) white Gaussian input; and (b) coloured Gaussian input.

partial-update LMS algorithms given in (2.63) the stochastic-partial-update LMS converges five times slower than the full-update LMS. Thus, as can be seen from Figure 2.12, a possibility exists for the sequential-partial-update LMS algorithm to outperform the stochastic-partial-update LMS algorithm for certain non-stationary input signals. In Figure 2.13 the learning curves of the full-update LMS and the stochastic-partial-update LMS algorithms are shown for the cyclostationary input signal (2.46) that causes the sequential-partial-update LMS to diverge (see Figure 2.8). As expected the stochastic-partial-update LMS exhibits no divergent behaviour.

In summary, the method stochastic partial updates is inherently stable for input signals that do not lead to stability problems for the full-update LMS. Despite its stability advantage over deterministic partial-update techniques (periodic and sequential partial updates), it may be outperformed by sequential partial updates for certain input signals.

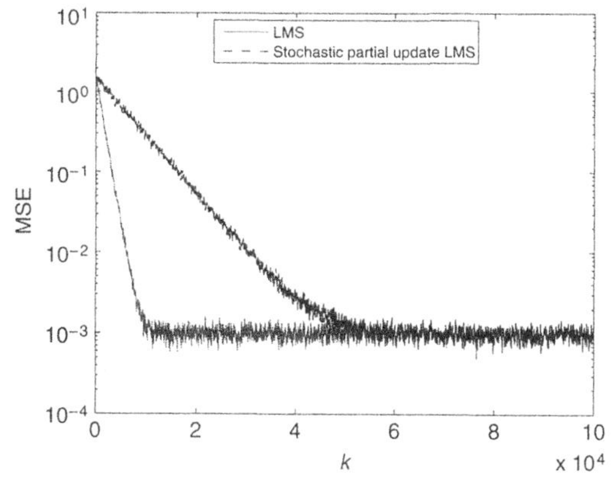

Figure 2.11 Time-averaged MSE curves for stochastic-partial-update LMS and full-update LMS for a cyclostationary input.

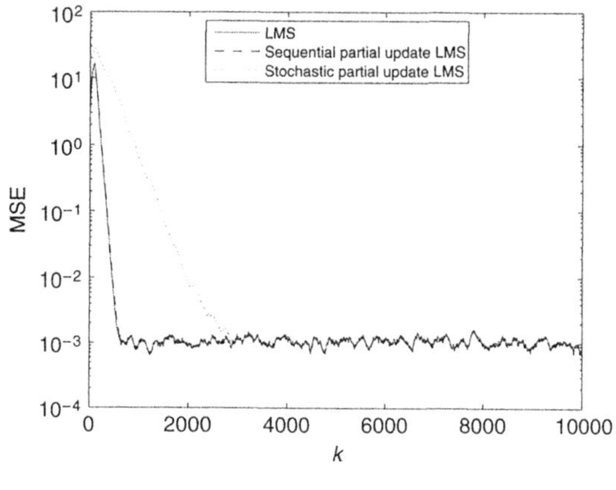

Figure 2.12 Time-averaged MSE curves for stochastic-partial-update LMS and full-update LMS for a cyclostationary input with $a_1 = 10$.

2.5 M-MAX UPDATES

Complexity reduction by M-max updates is a data-dependent partial update technique which is based on finding the M largest magnitude update vector entries. In the context of the NLMS algorithm, it was originally proposed as a generalization of max updates (Aboulnasr and Mayya, 1999; Douglas, 1995), which will be discussed in Section 2.6. Specifically, for the generic adaptive filter in (2.1), the method of M-max

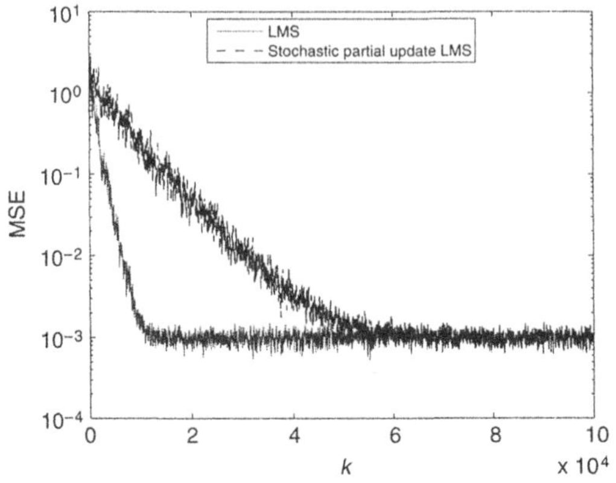

Figure 2.13 Time-averaged MSE curves for stochastic-partial-update LMS and full-update LMS for a cyclostationary input that causes sequential-partial-update LMS to diverge.

updates yields:

$$\begin{aligned} \boldsymbol{w}(k+1) &= \boldsymbol{w}(k) + \boldsymbol{I}_M(k)\boldsymbol{f}(k), \quad k = 0, 1, 2, \ldots \\ y(k) &= h(\boldsymbol{w}(k), \boldsymbol{x}(k)) \end{aligned} \tag{2.64}$$

where the coefficient selection matrix $\boldsymbol{I}_M(k)$ defined in (2.31) takes the form:

$$\boldsymbol{I}_M(k) = \begin{bmatrix} i_1(k) & 0 & \cdots & 0 \\ 0 & i_2(k) & \ddots & \vdots \\ \vdots & \ddots & \ddots & 0 \\ 0 & \cdots & 0 & i_N(k) \end{bmatrix},$$

$$i_j(k) = \begin{cases} 1 & \text{if } |f_j(k)| \in \max\limits_{1 \le l \le N}(|f_l(k)|, M) \\ 0 & \text{otherwise} \end{cases} \tag{2.65}$$

Here $\max_j(w_j, M)$ denotes the set of M maxima of the w_j (e.g. if $w_1 = 0.8$, $w_2 = 0.3$, $w_3 = 0.7$, $w_4 = 0.9$, then $\max_j(w_j, 2) = \{0.8, 0.9\}$).

The method of M-max updates is similar to sequential partial updates in that both approaches 'decimate' the update vector. However, the main difference between the two approaches lies in the way coefficient subsets are selected. In M-max updates the coefficient selection criterion requires the magnitude of update vector entries to be ranked. As different from deterministic round-robin selection, in each iteration the M-subset of adaptive filter coefficients corresponding to the M largest magnitude update vector entries are updated. This coefficient selection scheme finds the subset of update vector entries which is deemed to make the most contribution to the convergence of

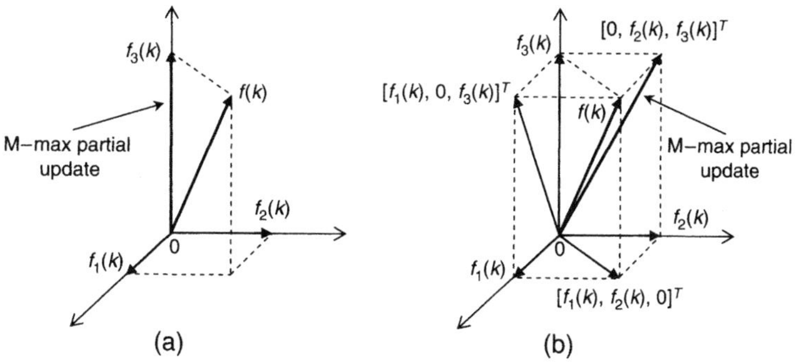

Figure 2.14 Projection of three-dimensional update vector $\boldsymbol{f}(k)$ ($N = 3$) with $|f_1(k)| < |f_2(k)| < |f_3(k)|$ to M-dimensional subspaces. The method of M-max updates selects the hyperplane that yields the closest projection: (a) for $M = 1$ M-max partial update vector is $[0, 0, f_3(k)]^T$ since this projection has the shortest distance from $\boldsymbol{f}(k)$; and (b) for $M = 2$ M-max partial update vector is $[0, f_2(k), f_3(k)]^T$.

the adaptive filter. The number of coefficients M is chosen to remain within the bounds of affordable complexity.

Selection of the M-subset of coefficients with M largest magnitude update entries is equivalent to finding an M-subset in $\mathcal{I}_1, \mathcal{I}_2, \ldots, \mathcal{I}_C$ that produces a partial update with the shortest distance from the full update vector $\boldsymbol{f}(k)$. In other words, $\boldsymbol{I}_M(k)$ for M-max updates solves the following optimization problem at every k:

$$\min_{\boldsymbol{D} \in \{\boldsymbol{I}_1, \ldots, \boldsymbol{I}_C\}} \|(\boldsymbol{D} - \boldsymbol{I})\boldsymbol{f}(k)\|_2 \tag{2.66a}$$

$$\boldsymbol{I}_l = \begin{bmatrix} i_1 & 0 & \cdots & 0 \\ 0 & i_2 & \ddots & \vdots \\ \vdots & \ddots & \ddots & 0 \\ 0 & \cdots & 0 & i_N \end{bmatrix}, \quad i_j = \begin{cases} 1 & \text{if } j \in \mathcal{I}_l \\ 0 & \text{otherwise} \end{cases} \tag{2.66b}$$

where $\|.\|_2$ denotes the Euclidean norm.

Another interpretation of M-max updates is a projection of the N-dimensional update vector $\boldsymbol{f}(k)$ onto an M-dimensional subspace spanned by an M-subset of basis vectors that define the original N-dimensional space. Out of all M-dimensional projections (and there are C of them), the one that has the shortest distance from $\boldsymbol{f}(k)$ is selected by M-max updates. Figure 2.14 provides an illustration of the projection concept in three-dimensional space ($N = 3$) for $M = 1$ and $M = 2$. The selected M-dimensional projection has the largest Euclidean norm among all projections.

The selection of coefficient subsets requires ranking of update entry magnitudes in order to identify the M-subset of coefficients with the largest Euclidean norm.

This assumes that the update vector entries have already been computed, and therefore yields no complexity savings. Consequently, the method of M-max updates can provide complexity reduction in the update process only if the required ranking can be performed without explicit computation of each entry of the update vector. The ranking operation introduces additional complexity overheads, increasing the computational complexity of M-max updates beyond that of sequential partial updates. Computationally efficient ranking algorithms are available. An overview of some of these algorithms is provided in Appendix A. When determining the number of coefficients to be updated per iteration, M, the complexity overheads introduced by ranking, needs to be taken into consideration. A major advantage of M-max partial updates is its capability to reduce convergence penalty incurred by other data independent approaches, viz. periodic, sequential and stochastic partial updates. In some applications it may even lead to a faster convergence than the full-update adaptive filter. A unique feature of M-max updates in this context is its ability to partially decorrelate some input signals by decreasing the eigenvalue spread of the input correlation matrix.

Let us consider the application of M-max updates to the LMS algorithm:

$$\boldsymbol{w}(k+1) = \boldsymbol{w}(k) + \mu e(k)\boldsymbol{I}_M(k)\boldsymbol{x}(k), \quad k = 0, 1, 2, \ldots \tag{2.67}$$

where $\boldsymbol{I}_M(k)$ is the M-max coefficient selection matrix in (2.65) with $\boldsymbol{f}(k) = e(k)\boldsymbol{x}(k)$. Using the LMS update vector, the diagonal entries of $\boldsymbol{I}_M(k)$ defined in (2.65) can be rewritten as

$$i_j(k) = \begin{cases} 1 & \text{if } |x(k-j+1)| \in \max\limits_{1 \le l \le N}(|x(k-l+1)|, M) \\ 0 & \text{otherwise} \end{cases} \tag{2.68}$$

We observe that ranking $|f_i(k)|, i = 1, \ldots, N$, is equivalently to ranking $|x(k-i+1)|$. Thus, the M-max updates are simply given by the M maxima of the magnitude of the input regressor vector entries, which does not require computation of the full update vector. This results in complexity reduction for the M-max LMS algorithm roughly by a factor of $B = N/M$.

Using the small step-size approximation, the M-max LMS algorithm can be replaced by the corresponding averaged system. Assuming a stationary input signal, the coefficient update correlation matrix is given by $\boldsymbol{R}_M = E\{\boldsymbol{I}_M(k)\boldsymbol{x}(k)\boldsymbol{x}^T(k)\}$. Suppose that the input signal $x(k)$ is an AR(1) process defined in (2.49). We next investigate how $\boldsymbol{R}_M$ is affected by an AR(1) input signal. The analysis will shed light on the decorrelation property of M-max updates.

Consider the case of $M = 1$. For notational convenience, let:

$$\boldsymbol{x}(k) = \begin{bmatrix} x(k) \\ x(k-1) \\ \vdots \\ x(k-N+1) \end{bmatrix} = \begin{bmatrix} x_1(k) \\ x_2(k) \\ \vdots \\ x_N(k) \end{bmatrix} \tag{2.69}$$

Define the events $A_i, i = 1, \ldots, N$ as:

$$A_i : |x_i(k)| \text{ is the maximum of } |x_1(k)|, |x_2(k)|, \ldots, |x_N(k)| \tag{2.70a}$$

Then we have

$$\boldsymbol{R}_M = \begin{bmatrix} E\{x_1^2(k) \mid A_1\} & E\{x_1(k)x_2(k) \mid A_1\} & \cdots & E\{x_1(k)x_N(k) \mid A_1\} \\ E\{x_2(k)x_1(k) \mid A_2\} & E\{x_2^2(k) \mid A_2\} & \cdots & E\{x_2(k)x_N(k) \mid A_2\} \\ \vdots & \vdots & & \vdots \\ E\{x_N(k)x_1(k) \mid A_N\} & E\{x_N(k)x_2(k) \mid A_N\} & \cdots & E\{x_N^2(k) \mid A_N\} \end{bmatrix} \tag{2.70b}$$

The joint probability density function of $\boldsymbol{x}(k)$ is N-dimensional zero-mean Gaussian:

$$p(\boldsymbol{x}(k)) = \frac{1}{(2\pi)^{N/2}|\boldsymbol{R}|^{1/2}} \exp\left(-\frac{1}{2}\boldsymbol{x}^T(k)\boldsymbol{R}^{-1}\boldsymbol{x}(k)\right) \tag{2.71}$$

where $\boldsymbol{R}$ is the $N \times N$ autocorrelation matrix of $\boldsymbol{x}(k)$, which is Toeplitz:

$$\boldsymbol{R} = \begin{bmatrix} r_x(0) & r_x(1) & \cdots & r_x(N-1) \\ r_x(1) & r_x(0) & \ddots & \vdots \\ \vdots & \ddots & \ddots & r_x(1) \\ r_x(N-1) & \cdots & r_x(1) & r_x(0) \end{bmatrix} \tag{2.72}$$

Here $r_x(l)$ is the autocorrelation function of $x(k)$ given by (2.50) for the AR(1) input.

Using the joint probability density function of $\boldsymbol{x}(k)$, the conditional expectations in (2.70b) can be written as:

$$\begin{aligned} E\{x_i(k)x_j(k) \mid A_i\} = {} & \frac{1}{(2\pi)^{N/2}|\boldsymbol{R}|^{1/2}} \int_{x_1=-|x_i|}^{|x_i|} \cdots \int_{x_i=-\infty}^{\infty} \cdots \int_{x_N=-|x_i|}^{|x_i|} x_i x_j \\ & \times \exp\left(-\frac{1}{2}[x_1, x_2, \ldots, x_N]\boldsymbol{R}^{-1}[x_1, x_2, \ldots, x_N]^T\right) \\ & \times \mathrm{d}x_1 \mathrm{d}x_2 \cdots \mathrm{d}x_N, \quad i, j = 1, \ldots, N \end{aligned} \tag{2.73}$$

For a given i, $i \in \{1, 2, \ldots, N\}$, all variables of integration $x_1, x_2, \ldots, x_N$ have the lower limit $-|x_i|$ and upper limit $|x_i|$ except for x_i itself which ranges from $-\infty$ to ∞. Equation (2.73) does not have a closed-form expression. Numerical integration techniques may be employed to evaluate it.

The correlation analysis for $M = 1$ can be extended to $M \geq 2$ in a straightforward way. However, due to notational complications this is not pursued here. Suffice it to say that $\boldsymbol{R}_M$ is expected to get closer to $\boldsymbol{R}$ as M approaches N.

2.5.1 Example 1: Eigenvalue spread of $\boldsymbol{R}_M$

The coefficient update correlation matrix $\boldsymbol{R}_M$ has been calculated using Monte Carlo simulations for $N = 3$, $M = 1$, $\sigma_\nu^2 = 1$ and a ranging from -0.9 to -0.1 in steps of 0.1. The resulting eigenvalue spread for $\boldsymbol{R}$ and $\boldsymbol{R}_M$ is plotted in Figure 2.15 for zero-mean white Gaussian $\nu(k)$ and binary $\nu(k)$ taking on values ± 1 with equal probability. It is interesting to note that, for the Gaussian AR(1) input process, setting $M = 1$ provides a significant reduction in the eigenvalue spread of the input signal applied to the adaptive filter (see Figure 2.15(a)). For $M = 2$ the eigenvalue spread is still smaller than the full-update case, but not as small as the case of $M = 1$. This observation points to an important feature of M-max updates; viz. the ability to reduce the correlation of the input signal. For white input signals, however, the use of M-max updates results in no change to the eigenvalue spread. We further notice that the eigenvalue spread reduction is more pronounced for highly correlated input signals with a close to -1. A question then arises as to whether this partial decorrelation property of M-max updates applies to *all* correlated input signals. The answer is no. To see this, consider another AR(1) process as the input signal this time driven by a binary signal. The eigenvalue spread plots are shown in Figure 2.15(b). The resulting input signal $x(k)$ is no longer Gaussian. However, it may be approximated by a Gaussian distribution thanks to the central limit theorem if the AR process has an impulse response with a slowly decaying envelope. The impulse response of the AR(1) process satisfies this requirement for a close to -1. Thus, as a approaches -1, the eigenvalue spread of M-max partial updates is seen to be smaller than the full update case (see Figure 2.15(b)). This result is expected since the input signal $x(k)$ is an approximately Gaussian AR(1) process, which was considered previously. On the other hand, if a is in the vicinity of zero, which implies the input signal $x(k)$ is definitely non-Gaussian, the use of M-max partial updates increases the eigenvalue spread of the input signal. Thus, the decorrelation property of M-max updates only applies to certain input distributions.

2.5.2 Example 2: Convergence performance

In this simulation example we illustrate the properties of M-max updates using the system identification problem in Section 2.3.1. The input signal $x(k)$ is a Gaussian

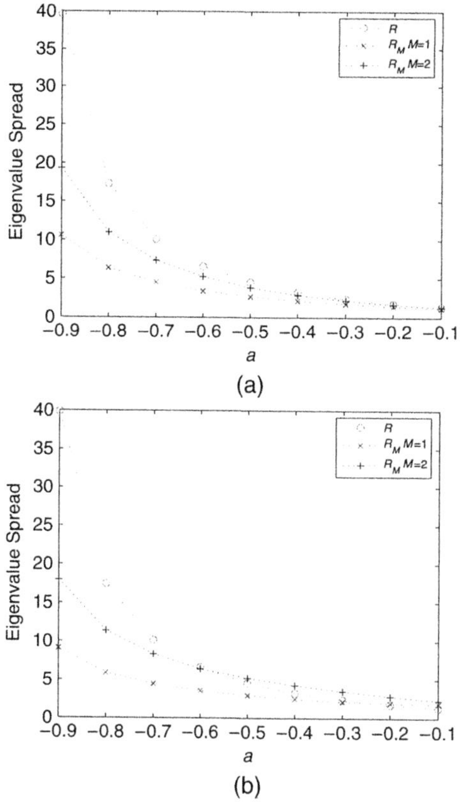

Figure 2.15 Eigenvalue spread for $\boldsymbol{R}$ and $\boldsymbol{R}_M$ (M-max updates) as a function AR(1) parameter a for: (a) white Gaussian $\nu(k)$; and (b) binary $\nu(k)$. The length of the adaptive filter is $N = 3$.

AR(1) process. The step-size parameter is set to $\mu = 5 \times 10^{-4}$, which is sufficiently small to allow approximation of the adaptive filters by their averaged counterparts. For $a = 0$ (white Gaussian input), $a = -0.5$ and $a = -0.9$ the time-averaged learning curves of the full-update LMS and the M-max LMS are shown in Figure 2.16. In order to make the effect of M on the convergence rate clearly visible, the time-averaging window length was set to $L = 5000$ in ((1.15)), which provides a good level of smoothing. The number of coefficients to be updated is varied in the range $1 \leq M \leq 4$.

For white input and AR(1) input with $a = -0.5$, the convergence rate of the M-max LMS improves as M is increased from 1 to 4 (see Figures 2.16(a)–(b)). This observation is intuitive since we expect that with larger M the M-max LMS will perform more like the full-update LMS algorithm. However, as we shall see below,

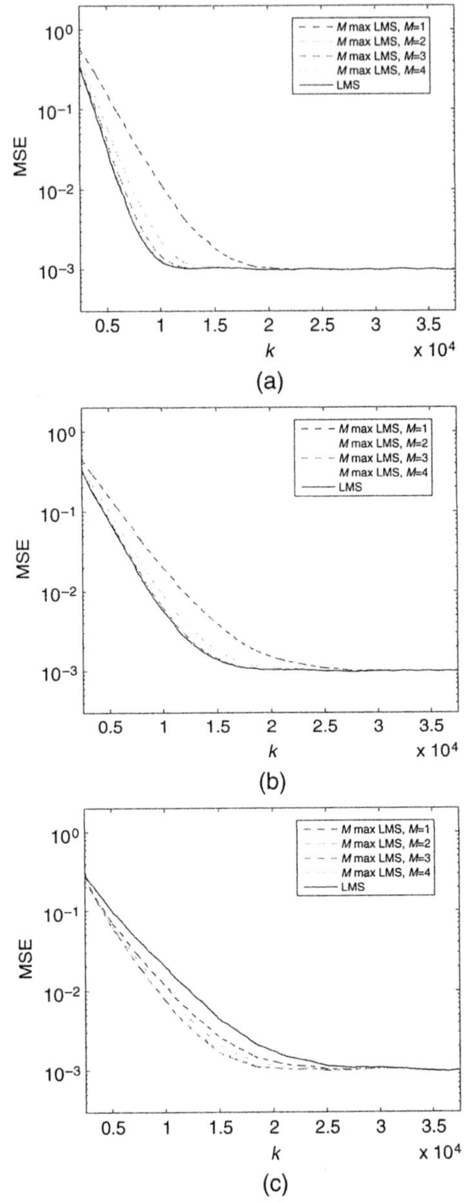

Figure 2.16 Time-averaged MSE curves for full-update and M-max LMS for Gaussian AR(1) input with: (a) $a = 0$ (white); (b) $a = -0.5$; and (c) $a = -0.9$.

increasing M does not always improve the convergence rate. When $M = 4$ the difference between the convergence rates of the full-update LMS and the M-max LMS is almost indiscernible. The M-max LMS incurs some convergence penalty for other values of M. In comparison with the periodic, sequential or stochastic partial updates, the reduction in convergence rate is relatively small. For example, for $M = 1$, the data-independent partial update techniques would yield a converge rate five times slower than the full-update LMS, whereas the M-max updates slow the convergence rate only by a factor of approximately 1.5. This is the major advantage of M-max updates. For the highly correlated input signal ($a = -0.9$), the M-max LMS algorithm exhibits a faster convergence rate than the full-update LMS algorithm for all values of M in the range $1 \leq M \leq 4$ (see Figure 2.16(c)). The fastest convergence is attained when $M = 3$ followed by $M = 2$, $M = 4$, $M = 1$ and $M = 5$ (full-update). In this case increasing M does not lead to a faster convergence.

2.5.3 Example 3: Convergence rate and eigenvalues of $\boldsymbol{R}_M$

In an attempt to elucidate the convergence behaviour of the M-max LMS observed in Figure 2.16, we will analyse the eigenvalues of $\boldsymbol{R}$ and $\boldsymbol{R}_M$. Before we do so, let us review the significance of the eigenvalues of $\boldsymbol{R}_M$ in relation to the convergence rate. Referring to (2.17), we see that the convergence rate of the averaged system is controlled by the smallest eigenvalues since their time constant dominates how fast the coefficient errors die away. In this sense the smallest eigenvalue of $\boldsymbol{R}_M$, $\lambda_{\min}$, gives an indication of the convergence rate. The larger $\lambda_{\min}$, the faster the convergence rate will be. While the eigenvalue spread is related to $\lambda_{\min}$, it does not provide an actual measure of how small or large $\lambda_{\min}$ is. While one adaptive algorithm may have a smaller eigenvalue spread than another one, it may still converge slower unless its $\lambda_{\min}$ is larger. Figures 2.17 and 2.18 show the eigenvalue spread and minimum eigenvalues of the correlation matrices for comparison purposes. The convergence rate of the M-max LMS algorithm with respect to the full-update LMS is accurately predicted by how its $\lambda_{\min}$ compares with that of $\boldsymbol{R}$. For example, in Figure 2.16(c), the converge rate is ranked from fastest to slowest as $M = 3$, $M = 2$, $M = 4$, $M = 1$ and $M = 5$. The minimum eigenvalue of $\boldsymbol{R}_M$ is also ranked in the same order as the convergence rate as shown in Figure 2.18(c). The eigenvalue spread of $\boldsymbol{R}_M$ is smaller than that of $\boldsymbol{R}$ in all cases except for $a = 0$ (white input) (see Figure 2.17). The reduced input signal correlation, as indicated by relatively smaller eigenvalue spread, does not always translate into a faster convergence rate.

If the update coefficient correlation matrices for full-update and partial-update adaptive filtering algorithms are related by simple scaling, the same scaling factor is also applied to the eigenvalues. For example, if $\boldsymbol{R}_M = \frac{1}{B}\boldsymbol{R}$, then the minimum eigenvalue of $\boldsymbol{R}_M$ will be $1/B$th that of $\boldsymbol{R}$. This explains the slow-down in convergence rate for the partial-update algorithm with correlation matrix $\boldsymbol{R}_M$. We have already observed this effect when discussing the data-independent partial update techniques early in this chapter.

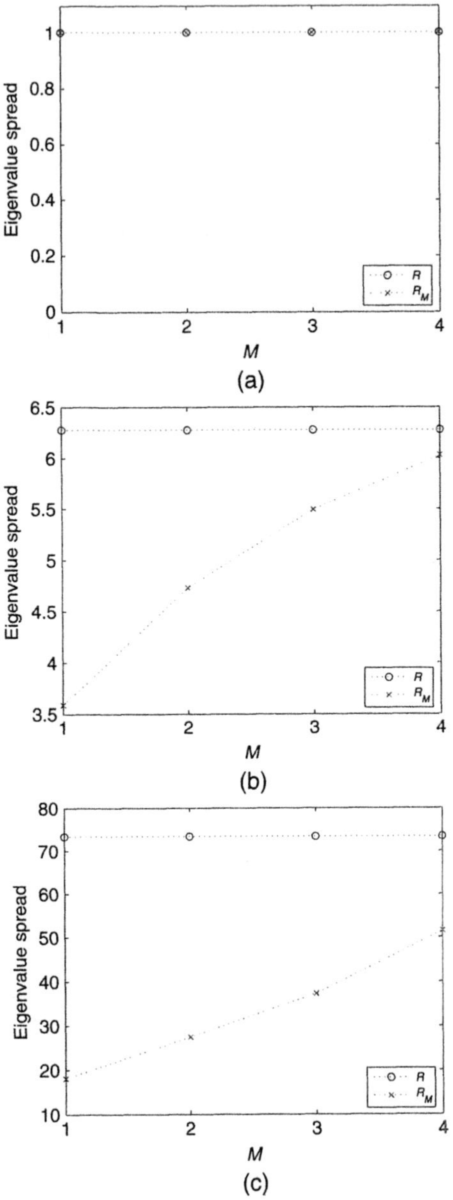

Figure 2.17 Eigenvalue spread for $\boldsymbol{R}$ and $\boldsymbol{R}_M$ versus M for: (a) $a = 0$ (white Gaussian input); (b) $a = -0.5$; and (c) $a = -0.9$.

The convergence analysis presented above assumes a very small step-size parameter μ so that the actual adaptive filter can be approximated by the corresponding averaged

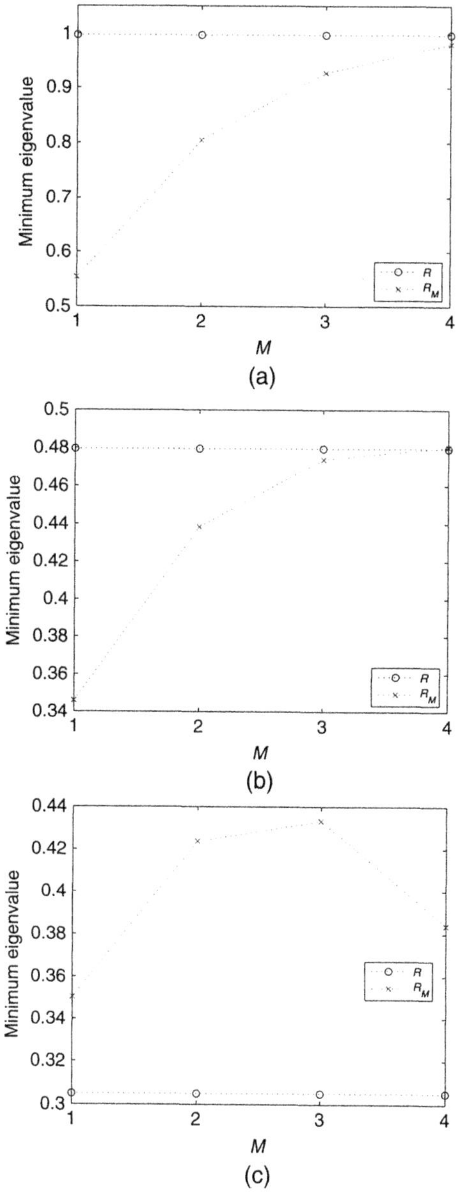

Figure 2.18 Minimum eigenvalue of $\boldsymbol{R}$ and $\boldsymbol{R}_M$ versus M for: (a) $a = 0$ (white Gaussian input); (b) $a = -0.5$; and (c) $a = -0.9$.

system, simplifying the analysis tremendously. The M-max LMS algorithm can still outperform the full-update LMS algorithm for large step-sizes that would rule out

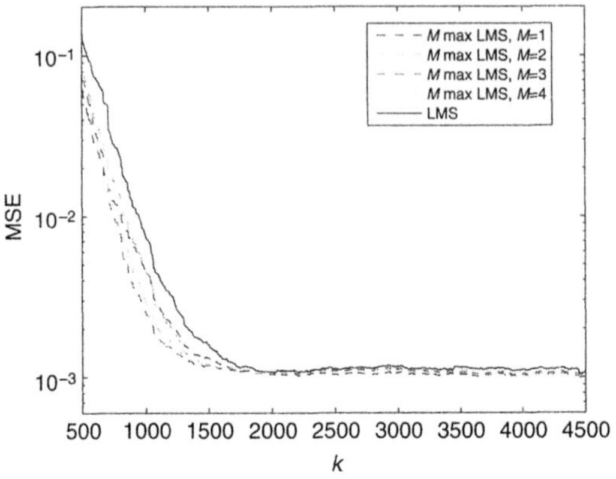

Figure 2.19 Time-averaged MSE curves for the M-max and full-update LMS algorithms for the AR(1) input with $a = -0.9$ and $\mu = 8 \times 10^{-3}$. The M-max LMS algorithm still out-performs the full-update LMS algorithm despite a large μ close to the stability bound.

averaging analysis. Figure 2.19 shows the time-averaged learning curves for the M-max LMS and the full-update LMS algorithms for the Gaussian AR(1) input with $a = -0.9$ and $\mu = 8 \times 10^{-3}$, which is very close to the upper bound for mean-squared stability of LMS. The time-averaging window length is $L = 1000$. Even for this relatively large step-size parameter, the M-max LMS algorithm still converges faster than the full-update LMS. We also note that the steady-state MSE for the M-max LMS algorithm is slightly smaller than that of the full-update LMS, which suggests that even faster convergence can be achieved for the M-max LMS if its step-size is adjusted so as to match its steady-state MSE with that of the full-update LMS.

2.5.4 Example 4: Convergence difficulties

The M-max LMS is susceptible to convergence problems if the eigenvalue spread of $\boldsymbol{R}_M$ is large, with some eigenvalues almost equal to zero, or $\boldsymbol{R}_M$ has some negative eigenvalues. We will illustrate such cases with simulation examples. Consider the cyclostationary input signal defined by:

$$x(k) = a_i u(k), \quad i = (k \bmod P) + 1 \tag{2.74}$$

where the a_i are constants obeying $|a_1| < |a_2| < \cdots < |a_P|$, P is the cyclic period with $P > N$, and $u(k)$ is an independent binary signal taking on values ± 1 with equal probability. For $M = 1$ the coefficient update correlation matrix is:

$$\boldsymbol{R}_M = \lim_{K \to \infty} \frac{1}{K} \sum_{k=0}^{K-1} \boldsymbol{I}_M(k)\boldsymbol{x}(k)\boldsymbol{x}^T(k)$$
$$= \frac{1}{P} \begin{bmatrix} \sum_{i=N}^{P} a_i^2 & & & \mathbf{0} \\ & a_P^2 & & \\ & & \ddots & \\ \mathbf{0} & & & a_P^2 \end{bmatrix}_{N \times N} \tag{2.75}$$

On the other hand, the autocorrelation matrix for the full-update LMS algorithm is:

$$\boldsymbol{R} = \frac{1}{P} \sum_{i=1}^{P} a_i^2 \boldsymbol{I} \tag{2.76}$$

The eigenvalue spreads for $\boldsymbol{R}_M$ and $\boldsymbol{R}$ are:

$$\chi(\boldsymbol{R}_M) = \frac{\sum_{i=N}^{P} a_i^2}{a_P^2} \quad \text{and} \quad \chi(\boldsymbol{R}) = 1 \tag{2.77}$$

We have simulated the full-update LMS and the M-max LMS algorithms for the input signal in (2.74) with $P = 500$ and:

$$a_i = 1 + \frac{i-1}{10P} \tag{2.78}$$

The eigenvalue spread for the selected input signal parameters is $\chi(\boldsymbol{R}_M) = 452.6929$ and $\chi(\boldsymbol{R}) = 1$. Figure 2.20 shows the time-averaged learning curves for $L = 1000$. The M-max LMS algorithm clearly fails to converge while the full-update LMS algorithm has no convergence difficulties. This observation is in line with the eigenvalue spread for the two algorithms. The simulated input signal does not yield any convergence problem for the data-independent partial update techniques, except for the usual convergence slow-down by a factor of $B = N/M = 5$.

2.5.5 Example 5: Instability

Consider a deterministic periodic signal with period $N = 5$ and one period given by $\{1, -0.4, -0.4, -0.4, -0.4\}$. The signal may be shifted in time without affecting the following analysis. It is easy to show that for this input signal and $M = 1$ the update

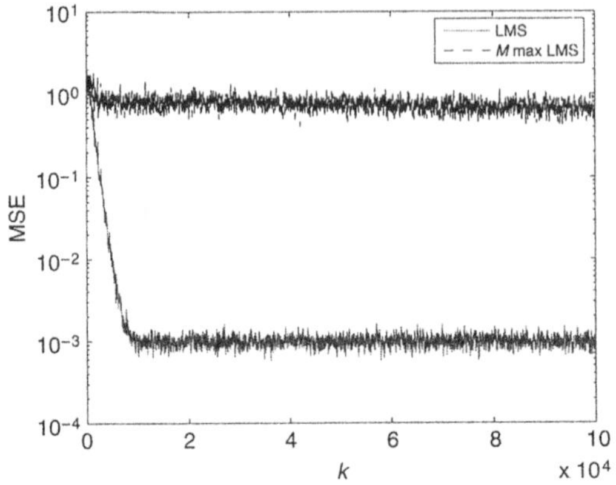

Figure 2.20 Time-averaged MSE curves for the M-max and full-update LMS algorithms for a cyclostationary input.

coefficient correlation matrix is:

$$\boldsymbol{R}_M = \frac{1}{5}\begin{bmatrix} 1 & -0.4 & -0.4 & -0.4 & -0.4 \\ -0.4 & 1 & -0.4 & -0.4 & -0.4 \\ -0.4 & -0.4 & 1 & -0.4 & -0.4 \\ -0.4 & -0.4 & -0.4 & 1 & -0.4 \\ -0.4 & -0.4 & -0.4 & -0.4 & 1 \end{bmatrix} \tag{2.79}$$

with eigenvalues:

$$\begin{aligned} \lambda_1 &= -0.12 \\ \lambda_2 &= \lambda_3 = \lambda_4 = \lambda_5 = 0.28. \end{aligned} \tag{2.80}$$

Since $\lambda_1 < 0$, the M-max LMS algorithm diverges for this input signal. The input autocorrelation matrix of the full-update LMS algorithm is:

$$\boldsymbol{R} = \begin{bmatrix} 0.328 & -0.064 & -0.064 & -0.064 & -0.064 \\ -0.064 & 0.328 & -0.064 & -0.064 & -0.064 \\ -0.064 & -0.064 & 0.328 & -0.064 & -0.064 \\ -0.064 & -0.064 & -0.064 & 0.328 & -0.064 \\ -0.064 & -0.064 & -0.064 & -0.064 & 0.328 \end{bmatrix} \tag{2.81}$$

which has positive eigenvalues. Figure 2.21 confirms the instability of the M-max LMS algorithm.

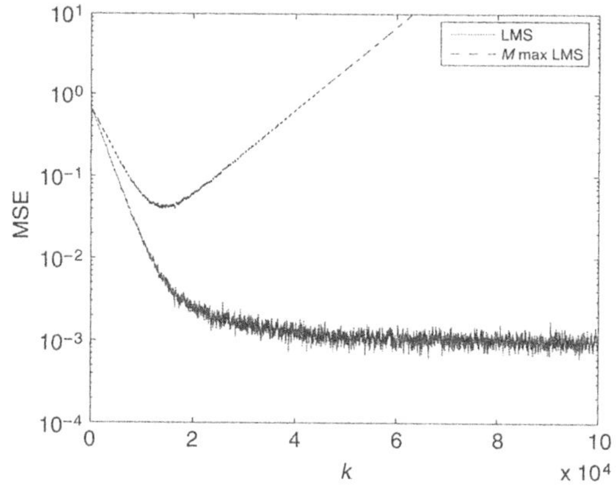

Figure 2.21 Time-averaged MSE curves for the M-max and full-update LMS algorithms for a periodic input signal for which $\boldsymbol{R}_M$ has a negative eigenvalue.

2.6 SELECTIVE PARTIAL UPDATES

The M-max update technique is applicable to any adaptive filtering algorithm with a generic update equation given by (2.1). What makes M-max updates universally applicable is that it searches for a subset among all possible coefficient update subsets that is closest to the full-update vector in the Euclidean norm sense [see (2.66)]. The method of selective partial updates, on the other hand, requires more insight into the application under consideration, and is usually applicable to 'normalized' adaptive filtering algorithms with time-varying step-sizes (Doğançay and Tanrıkulu, 2001a). In this sense it does not always lead to a simple modification of the update vector of a generic adaptive filter. There are two alternative methods that can be utilized to develop selective-partial-update algorithms; viz. application of the *principle of minimum disturbance* (Widrow and Lehr, 1990) and approximation of Newton's method. In this section we will develop the method of selective partial updates for the NLMS algorithm.

2.6.1 Constrained optimization

Let us assume a linear adaptive filter with output signal $y(k) = \boldsymbol{w}^T(k)\boldsymbol{x}(k)$. Then we have:

$$\begin{aligned} e(k) &= d(k) - y(k) \qquad \textit{a priori} \text{ error} \\ e_p(k) &= d(k) - \boldsymbol{w}^T(k+1)\boldsymbol{x}(k) \qquad \textit{a posteriori} \text{ error} \end{aligned} \tag{2.82}$$

The *a posteriori* error gives the error signal at the adaptive filter output when the current adaptive filter coefficients are replaced with the updated coefficients. Using the principle of minimum disturbance, a selective-partial-update adaptive filtering algorithm is derived from the solution of the following constrained optimization problem:

$$\min_{\boldsymbol{I}_M(k)} \min_{\boldsymbol{w}_{\mathcal{M}}(k+1)} \|\boldsymbol{w}_{\mathcal{M}}(k+1) - \boldsymbol{w}_{\mathcal{M}}(k))\|_2^2 \tag{2.83a}$$

subject to:

$$e_p(k) - \left(1 - \frac{\mu\|\boldsymbol{x}_{\mathcal{M}}(k)\|_2^2}{\epsilon + \|\boldsymbol{x}_{\mathcal{M}}(k)\|_2^2}\right) e(k) = 0 \tag{2.83b}$$

where ϵ is a small positive regularization constant which also serves the purpose of avoiding division by zero, $\boldsymbol{w}_{\mathcal{M}}(k)$ is an $M \times 1$ subvector of $\boldsymbol{w}(k)$ comprised of entries given by the set $\{w_j(k) : i_j(k) = 1\}$, and $\boldsymbol{x}_{\mathcal{M}}(k)$ is an $M \times 1$ subvector of $\boldsymbol{x}(k)$ defined similarly to $\boldsymbol{w}_{\mathcal{M}}(k)$. Here the $i_j(k)$ are the diagonal elements of the coefficient selection matrix $\boldsymbol{I}_M(k)$. Note that $\boldsymbol{w}^T(k)\boldsymbol{I}_M(k)\boldsymbol{w}(k) = \|\boldsymbol{w}_{\mathcal{M}}(k)\|_2^2$ and $\boldsymbol{x}^T(k)\boldsymbol{I}_M(k)\boldsymbol{x}(k) = \|\boldsymbol{x}_{\mathcal{M}}(k)\|_2^2$. Thus $\boldsymbol{w}_{\mathcal{M}}(k)$ is the subset of adaptive filter coefficients selected for update according to $\boldsymbol{I}_M(k)$. For example, if $N = 3$ and $M = 2$, $\boldsymbol{I}_M(k)$ and $\boldsymbol{w}_{\mathcal{M}}(k)$ are related as follows:

$$\begin{aligned}
\boldsymbol{I}_M(k) &= \mathrm{diag}(1,1,0), & \boldsymbol{w}_{\mathcal{M}}(k) &= [w_1(k), w_2(k)]^T \\
\boldsymbol{I}_M(k) &= \mathrm{diag}(1,0,1), & \boldsymbol{w}_{\mathcal{M}}(k) &= [w_1(k), w_3(k)]^T \\
\boldsymbol{I}_M(k) &= \mathrm{diag}(0,1,1), & \boldsymbol{w}_{\mathcal{M}}(k) &= [w_2(k), w_3(k)]^T.
\end{aligned}$$

Using this relationship between $\boldsymbol{I}_M(k)$ and $\boldsymbol{w}_{\mathcal{M}}(k)$, we have $\|\boldsymbol{w}_{\mathcal{M}}(k+1) - \boldsymbol{w}_{\mathcal{M}}(k))\|_2^2 = \|\boldsymbol{I}_M(k)(\boldsymbol{w}(k+1) - \boldsymbol{w}(k))\|_2^2$ in (2.83). Since only the coefficients identified by $\boldsymbol{I}_M(k)$ are updated, an implicit constraint of any partial-update technique, including (2.83), is that:

$$\boldsymbol{w}(k+1) - \boldsymbol{w}(k) = \boldsymbol{I}_M(k)(\boldsymbol{w}(k+1) - \boldsymbol{w}(k)) \tag{2.84}$$

We will have occasion to resort to this partial-update 'constraint' in the sequel.

The constrained optimization problem in (2.83) may be solved either by using the method of Lagrange multipliers as we shall see next or by formulating it as an underdetermined minimum-norm estimation problem. We will use the latter approach when we discuss the ℓ_1 and ℓ_∞-norm constrained optimization problems later in the section. Suppose that $\boldsymbol{I}_M(k)$ is fixed for the time being. Then we have:

$$\min_{\boldsymbol{w}_{\mathcal{M}}(k+1)} \|\boldsymbol{w}_{\mathcal{M}}(k+1) - \boldsymbol{w}_{\mathcal{M}}(k))\|_2^2 \tag{2.85a}$$

subject to:

$$e_p(k) - \left(1 - \frac{\mu \|\boldsymbol{x}_{\mathcal{M}}(k)\|_2^2}{\epsilon + \|\boldsymbol{x}_{\mathcal{M}}(k)\|_2^2}\right) e(k) = 0. \tag{2.85b}$$

The cost function to be minimized is:

$$\begin{aligned} J(\boldsymbol{w}_{\mathcal{M}}(k+1), \lambda) &= \|\boldsymbol{w}_{\mathcal{M}}(k+1) - \boldsymbol{w}_{\mathcal{M}}(k))\|_2^2 \\ &\quad + \lambda \left(e_p(k) - \left(1 - \frac{\mu \|\boldsymbol{x}_{\mathcal{M}}(k)\|_2^2}{\epsilon + \|\boldsymbol{x}_{\mathcal{M}}(k)\|_2^2}\right) e(k) \right) \end{aligned} \tag{2.86}$$

where λ is a Lagrange multiplier. Setting:

$$\frac{\partial J(\boldsymbol{w}_{\mathcal{M}}(k+1), \lambda)}{\partial \boldsymbol{w}_{\mathcal{M}}(k+1)} = \mathbf{0} \quad \text{and} \quad \frac{\partial J(\boldsymbol{w}_{\mathcal{M}}(k+1), \lambda)}{\partial \lambda} = 0 \tag{2.87}$$

yields:

$$\boldsymbol{w}_{\mathcal{M}}(k+1) - \boldsymbol{w}_{\mathcal{M}}(k) - \frac{\lambda}{2} \boldsymbol{x}_{\mathcal{M}}(k) = \mathbf{0} \tag{2.88a}$$

$$(\boldsymbol{w}(k+1) - \boldsymbol{w}(k))^T \boldsymbol{x}(k) - \mu e(k) \frac{\|\boldsymbol{x}_{\mathcal{M}}(k)\|_2^2}{\epsilon + \|\boldsymbol{x}_{\mathcal{M}}(k)\|_2^2} = 0 \tag{2.88b}$$

From (2.88a) we have:

$$\begin{aligned} (\boldsymbol{w}_{\mathcal{M}}(k+1) - \boldsymbol{w}_{\mathcal{M}}(k))^T \boldsymbol{x}_{\mathcal{M}}(k) &= \frac{\lambda}{2} \boldsymbol{x}_{\mathcal{M}}^T(k) \boldsymbol{x}_{\mathcal{M}}(k) && \text{(2.89a)} \\ &= (\boldsymbol{w}(k+1) - \boldsymbol{w}(k))^T \boldsymbol{x}(k) && \text{(2.89b)} \end{aligned}$$

since only the adaptive filter coefficients selected by $\boldsymbol{I}_M(k)$ are updated [see (2.84)]. Substituting (2.89) into (2.88b) yields:

$$\lambda = 2\mu e(k) \frac{1}{\epsilon + \|\boldsymbol{x}_{\mathcal{M}}(k)\|_2^2} \tag{2.90}$$

Using this result in (2.88a) gives the desired adaptation equation for fixed $\boldsymbol{I}_M(k)$:

$$\boldsymbol{w}_{\mathcal{M}}(k+1) = \boldsymbol{w}_{\mathcal{M}}(k) + \frac{\mu}{\epsilon + \|\boldsymbol{x}_{\mathcal{M}}(k)\|_2^2} e(k) \boldsymbol{x}_{\mathcal{M}}(k) \tag{2.91}$$

What remains to be done is to find $\boldsymbol{I}_M(k)$ that results in the minimum-norm coefficient update. This is done by solving:

$$\min_{\boldsymbol{I}_M(k)} \left\| \frac{\mu}{\epsilon + \|\boldsymbol{x}_{\mathcal{M}}(k)\|_2^2} e(k) \boldsymbol{x}_{\mathcal{M}}(k) \right\|_2^2 \tag{2.92}$$

which is equivalent to:

$$\max_{\boldsymbol{I}_M(k)} \|\boldsymbol{x}_{\mathcal{M}}(k)\|_2^2 + \frac{\epsilon^2}{\|\boldsymbol{x}_{\mathcal{M}}(k)\|_2^2} \tag{2.93}$$

If any M entries of the regressor vector were zero, then the solution would be:

$$\min_{\boldsymbol{I}_M(k)} \|\boldsymbol{x}_{\mathcal{M}}(k)\|_2^2 \tag{2.94}$$

which would select the adaptive filter coefficients corresponding to the M zero input samples. For such an input (2.83) would yield the trivial solution $\boldsymbol{w}_{\mathcal{M}}(k+1) = \boldsymbol{w}_{\mathcal{M}}(k)$. However, this solution is not what we wanted. We get zero update and therefore the adaptation process freezes while we could have selected some non-zero subset of the regressor vector yielding a non-zero update. This example highlights an important consideration when applying the principle of minimum disturbance, viz. the requirement to avoid the trivial solution by ruling out zero regressor subsets from consideration. Letting $\epsilon \to 0$ does the trick by replacing (2.93) with:

$$\max_{\boldsymbol{I}_M(k)} \|\boldsymbol{x}_{\mathcal{M}}(k)\|_2^2 \tag{2.95}$$

The solution to this selection criterion is identical to the coefficient selection matrix of the M-max LMS algorithm:

$$\boldsymbol{I}_M(k) = \begin{bmatrix} i_1(k) & 0 & \cdots & 0 \\ 0 & i_2(k) & \ddots & \vdots \\ \vdots & \ddots & \ddots & 0 \\ 0 & \cdots & 0 & i_N(k) \end{bmatrix},$$
$$i_j(k) = \begin{cases} 1 & \text{if } |x(k-j+1)| \in \max_{1 \le l \le N}(|x(k-l+1)|, M) \\ 0 & \text{otherwise} \end{cases} \tag{2.96}$$

If the step-size parameter μ in the optimization constraint obeys the inequality:

$$\left| 1 - \frac{\mu \|\boldsymbol{x}_{\mathcal{M}}(k)\|_2^2}{\epsilon + \|\boldsymbol{x}_{\mathcal{M}}(k)\|_2^2} \right| < 1 \tag{2.97}$$

then the *a posteriori* error magnitude is always smaller than the *a priori* error magnitude for non-zero $e(k)$. This condition is readily met if $0 < \mu < 2$ with $\mu = 1$ yielding $e_p(k) \approx 0$.

Combining (2.91) and (2.96) we obtain the solution to (2.83) as:

$$\boldsymbol{w}(k+1) = \boldsymbol{w}(k) + \frac{\mu}{\epsilon + \|\boldsymbol{I}_M(k)\boldsymbol{x}(k)\|_2^2} e(k)\boldsymbol{I}_M(k)\boldsymbol{x}(k) \tag{2.98}$$

This algorithm is known as the selective-partial-update NLMS algorithm. When $M = N$ (i.e. all coefficients are updated) it reduces to the conventional NLMS algorithm. The M-max NLMS algorithm is defined by:

$$\boldsymbol{w}(k+1) = \boldsymbol{w}(k) + \frac{\mu}{\epsilon + \|\boldsymbol{x}(k)\|_2^2} e(k)\boldsymbol{I}_M(k)\boldsymbol{x}(k) \tag{2.99}$$

The selective-partial-update NLMS algorithm differs from the M-max NLMS algorithm in the way the step-size parameter μ is normalized. The following discussion will show that the selective-partial-update NLMS algorithm is a natural extension of Newton's method.

2.6.2 Instantaneous approximation of Newton's method

The selective-partial-update NLMS can also be developed by considering an instantaneous approximation to Newton's method. Newton's method is a fast-converging iterative estimation method and for the quadratic cost function:

$$J(\mathbf{w}) = E\{(x^T(k)\mathbf{w} - d(k))^2\} \tag{2.100}$$

it is given by the recursion:

$$\boldsymbol{w}(k+1) = \boldsymbol{w}(k) + \mu \boldsymbol{R}^{-1}(\boldsymbol{p} - \boldsymbol{R}\boldsymbol{w}(k)), \quad k = 0, 1, 2, \ldots \tag{2.101}$$

Compared with the steepest-descent algorithm:

$$\boldsymbol{w}(k+1) = \boldsymbol{w}(k) + \mu(\boldsymbol{p} - \boldsymbol{R}\boldsymbol{w}(k)), \quad k = 0, 1, 2, \ldots \tag{2.102}$$

Newton's method uses a matrix step-size parameter $\mu \boldsymbol{R}^{-1}$. For $\mu = 1$ Newton's method converges after one iteration:

$$\boldsymbol{w}(1) = \boldsymbol{w}(0) + \boldsymbol{R}^{-1}(\boldsymbol{p} - \boldsymbol{R}\boldsymbol{w}(0)) \tag{2.103a}$$
$$= \boldsymbol{R}^{-1}\boldsymbol{p}. \tag{2.103b}$$

Thus the NLMS algorithm, which is an instantaneous approximation of Newton's method, is expected to converge faster than the LMS algorithm obtained from an instantaneous approximation of the steepest-descent algorithm. Partial updating of the adaptive filter coefficients replaces $\boldsymbol{R} = E\{\boldsymbol{x}(k)\boldsymbol{x}^T(k)\}$ with $\boldsymbol{R}_M = E\{\boldsymbol{I}_M(k)\boldsymbol{x}(k)\boldsymbol{x}^T(k)\}$ and $\boldsymbol{p} = E\{\boldsymbol{x}(k)d(k)\}$ with $\boldsymbol{p}_M = E\{\boldsymbol{I}_M(k)\boldsymbol{x}(k)d(k)\}$. In practical applications, Newton's method may require a small regularization parameter to improve the conditioning of $\boldsymbol{R}$ or $\boldsymbol{R}_M$. Thus a regularized partial-update version of Newton's method is given by:

$$\boldsymbol{w}(k+1) = \boldsymbol{w}(k) + \mu(\epsilon \boldsymbol{I} + \boldsymbol{R}_M)^{-1}(\boldsymbol{p}_M - \boldsymbol{R}_M \boldsymbol{w}(k)), \quad k = 0, 1, 2, \ldots \tag{2.104}$$

Here ϵ is a small positive regularization parameter.

The instantaneous approximation of the regularized partial-update Newton's method (2.104) is:

$$\boldsymbol{w}(k+1) = \boldsymbol{w}(k) + \mu(\epsilon \boldsymbol{I} + \boldsymbol{I}_M(k)\boldsymbol{x}(k)\boldsymbol{x}^T(k))^{-1} e(k)\boldsymbol{I}_M(k)\boldsymbol{x}(k), \quad k = 0, 1, 2, \ldots \tag{2.105}$$

where we simply stripped the expectation operator off the correlation matrix and cross-correlation vector. According to the matrix inversion lemma (Sayed, 2003), we have:

$$(\boldsymbol{A} + \boldsymbol{BCD})^{-1} = \boldsymbol{A}^{-1} - \boldsymbol{A}^{-1}\boldsymbol{B}(\boldsymbol{C}^{-1} + \boldsymbol{D}\boldsymbol{A}^{-1}\boldsymbol{B})^{-1}\boldsymbol{D}\boldsymbol{A}^{-1} \tag{2.106}$$

where all matrix inverses are assumed to exist. Applying the matrix inversion lemma to $(\epsilon \boldsymbol{I} + \boldsymbol{I}_M(k)\boldsymbol{x}(k)\boldsymbol{x}^T(k))^{-1}$ in (2.105) with $\boldsymbol{A} = \epsilon \boldsymbol{I}$, $\boldsymbol{B} = \boldsymbol{I}_M(k)\boldsymbol{x}(k)$, $\boldsymbol{C} = 1$ and $\boldsymbol{D} = \boldsymbol{x}^T(k)$, we obtain:

$$(\epsilon \boldsymbol{I} + \boldsymbol{I}_M(k)\boldsymbol{x}(k)\boldsymbol{x}^T(k))^{-1} = \frac{1}{\epsilon}\boldsymbol{I} - \frac{1}{\epsilon}\boldsymbol{I}\frac{\boldsymbol{I}_M(k)\boldsymbol{x}(k)\boldsymbol{x}^T(k)}{\epsilon + \boldsymbol{x}^T(k)\boldsymbol{I}_M(k)\boldsymbol{x}(k)} \tag{2.107}$$

and:

$$\begin{aligned}
(\epsilon \boldsymbol{I} &+ \boldsymbol{I}_M(k)\boldsymbol{x}(k)\boldsymbol{x}^T(k))^{-1}\boldsymbol{I}_M(k)\boldsymbol{x}(k) \\
&= \frac{1}{\epsilon}\boldsymbol{I}_M(k)\boldsymbol{x}(k)\left(1 - \frac{\boldsymbol{x}^T(k)\boldsymbol{I}_M(k)\boldsymbol{x}(k)}{\epsilon + \boldsymbol{x}^T(k)\boldsymbol{I}_M(k)\boldsymbol{x}(k)}\right) && \text{(2.108a)} \\
&= \frac{\boldsymbol{I}_M(k)\boldsymbol{x}(k)}{\epsilon + \boldsymbol{x}^T(k)\boldsymbol{I}_M(k)\boldsymbol{x}(k)} && \text{(2.108b)}
\end{aligned}$$

Substituting this into (2.105) yields:

$$\boldsymbol{w}_{\mathcal{M}}(k+1) = \boldsymbol{w}_{\mathcal{M}}(k) + \frac{\mu}{\epsilon + \|\boldsymbol{x}_{\mathcal{M}}(k)\|_2^2} e(k)\boldsymbol{x}_{\mathcal{M}}(k) \tag{2.109}$$

which is identical to (2.91). Incorporating the partial-update selection criterion (2.96) into the above recursion finally gives the selective-partial-update NLMS algorithm in (2.98).

2.6.3 q-Norm constrained optimization

It is possible to extend the method of selective partial updates to arbitrary vector norms different to the ℓ_2 norm (the Euclidean norm) used in (2.83). Consider the general q-norm constrained optimization problem:

$$\min_{\boldsymbol{w}(k+1)} \|\boldsymbol{w}(k+1) - \boldsymbol{w}(k))\|_q^2 \tag{2.110a}$$

subject to:

$$e_p(k) - \left(1 - \frac{\mu \|\boldsymbol{x}(k)\|_r^r}{\epsilon + \|\boldsymbol{x}(k)\|_r^r}\right) e(k) = 0 \tag{2.110b}$$

where $r = \frac{q}{q-1}$ and:

$$\|\boldsymbol{x}(k)\|_r = \left(|x_1(k)|^r + |x_2(k)|^r + \cdots + |x_N(k)|^r\right)^{1/r} \tag{2.111}$$

Here we use the simplified notation for the regressor vector entries given in (2.69). The constraint in (2.110b) can be rewritten as:

$$\boldsymbol{x}^T(k)\, \delta\boldsymbol{w}(k+1) = \frac{\mu \|\boldsymbol{x}(k)\|_r^r}{\epsilon + \|\boldsymbol{x}(k)\|_r^r} e(k) \tag{2.112}$$

where $\delta\boldsymbol{w}(k+1) = \boldsymbol{w}(k+1) - \boldsymbol{w}(k)$. Thus the constrained optimization problem in (2.110) is equivalent to finding the minimum ℓ_q-norm solution $\delta\boldsymbol{w}(k+1)$ to the under-determined linear equation in (2.112).

The minimum ℓ_q-norm solution for $1 \leq r < \infty$ is given by (Douglas, 1994)

$$\delta\boldsymbol{w}(k+1) = \frac{\mu}{\epsilon + \|\boldsymbol{x}(k)\|_r^r} e(k)\boldsymbol{f}_r(k) \tag{2.113a}$$

or:

$$\boldsymbol{w}(k+1) = \boldsymbol{w}(k) + \frac{\mu}{\epsilon + \|\boldsymbol{x}(k)\|_r^r} e(k)\boldsymbol{f}_r(k) \tag{2.113b}$$

where:

$$\boldsymbol{f}_r(k) = \begin{bmatrix} \text{sign}(x_1(k))|x_1(k)|^{r-1} \\ \text{sign}(x_2(k))|x_2(k)|^{r-1} \\ \vdots \\ \text{sign}(x_N(k))|x_N(k)|^{r-1} \end{bmatrix} \tag{2.114}$$

The signum function sign(·) is defined by:

$$\text{sign}(x) = \begin{cases} 1 & \text{if } x > 0 \\ 0 & \text{if } x = 0 \\ -1 & \text{if } x < 0 \end{cases} \tag{2.115}$$

The minimum ℓ_1-norm solution to (2.110) takes the following simple form (Douglas, 1994)

$$\begin{aligned} w_i(k+1) &= w_i(k) + f_i(k), \quad i = 1, \ldots, N, \\ f_i(k) &= \begin{cases} \dfrac{\mu}{x_i(k)} e(k) & \text{if } i = \arg\max_j |x_j(k)| \\ 0 & \text{otherwise} \end{cases} \end{aligned} \tag{2.116}$$

which is referred to as the *max-NLMS algorithm* (Douglas, 1995). In deriving the max-NLMS algorithm we assumed no regularization (i.e. $\epsilon = 0$). Here the ℓ_∞ norm is defined by:

$$\|\boldsymbol{x}(k)\|_\infty = \max_i |x_i(k)| \tag{2.117}$$

An important feature of the max-NLMS algorithm is that it employs partial coefficient updates. Indeed, comparison of (2.116) with the selective-partial-update NLMS algorithm in (2.98), reveals that the max-NLMS algorithm is a special case of the selective-partial-update NLMS algorithm with $M = 1$ and $\epsilon = 0$. This is easily seen by noting that for $M = 1$ we have:

$$\|\boldsymbol{I}_M(k)\boldsymbol{x}(k)\|_2^2 = x_m^2(k), \quad m = \arg\max_i |x_i(k)| \tag{2.118a}$$

$$\boldsymbol{I}_M(k)\boldsymbol{x}(k) = \begin{bmatrix} i_1(k) & 0 & \cdots & 0 \\ 0 & i_2(k) & \ddots & \vdots \\ \vdots & \ddots & \ddots & 0 \\ 0 & \cdots & 0 & i_N(k) \end{bmatrix} \boldsymbol{x}(k), \quad i_j(k) = \begin{cases} 1 & \text{if } j = m \\ 0 & \text{otherwise} \end{cases} \tag{2.118b}$$

The above relations reduce the selective-partial-update NLMS algorithm in (2.98) to the max-NLMS algorithm in (2.116).

Another important class of adaptive filtering algorithms with selective partial updates is given by the solution of the following minimum ℓ_∞-norm constrained optimization problem:

$$\min_{\boldsymbol{w}_{\mathcal{M}}(k+1)} \|\boldsymbol{w}_{\mathcal{M}}(k+1) - \boldsymbol{w}_{\mathcal{M}}(k)\|_\infty^2 \tag{2.119a}$$

subject to:

$$e_p(k) - \left(1 - \frac{\mu \|\boldsymbol{x}_{\mathcal{M}}(k)\|_1}{\epsilon + \|\boldsymbol{x}_{\mathcal{M}}(k)\|_1}\right) e(k) = 0 \tag{2.119b}$$

For a given $\boldsymbol{I}_M(k)$, Eq. (2.119b) can be expressed as:

$$\boldsymbol{x}_{\mathcal{M}}^T(k)\, \delta \boldsymbol{w}_{\mathcal{M}}(k+1) = \frac{\mu \|\boldsymbol{x}_{\mathcal{M}}(k)\|_1}{\epsilon + \|\boldsymbol{x}_{\mathcal{M}}(k)\|_1} e(k) \tag{2.120}$$

According to (2.113), the minimum ℓ_∞-norm solution $\delta \boldsymbol{w}_{\mathcal{M}}(k+1)$, which also solves (2.119), is given by:

$$\boldsymbol{w}_{\mathcal{M}}(k+1) = \boldsymbol{w}_{\mathcal{M}}(k) + \frac{\mu}{\epsilon + \|\boldsymbol{x}_{\mathcal{M}}(k)\|_1} e(k) \operatorname{sign}(\boldsymbol{x}_{\mathcal{M}}(k)) \tag{2.121}$$

where $\operatorname{sign}(\boldsymbol{x}(k)) = [\operatorname{sign}(x_1(k)), \operatorname{sign}(x_2(k)), \ldots, \operatorname{sign}(x_N(k))]^T$.

The optimum coefficient subset has the minimum Euclidean norm and is obtained from:

$$\min_{\boldsymbol{I}_M(k)} \|\delta \boldsymbol{w}_{\mathcal{M}}(k+1)\|_2^2 \tag{2.122}$$

This optimization problem can be equivalently written as:

$$\min_{\boldsymbol{I}_M(k)} \left\| \frac{\operatorname{sign}(\boldsymbol{x}_{\mathcal{M}}(k))}{\epsilon + \|\boldsymbol{x}_{\mathcal{M}}(k)\|_1} \right\|_2^2 \tag{2.123a}$$

$$\min_{\boldsymbol{I}_M(k)} \frac{M}{(\epsilon + \|\boldsymbol{x}_{\mathcal{M}}(k)\|_1)^2} \tag{2.123b}$$

$$\max_{\boldsymbol{I}_M(k)} \|\boldsymbol{x}_{\mathcal{M}}(k)\|_1^2 \tag{2.123c}$$

$$\max_{\boldsymbol{I}_M(k)} \|\boldsymbol{x}_{\mathcal{M}}(k)\|_2^2 \tag{2.123d}$$

where we assume that $x_i(k) \neq 0$, $i = 1, \ldots, N$. The remarks made about zero regressor subvectors following (2.93) also apply here. The coefficient selection criterion given above is exactly the same as that for M-max and selective partial updates. Therefore the coefficient selection matrix $\boldsymbol{I}_M(k)$ is given by (2.96). Incorporating the optimum coefficient selection matrix $\boldsymbol{I}_M(k)$ into (2.121), we obtain:

$$\boldsymbol{w}(k+1) = \boldsymbol{w}(k) + \frac{\mu}{\epsilon + \|\boldsymbol{I}_M(k)\boldsymbol{x}(k)\|_1} e(k)\boldsymbol{I}_M(k)\text{sign}(\boldsymbol{x}(k)) \qquad (2.124)$$

which is referred to as the *selective-partial-update sign-NLMS algorithm*. For $M = N$, the selective-partial-update sign-NLMS algorithm reduces to the conventional sign-NLMS algorithm (Nagumo and Noda, 1967). Again we note that the selective-partial-update sign-NLMS algorithm and the M-max sign-NLMS algorithm are different with respect to the normalization term they use. The chief advantage of the sign-NLMS algorithm is its reduced update complexity thanks to the use of signum operator and ℓ_1 norm.

2.6.4 Averaged system

The averaged system approximating the selective-partial-update NLMS for sufficiently small μ is:

$$\begin{aligned}\boldsymbol{w}^a(k+1) &= \boldsymbol{w}^a(k) + \mu E\left\{\frac{1}{\epsilon + \boldsymbol{x}^T(k)\boldsymbol{I}_M(k)\boldsymbol{x}(k)} e(k)\boldsymbol{I}_M(k)\boldsymbol{x}(k)\right\} \\ &= \boldsymbol{w}^a(k) + \mu(\boldsymbol{\varphi}_M - \boldsymbol{\Phi}_M \boldsymbol{w}^a(k)) \qquad (2.125)\end{aligned}$$

where:

$$\begin{aligned}\boldsymbol{\Phi}_M &= E\left\{\frac{\boldsymbol{I}_M(k)\boldsymbol{x}(k)\boldsymbol{x}^T(k)}{\epsilon + \boldsymbol{x}^T(k)\boldsymbol{I}_M(k)\boldsymbol{x}(k)}\right\} \\ &= E\{(\epsilon\boldsymbol{I} + \boldsymbol{I}_M(k)\boldsymbol{x}(k)\boldsymbol{x}^T(k))^{-1}\boldsymbol{I}_M(k)\boldsymbol{x}(k)\boldsymbol{x}^T(k)\} \qquad (2.126a)\\ \boldsymbol{\varphi}_M &= E\left\{\frac{\boldsymbol{I}_M(k)\boldsymbol{x}(k)d(k)}{\epsilon + \boldsymbol{x}^T(k)\boldsymbol{I}_M(k)\boldsymbol{x}(k)}\right\} \\ &= E\{(\epsilon\boldsymbol{I} + \boldsymbol{I}_M(k)\boldsymbol{x}(k)\boldsymbol{x}^T(k))^{-1}\boldsymbol{I}_M(k)\boldsymbol{x}(k)d(k)\} \qquad (2.126b)\end{aligned}$$

The corresponding $\boldsymbol{\Phi}_M$ and $\boldsymbol{\varphi}_M$ for the conventional NLMS algorithm is obtained by setting $M = N$. For M-max updates $\boldsymbol{\Phi}_M$ and $\boldsymbol{\varphi}_M$ are modified to:

$$\boldsymbol{\Phi}_M = E\left\{\frac{\boldsymbol{I}_M(k)\boldsymbol{x}(k)\boldsymbol{x}^T(k)}{\epsilon + \boldsymbol{x}^T(k)\boldsymbol{x}(k)}\right\} \qquad (2.127a)$$

$$\boldsymbol{\varphi}_M = E\left\{\frac{\boldsymbol{I}_M(k)\boldsymbol{x}(k)d(k)}{\epsilon + \boldsymbol{x}^T(k)\boldsymbol{x}(k)}\right\} \qquad (2.127b)$$

2.6.5 Example 1: Eigenanalysis

The eigenanalysis of the correlation matrix $\mathbf{\Phi}_M$ reveals the convergence properties of the selective-partial-update NLMS. In Figures 2.22 and 2.23 the eigenvalue spread and minimum eigenvalues of $\mathbf{\Phi}_M$ are shown for the selective-partial-update and M-max NLMS algorithms. The input signal is a Gaussian AR(1) process with the AR parameter a set to 0 (white input), -0.5 and -0.9 to generate increasingly correlated input signals. The adaptive filter has $N = 5$ coefficients, and the regularization parameter is $\epsilon = 10^{-5}$. In Figures 2.22 and 2.23, $M = 5$ corresponds to the conventional (full-update) NLMS algorithm. Figure 2.22 shows that for the input signals considered the method of M-max updates results in a smaller eigenvalue spread than selective partial updates except for the white input. In Figure 2.23(a) we observe an interesting property of selective partial updates. For white input the minimum eigenvalue of $\mathbf{\Phi}_M$ is the same as that for the full-update NLMS irrespective of M. In other words, the eigenvalues of $\mathbf{\Phi}_M$ for selective partial updates are identical to those for full coefficient updates. Thus we expect the same convergence rate for selective partial updates as the full-update NLMS. The method of M-max updates does not share this property. In fact, for white input its convergence rate slows down with decreasing M. Figure 2.23(b) and (c) shows that for the correlated input signals considered, the selective-partial-update NLMS always outperforms the conventional NLMS for $M < 5$ because it has a larger minimum eigenvalue. Another interesting property of selective partial updates is that the minimum eigenvalue of $\mathbf{\Phi}_M$ increases with decreasing M and $M = 1$ gives the fastest convergence for the AR(1) inputs considered. Obviously, all these conclusions are subject to the caveat of having a sufficiently small step-size parameter. Therefore, there is no guarantee that they will carry over to large step-size parameters.

2.6.6 Example 2: Convergence performance

We use the system identification problem in Section 2.3.1 to demonstrate convergence properties of the selective-partial-update NLMS algorithm. The input signal $x(k)$ is a Gaussian AR(1) process. The step-size parameter is set to $\mu = 0.1$. For $a = 0$ (white Gaussian input), $a = -0.5$ and $a = -0.9$, the estimated learning curves of the full-update NLMS and the selective-partial-update NLMS are shown in Figure 2.24. To obtain smooth MSE curves, a combination of ensemble and time averaging has been employed. We observe that the convergence behaviour of the selective-partial-update NLMS algorithm is, in most cases, accurately predicted by the convergence analysis based on the eigenanalysis of $\mathbf{\Phi}_M$ (see Figures 2.22 and 2.23). However, for $M = 1$ the steady-state MSE is slightly larger than the other selective-partial-update implementations with $M > 1$. This suggests that the step-size for $M = 1$ should be decreased to bring its steady-state MSE to the same level as other cases. Reducing the step-size parameter would slow down the convergence rate. Therefore, we should not read too much into the seemingly fast convergence rate obtained for $M = 1$.

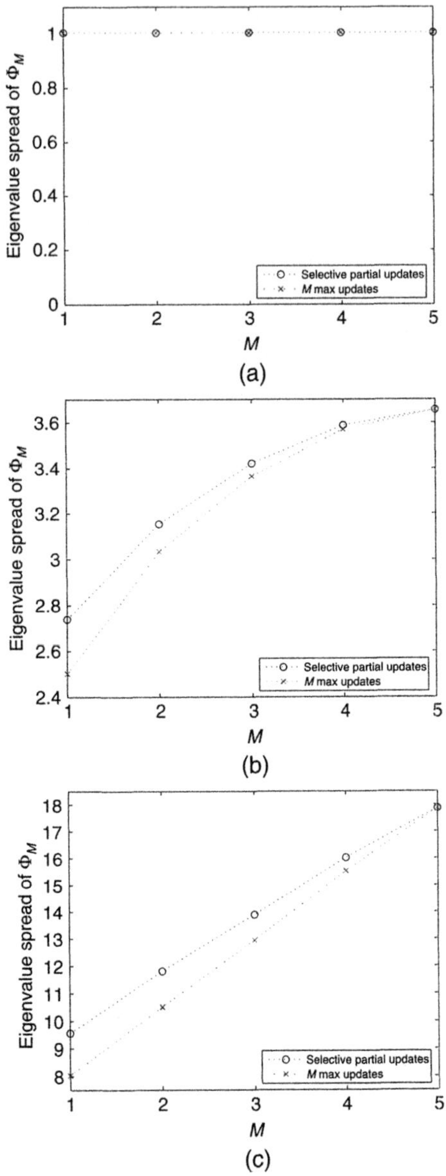

Figure 2.22 Eigenvalue spread of $\mathbf{\Phi}_M$ versus M for: (a) $a = 0$ (white Gaussian input); (b) $a = -0.5$; and (c) $a = -0.9$.

2.6.7 Example 3: Instability

For certain input signals the method of selective partial updates can suffer from slow convergence or even become unstable, no matter how small μ is. The convergence

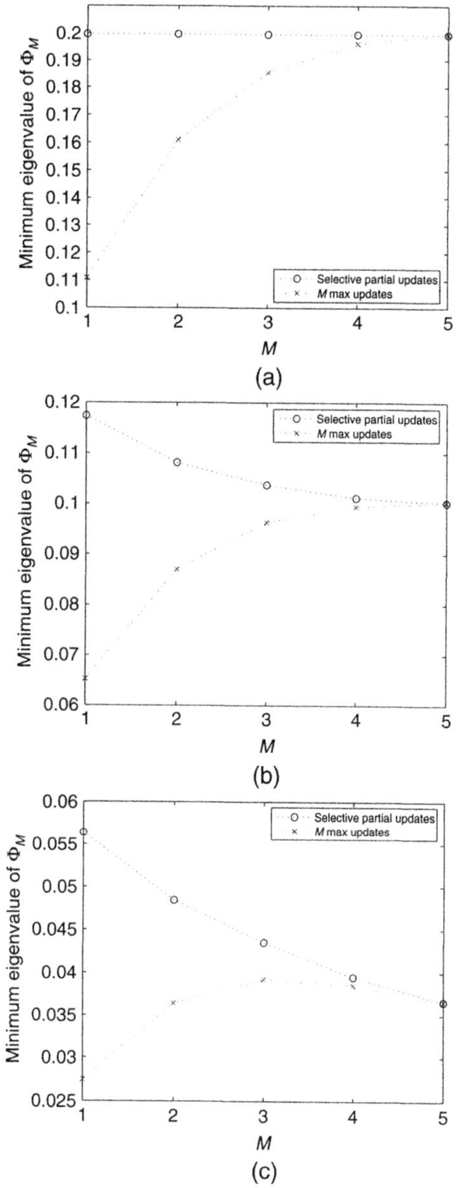

Figure 2.23 Minimum eigenvalue of $\mathbf{\Phi}_M$ versus M for: (a) $a = 0$ (white Gaussian input); (b) $a = -0.5$; and (c) $a = -0.9$.

problems arise if the input signal causes some eigenvalues of the correlation matrix $\mathbf{\Phi}_M$ to be almost zero or negative. The cyclostationary and periodic input signals illustrated in Figures 2.20 and 2.21 are two such signals. In this simulation example we illustrate the instability of both selective partial updates and M-max updates to emphasize the fact that the M-max NLMS algorithm is not immune from stability problems. The estimated learning curves of the full-update, M-max and selective-partial-update NLMS algorithms for $M = 1$ are shown in Figure 2.25. The step-size parameter is set to $\mu = 0.1$ for the cyclostationary input and to $\mu = 0.01$ for the deterministic periodic input.

2.7 SET MEMBERSHIP PARTIAL UPDATES

The method of set membership partial updates is an extension of selective partial updates employing a specific time-varying normalized step-size parameter that keeps the *a posteriori* error magnitude bounded. Similar to selective partial updates, it also requires insight into the adaptive filtering problem at hand and is best utilized when designing a partial-update filter from scratch. We first provide a brief review of set membership filtering (Gollamudi *et al.*, 1998) and then describe the method of set membership partial updates in the context of minimum disturbance principle.

Consider a linear adaptive filtering problem with the set of adaptive filter coefficients defined by:

$$\mathcal{H}_k = \{\boldsymbol{w} \in \mathbb{R}^N : |d(k) - \boldsymbol{w}^T \boldsymbol{x}(k)| \leq \gamma\} \tag{2.128}$$

which is referred to as the constraint set. This set contains all adaptive filter coefficient vectors for which the *a priori* error $e(k)$ is bounded by γ in magnitude at time k. Taking the intersection of constraint sets up to time instant k, we obtain:

$$\psi_k = \bigcap_{i=0}^{k} \mathcal{H}_i = \{\boldsymbol{w} \in \mathbb{R}^N : |d(i) - \boldsymbol{w}^T \boldsymbol{x}(i)| \leq \gamma,\ i = 0, \ldots, k\} \tag{2.129}$$

which is called the exact membership set. In the limit as $k \to \infty$, ψ_k converges to the feasibility set:

$$\Theta = \{\boldsymbol{w} \in \mathbb{R}^N : |d(k) - \boldsymbol{w}^T \boldsymbol{x}(k)| \leq \gamma \ \forall k\} \tag{2.130}$$

This set contains all possible adaptive filter coefficients that guarantee $|e(k)|$ to be bounded by γ for any input signal and corresponding desired response. Central to set membership adaptive filtering is the concept of finding an estimate of adaptive filter coefficients that is a member of the feasibility set for a given error bound γ. Several adaptive techniques are available to solve this estimation problem.

The method of set membership partial updates employs a modified version of the principle of minimum disturbance that underpins the derivation of selective partial updates (Werner *et al.*, 2004). The principle of minimum disturbance requires the

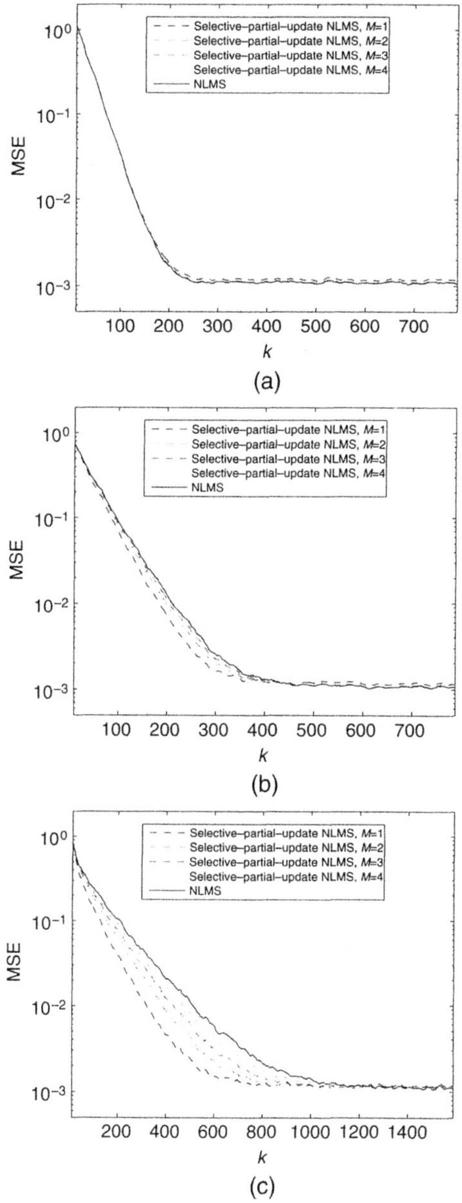

Figure 2.24 MSE curves for full-update and selective-partial-update NLMS algorithms for Gaussian AR(1) input with: (a) $a = 0$ (white); (b) $a = -0.5$; and (c) $a = -0.9$.

updated subset of adaptive filter coefficients to meet an equality constraint which aims to reduce the magnitude of the *a posteriori* error $e_p(k)$ with respect to the *a*

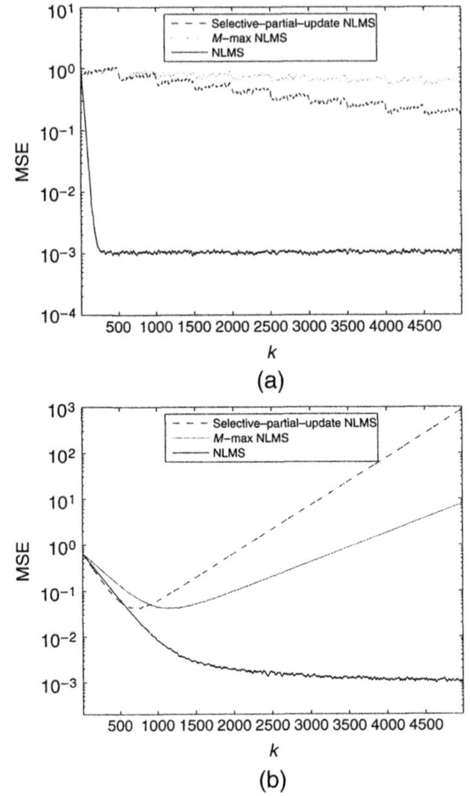

Figure 2.25 MSE curves for full-update, M-max and selective-partial-update NLMS algorithms for: (a) a cyclostationary input; and (b) a periodic input signal for which $\mathbf{\Phi}_M$ has a negative eigenvalue.

priori error $e(k)$. In set membership partial updates the equality constraint is relaxed by replacing it with an inequality constraint imposing a bound on the *a posteriori* error. Suppose that the adaptive filter coefficient vector $\boldsymbol{w}(k)$ is not a member of the constraint set $\mathcal{H}_k$, i.e., $|e(k)| > \gamma$. Then the method of set membership partial updates aims to find the updated coefficients $\boldsymbol{w}(k+1)$ satisfying the following requirements:

- $\boldsymbol{w}(k+1)$ differs from $\boldsymbol{w}(k)$ in at most M coefficients, i.e., it obeys (2.84)
- $\boldsymbol{w}(k+1)$ is a member of $\mathcal{H}_k$, i.e., $|d(k) - \boldsymbol{w}^T(k+1)\boldsymbol{x}(k)| \leq \gamma$
- $\|\boldsymbol{w}_{\mathcal{M}}(k+1) - \boldsymbol{w}_{\mathcal{M}}(k))\|_2^2$ is minimized over $\boldsymbol{I}_M(k)$

More formally this can be recast as a constrained optimization problem:

$$\min_{\boldsymbol{I}_M(k)} \min_{\boldsymbol{w}_{\mathcal{M}}(k+1)} \|\boldsymbol{w}_{\mathcal{M}}(k+1) - \boldsymbol{w}_{\mathcal{M}}(k))\|_2^2 \qquad (2.131a)$$

subject to:

$$e_p(k) - \left(1 - \alpha(k)\frac{\|\boldsymbol{x}_{\mathcal{M}}(k)\|_2^2}{\epsilon + \|\boldsymbol{x}_{\mathcal{M}}(k)\|_2^2}\right) e(k) = 0 \tag{2.131b}$$

where $\alpha(k)$ is a time-varying step-size defined by:

$$\alpha(k) = \begin{cases} \left(1 - \frac{\gamma}{|e(k)|}\right)\left(1 + \frac{\epsilon}{\|\boldsymbol{x}_{\mathcal{M}}(k)\|_2^2}\right) & \text{if } \boldsymbol{w}(k) \notin \mathcal{H}_k \text{ , i.e., } |e(k)| > \gamma \\ 0 & \text{otherwise} \end{cases} \tag{2.132}$$

As we have seen in the previous section, the solution to (2.131) is given by:

$$\boldsymbol{w}(k+1) = \boldsymbol{w}(k) + \frac{\alpha(k)}{\epsilon + \|\boldsymbol{I}_M(k)\boldsymbol{x}(k)\|_2^2} e(k)\boldsymbol{I}_M(k)\boldsymbol{x}(k) \tag{2.133}$$

where $\boldsymbol{I}_M(k)$ is defined in (2.96). We refer to (2.133) as the *set-membership partial-update NLMS algorithm*. This algorithm may be considered a time-varying step-size version of the selective-partial-update NLMS with a step-size given by (2.132). If $\boldsymbol{w}(k)$ is in $\mathcal{H}(k)$, then $\boldsymbol{w}(k+1) = \boldsymbol{w}(k)$ and $e_p(k) = e(k)$. Otherwise, $\boldsymbol{w}(k)$ is projected to the closest boundary of $\mathcal{H}_k$ to obtain:

$$e_p(k) = \gamma \frac{e(k)}{|e(k)|}. \tag{2.134}$$

If we set $\gamma = 0$, the constraint in (2.131b) reduces to $e_p(k) = 0$. Solving (2.131b) for this constraint leads to the selective-partial-update NLMS algorithm with $\epsilon = 0$ and $\mu = 1$. Thus for $\gamma = 0$, the set-membership partial-update NLMS algorithm reduces to the selective-partial-update NLMS algorithm with $\epsilon = 0$ and $\mu = 1$. On the other hand, replacing $\alpha(k)$ with the constant step-size μ, we go back to the selective-partial-update NLMS algorithm.

For small M stability problems may arise if $\alpha(k)$ falls outside the stability bound for the selective-partial-update NLMS algorithm. Therefore an appropriate selection of the threshold γ is an important design consideration. Methods for resolving stability problems were proposed in (Werner *et al.*, 2004), based on adjusting γ at each iteration.

An obvious advantage of set membership filtering is its sparse coefficient updates akin to periodic partial updates. However, the updates often occur at irregular intervals and cannot be predicted ahead of time. Consequently, the sparse updates of set membership filtering do not lead to any complexity reduction as the hardware needs to be equipped with sufficient resources to be able to update the entire coefficient set at any iteration. However, the average computational complexity can be relatively small as a result of sparse updates. This in turn has the potential to reduce the power consumption. Compared with the NLMS algorithm, the conventional set membership NLMS algorithm has some performance advantages (Gollamudi *et al.*, 1998). Combining set membership filtering with selective partial updates therefore leads to desirable performance improvement.

2.7.1 Example 1: Convergence performance

We use the system identification problem in Section 2.3.1 to simulate the set-membership partial-update NLMS algorithm. The input signal $x(k)$ is a Gaussian AR(1) process. The bound on the magnitude of *a posteriori* error is set to $\gamma = 0.01$ for $2 \leq M \leq 5$ and to $\gamma = 0.1$ for $M = 1$. A larger error bound is required for $M = 1$ to ensure stability. Referring to (2.132), we see that γ in effect controls the convergence rate of the set membership partial-update NLMS algorithm. In general, the larger γ, the slower the convergence rate will be. For $a = 0$ (white Gaussian input), $a = -0.5$ and $a = -0.9$, the learning curves of the NLMS, set-membership NLMS and set-membership partial-update NLMS algorithms are shown in Figure 2.26 for comparison purposes. We observe that the set-membership algorithms achieve a smaller steady-state MSE than the NLMS algorithm.

2.7.2 Example 2: Instability

Similar to the method of selective partial updates, the method of set membership partial updates is vulnerable to convergence difficulties for certain cyclostationary and periodic input signals, no matter how large γ is. The learning curves of the set-membership NLMS and set-membership partial-update NLMS algorithms are shown in Figure 2.27 for the two input signals considered in Figures 2.20 and 2.21. Only the case of $M = 1$ is illustrated. The error bound is set to $\gamma = 1$ for the cyclostationary input and to $\gamma = 0.2$ for the deterministic periodic input.

2.8 BLOCK PARTIAL UPDATES

The block partial updates method is another data-dependent partial-update technique in which the adaptive filter coefficients are partitioned into subvectors or blocks, in a way similar to sequential partial updates, and coefficient selection is performed over these subvectors (Schertler, 1998). In block partial updates the coefficient vector and the corresponding regressor vector are partitioned as:

$$\boldsymbol{w}(k) = \begin{bmatrix} w_1(k) \\ w_2(k) \\ \vdots \\ w_N(k) \end{bmatrix} = \begin{bmatrix} \boldsymbol{w}_1(k) \\ \boldsymbol{w}_2(k) \\ \vdots \\ \boldsymbol{w}_{N_b}(k) \end{bmatrix},$$

$$\boldsymbol{x}(k) = \begin{bmatrix} x(k) \\ x(k-1) \\ \vdots \\ x(k-N+1) \end{bmatrix} = \begin{bmatrix} \boldsymbol{x}_1(k) \\ \boldsymbol{x}_2(k) \\ \vdots \\ \boldsymbol{x}_{N_b}(k) \end{bmatrix} \tag{2.135}$$

where N_b is the number of blocks (subvectors) with each block given by:

$$\boldsymbol{w}_i(k) = \begin{bmatrix} w_{(i-1)L_b+1}(k) \\ w_{(i-1)L_b+2}(k) \\ \vdots \\ w_{iL_b}(k) \end{bmatrix}_{L_b \times 1}, \quad \boldsymbol{x}_i(k) = \begin{bmatrix} x(k-(i-1)L_b) \\ x(k-(i-1)L_b-1) \\ \vdots \\ x(k-iL_b+1) \end{bmatrix}_{L_b \times 1},$$
$$i = 1, \ldots, N_b \tag{2.136}$$

Here L_b is the block size, and it is assumed that $N/N_b = L_b$.

After partitioning the coefficient and regressor vectors as described above, data-dependent partial-update techniques can be applied to the partitioned vectors by replacing magnitude of vector entries with Euclidean norm of vector partitions in ranking operations. For example, the block-partial-update version of the selective-partial-update NLMS algorithm becomes:

$$\boldsymbol{w}(k+1) = \boldsymbol{w}(k) + \frac{\mu}{\epsilon + \|\boldsymbol{I}_M(k)\boldsymbol{x}(k)\|_2^2} e(k)\boldsymbol{I}_M(k)\boldsymbol{x}(k) \tag{2.137}$$

where:

$$\boldsymbol{I}_M(k) = \begin{bmatrix} \boldsymbol{i}_1(k) & & & \boldsymbol{0} \\ & \boldsymbol{i}_2(k) & & \\ & & \ddots & \\ \boldsymbol{0} & & & \boldsymbol{i}_{N_b}(k) \end{bmatrix},$$

$$\boldsymbol{i}_j(k) = \begin{cases} \boldsymbol{I}_{L_b} & \text{if } \|\boldsymbol{x}_j(k)\|_2^2 \in \max_{1 \le l \le N_b}(\|\boldsymbol{x}_l(k)\|_2^2, M) \\ \boldsymbol{0}_{L_b} & \text{otherwise} \end{cases} \tag{2.138}$$

The matrices $\boldsymbol{I}_{L_b}$ and $\boldsymbol{0}_{L_b}$ are $L_b \times L_b$ identity and zero matrices, respectively.

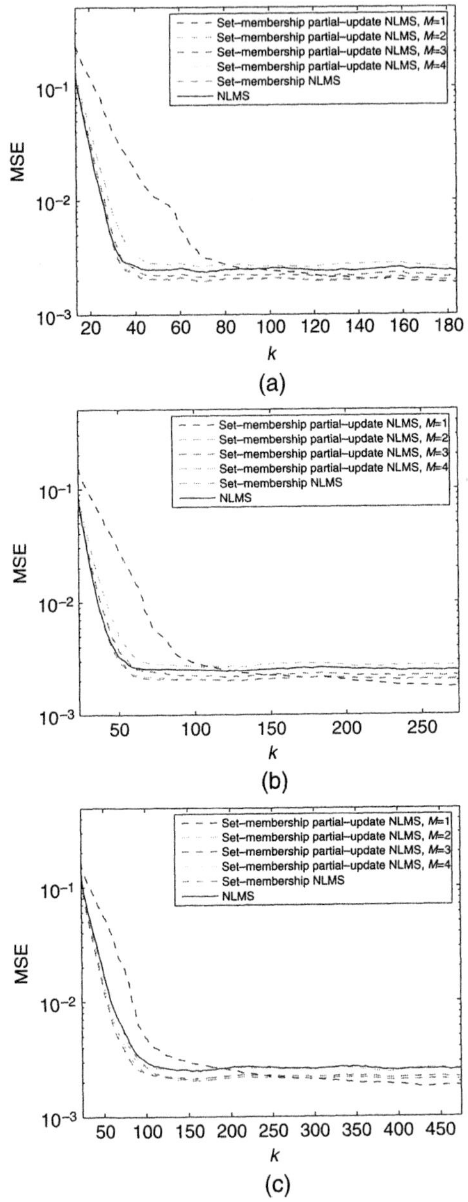

Figure 2.26 MSE curves for full-update and set-membership partial-update NLMS algorithms for Gaussian AR(1) input with: (a) $a = 0$ (white); (b) $a = -0.5$; and (c) $a = -0.9$.

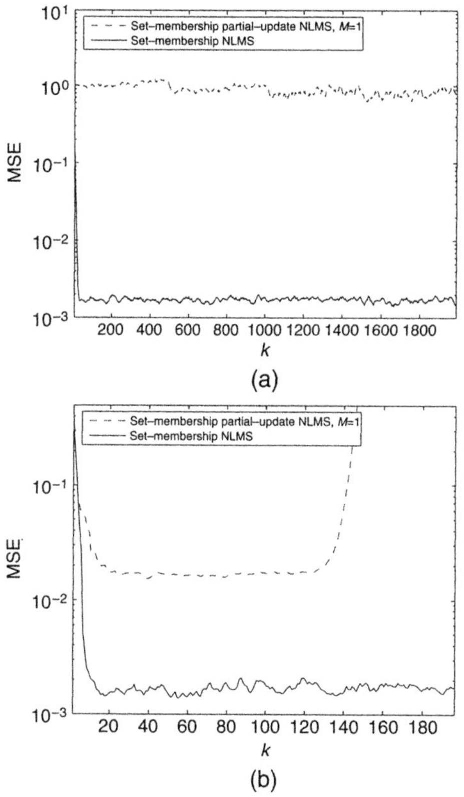

Figure 2.27 MSE curves for set-membership partial-update NLMS algorithms for: (a) a cyclostationary input; and (b) a periodic input signal.

The main motivation behind block partial updates is to reduce memory requirements associated with the sortline algorithm (see Appendix A). The partitioned coefficient vector reduces the length of data to be ranked. On the other hand, the shift structure exploited by sortline is lost in the process of partitioning the regressor vector. As a result, the heapsort algorithm is used to rank the data. The memory requirements of the ranking operation can also be reduced by using the heapsort algorithm without applying block partial updates. The downside of this approach is the increased comparison operations.

The use of block partial updates in adaptive filters with data-dependent coefficient selection usually results in slower convergence as a result of 'quantization' of coefficient selection. This so-called quantization effect gets worse as the block size L_b increases.

2.9 COMPLEXITY CONSIDERATIONS

The computational complexity analysis usually focuses on the adaptation complexity since filtering is part of the adaptation process (to compute the output error) and its complexity is independent of the adaptation process. For data-dependent partial update techniques (i.e. M-max, selective partial updates and set membership partial updates), the additional complexity overhead associated with sorting of the magnitude of regressor entries is very much dependent on the sorting algorithm used. An in-depth discussion of adaptation complexity for individual partial-update adaptive filters is provided in Chapter 4.

In most cases, practical implementation of adaptive filters will require further complexity reduction by way of approximation. For example, the division operation places a severe demand on the computational resources of digital signal processing hardware because most digital signal processors are optimized for linear filtering operations. To get around this, the division operation is often replaced by a series expansion or table look-up. The multiply operation can also be simplified by approximating constant parameters with powers of two, wherever possible, in order to substitute computationally inexpensive shift operations for multiply operations.

The filtering complexity of an adaptive filter can be significantly reduced by resorting to fast convolution methods based on FFT such as overlap-add and overlap-save methods. FFT-based block adaptive filters also provide reduced filtering complexity albeit at the expense of sometimes significant bulk delay.

Chapter | three

Convergence and stability analysis

3.1 INTRODUCTION

In Chapter 2 we utilized averaging theory to gain insight into the stability and convergence performance of different partial-update techniques. While averaging theory allows significant simplification in the analysis of adaptive filters, its application is only limited to cases where the step-size parameter is very small. Stochastic gradient algorithms derived from an instantaneous approximation of the steepest descent algorithm and its variants behave differently to their averaged approximations as the step-size gets larger. This difference in convergence behaviour is attributed to gradient noise caused by removal of the expectation operator.

This chapter provides an in-depth coverage of steady-state and convergence (transient) analysis of several partial-update LMS and NLMS algorithms. Exact performance analysis of adaptive filters is challenging. The same is also true for partial-update adaptive filters. Simplifying assumptions are often necessary to obtain closed-form analysis expressions. The chapter mostly draws on the concept of energy conservation (Sayed, 2003), which provides a unifying framework for convergence and stability analysis. The assumptions made by energy conservation arguments are not as stringent as independence assumptions traditionally employed in the convergence analysis of stochastic gradient algorithms.

3.2 CONVERGENCE PERFORMANCE

Consider the task of estimating a system from its input and noisy output observations $x(k)$, $d(k)$. The optimum solution in the sense of minimizing the MSE is given by:

$$\begin{aligned} \boldsymbol{w}_o &= \arg\min_{\boldsymbol{w}} E\{(d(k) - \boldsymbol{x}^T(k)\boldsymbol{w})^2\} \\ &= \boldsymbol{R}^{-1}\boldsymbol{p} \end{aligned} \tag{3.1}$$

where $\boldsymbol{R} = E\{\boldsymbol{x}(k)\boldsymbol{x}^T(k)\}$ is the input autocorrelation matrix and $\boldsymbol{p} = E\{\boldsymbol{x}(k)d(k)\}$ is the cross-correlation vector between the input signal and the desired response. The desired response (noisy system output) is related to the input via:

$$d(k) = \boldsymbol{x}^T(k)\boldsymbol{w}_o + v(k) \tag{3.2}$$

where $v(k)$ is the noise obtained by subtracting the output of the optimum filter from the desired response. We observe that $\boldsymbol{x}(k)$ and $v(k)$ are uncorrelated thanks to the orthogonality principle of the optimum solution $\boldsymbol{w}_o$:

$$\boldsymbol{p} - \boldsymbol{R}\boldsymbol{w}_o = \boldsymbol{0} \tag{3.3a}$$

$$E\{\boldsymbol{x}(k)(d(k) - \boldsymbol{x}^T(k)\boldsymbol{w}_o)\} = \boldsymbol{0} \tag{3.3b}$$

$$E\{\boldsymbol{x}(k)v(k)\} = \boldsymbol{0} \tag{3.3c}$$

The minimum MSE is:

$$\begin{aligned} J_{\min} &= \min_{\boldsymbol{w}} E\{(d(k) - \boldsymbol{x}^T(k)\boldsymbol{w})^2\} \\ &= E\{(d(k) - \boldsymbol{x}^T(k)\boldsymbol{w}_o)^2\} \\ &= \sigma_d^2 - \boldsymbol{p}^T\boldsymbol{w}_o \\ &= \sigma_v^2 \end{aligned} \tag{3.4}$$

where $\sigma_v^2 = E\{v^2(k)\}$.

In practice, prior knowledge of $\boldsymbol{R}$ and $\boldsymbol{p}$ is not available. We are therefore interested in estimating the unknown system by means of an adaptive filter which learns $\boldsymbol{R}$ and $\boldsymbol{p}$ iteratively. Supposing that complexity constraints do not permit full coefficient updates, we have the following partial-update stochastic gradient algorithm:

$$\boldsymbol{w}(k+1) = \boldsymbol{w}(k) + \frac{\mu}{f(\boldsymbol{x}(k))} e(k)\boldsymbol{I}_M(k)\boldsymbol{x}(k) \tag{3.5a}$$

$$\boldsymbol{I}_M(k)\boldsymbol{w}(k+1) = \boldsymbol{I}_M(k)\boldsymbol{w}(k) + \frac{\mu}{f(\boldsymbol{x}(k))} e(k)\boldsymbol{I}_M(k)\boldsymbol{x}(k) \tag{3.5b}$$

$$\boldsymbol{w}_{\mathcal{M}}(k+1) = \boldsymbol{w}_{\mathcal{M}}(k) + \frac{\mu}{f(\boldsymbol{x}(k))} e(k)\boldsymbol{x}_{\mathcal{M}}(k) \tag{3.5c}$$

All three recursions given above are identical with the last one representing a reduced-size adaptive filter in accordance with partial updates. In (3.5) $f(\cdot)$ is a scalar normalization function, which enables generalization of (3.5c) to normalized LMS. For example, for $f(\boldsymbol{x}(k)) = 1$ we have the LMS algorithm and setting $f(\boldsymbol{x}(k)) = \epsilon + \boldsymbol{x}^T(k)\boldsymbol{I}_M(k)\boldsymbol{x}(k)$ gives the selective-partial-update NLMS algorithm. Subtracting both sides of (3.5c) from the $M \times 1$ subvector of the optimum solution $\boldsymbol{w}^o_{\mathcal{M}}(k)$ which is obtained from $\boldsymbol{w}_o$ by extracting its M entries corresponding to the unity diagonal entries of $\boldsymbol{I}_M(k)$, we obtain:

$$\Delta \boldsymbol{w}_{\mathcal{M}}(k, k+1) = \Delta \boldsymbol{w}_{\mathcal{M}}(k, k) - \frac{\mu}{f(\boldsymbol{x}(k))} e(k) \boldsymbol{x}_{\mathcal{M}}(k) \quad (3.6)$$

where $\Delta \boldsymbol{w}_{\mathcal{M}}(k, l) = \boldsymbol{w}^o_{\mathcal{M}}(k) - \boldsymbol{w}_{\mathcal{M}}(l)$ is the partial-update coefficient error at iteration k. Since coefficient subsets selected by $\boldsymbol{I}_M(k)$ vary with k, the partial-update coefficient error is dependent on k.

In terms of $v(k)$ and coefficient error vector the *a priori* error can be expressed as:

$$\begin{aligned} e(k) &= d(k) - \boldsymbol{x}^T(k)\boldsymbol{w}(k) && (3.7a)\\ &= \boldsymbol{x}^T(k)\boldsymbol{w}_o + v(k) - \boldsymbol{x}^T(k)\boldsymbol{w}(k) && (3.7b)\\ &= e_a(k) + v(k) && (3.7c) \end{aligned}$$

where:

$$e_a(k) = \boldsymbol{x}^T(k)(\boldsymbol{w}_o - \boldsymbol{w}(k)) \quad (3.8)$$

is the so-called *a priori* estimation error. In convergence analysis of partial-update adaptive filters we are particularly interested in the time evolution of learning curve or MSE since it enables comparison of convergence performance for different partial-update techniques. Our main focus in this chapter is, therefore, on the analysis of mean-square convergence of partial-update adaptive filters. MSE is defined by:

$$\begin{aligned} J(k) &= E\{e^2(k)\}\\ &= E\{e_a^2(k)\} + \sigma_v^2 && (3.9) \end{aligned}$$

where $v(k)$ is assumed to be zero-mean and statistically independent of $e_a(k)$. The independence assumption is justified if $v(k)$ is independent of $\boldsymbol{x}(k)$.

The gradient noise present in stochastic gradient algorithms leads to a larger MSE than the minimum MSE obtained by the optimum solution. The difference between the MSE attained by an adaptive filter and the minimum MSE is called the excess MSE (EMSE):

$$\begin{aligned} J_E(k) &= J(k) - J_{\min}\\ &= E\{e_a^2(k)\} && (3.10) \end{aligned}$$

A closely related performance criterion to the excess MSE is misadjustment which is defined as:

$$\begin{aligned} J_A(k) &= \frac{J_E(k)}{J_{\min}}\\ &= \frac{E\{e_a^2(k)\}}{\sigma_v^2} && (3.11) \end{aligned}$$

The mean-square value of the coefficient error is called the mean-square deviation (MSD) and is given by:

$$J_D(k) = E\{\|\boldsymbol{w}_o - \boldsymbol{w}(k)\|_2^2\}. \tag{3.12}$$

Steady-state values of the aforementioned performance criteria are also of interest in terms of performance comparison upon convergence of the adaptive filter coefficients. The limiting values of the performance criteria as $k \to \infty$ give the steady-state values. For example, the steady-state MSE and excess MSE are defined by:

$$\text{Steady-state MSE} \;\; = \lim_{k\to\infty} J(k) \tag{3.13a}$$

$$\text{Steady-state EMSE} = \lim_{k\to\infty} J_E(k). \tag{3.13b}$$

The aim of stability analysis is to determine a bound on the step-size parameter μ so that $J(k)$ is guaranteed to be bounded (i.e. $J(k) < \infty$ for all k). In this context, analysis of the time evolution of MSE plays a central role in characterizing the necessary conditions on the step-size parameter in order to ensure that $J(k)$ does not grow without bound. The key challenge in convergence and stability analysis is to derive closed-form expressions relating readily computable quantities to time evolution of MSE or EMSE.

3.3 STEADY-STATE ANALYSIS

Premultiplying both sides of (3.6) with $\boldsymbol{x}_{\mathcal{M}}^T(k)$ we obtain:

$$\underbrace{\boldsymbol{x}_{\mathcal{M}}^T(k)\Delta\boldsymbol{w}_{\mathcal{M}}(k, k+1)}_{\varepsilon_p(k)} = \underbrace{\boldsymbol{x}_{\mathcal{M}}^T(k)\Delta\boldsymbol{w}_{\mathcal{M}}(k, k)}_{\varepsilon_a(k)} - \frac{\mu}{f(\boldsymbol{x}(k))} e(k)\|\boldsymbol{x}_{\mathcal{M}}(k)\|_2^2 \tag{3.14}$$

where $\varepsilon_a(k)$ is the partial-update *a priori* estimation error and $\varepsilon_p(k)$ is the partial-update *a posteriori* estimation error. Solving for $e(k)$ we get:

$$e(k) = \frac{f(\boldsymbol{x}(k))}{\mu\|\boldsymbol{x}_{\mathcal{M}}(k)\|_2^2}(\varepsilon_a(k) - \varepsilon_p(k)) \tag{3.15}$$

Substituting this back into (3.6) yields:

$$\Delta\boldsymbol{w}_{\mathcal{M}}(k, k+1) + \frac{\boldsymbol{x}_{\mathcal{M}}(k)}{\|\boldsymbol{x}_{\mathcal{M}}(k)\|_2^2}\varepsilon_a(k) = \Delta\boldsymbol{w}_{\mathcal{M}}(k, k) + \frac{\boldsymbol{x}_{\mathcal{M}}(k)}{\|\boldsymbol{x}_{\mathcal{M}}(k)\|_2^2}\varepsilon_p(k) \tag{3.16}$$

Taking the squared Euclidean norm of both sides we get:

$$\|\Delta\boldsymbol{w}_{\mathcal{M}}(k, k+1)\|_2^2 + \frac{\varepsilon_a^2(k)}{\|\boldsymbol{x}_{\mathcal{M}}(k)\|_2^2} = \|\Delta\boldsymbol{w}_{\mathcal{M}}(k, k)\|_2^2 + \frac{\varepsilon_p^2(k)}{\|\boldsymbol{x}_{\mathcal{M}}(k)\|_2^2} \tag{3.17}$$

This expression provides an *exact* relation between partial-update coefficient errors and partial-update *a priori* and *a posteriori* estimation errors. Furthermore, it makes no assumptions about the signal statistics.

Taking the expectation of both sides of (3.17), we have:

$$E\{\|\Delta \boldsymbol{w}_{\mathcal{M}}(k,k+1)\|_2^2\} + E\left\{\frac{\varepsilon_a^2(k)}{\|\boldsymbol{x}_{\mathcal{M}}(k)\|_2^2}\right\} = E\{\|\Delta \boldsymbol{w}_{\mathcal{M}}(k,k)\|_2^2\} + E\left\{\frac{\varepsilon_p^2(k)}{\|\boldsymbol{x}_{\mathcal{M}}(k)\|_2^2}\right\} \quad (3.18)$$

The steady-state behaviour of an adaptive filter is characterized by:

$$\lim_{k\to\infty} E\{\|\boldsymbol{w}_o - \boldsymbol{w}(k)\|_2^2\} = c_1 < \infty \quad (3.19)$$

This also implies that:

$$E\{\|\Delta \boldsymbol{w}_{\mathcal{M}}(k,k+1)\|_2^2\} = E\{\|\Delta \boldsymbol{w}_{\mathcal{M}}(k,k)\|_2^2\} = c_2 < \infty \quad \text{as } k \to \infty \quad (3.20)$$

Thus at steady-state (3.18) becomes:

$$E\left\{\frac{\varepsilon_a^2(k)}{\|\boldsymbol{x}_{\mathcal{M}}(k)\|_2^2}\right\} = E\left\{\frac{\varepsilon_p^2(k)}{\|\boldsymbol{x}_{\mathcal{M}}(k)\|_2^2}\right\} \quad \text{as } k \to \infty \quad (3.21)$$

Substituting (3.14) into the above expression we get:

$$\mu E\left\{\left(\frac{e(k)}{f(\boldsymbol{x}(k))}\right)^2 \|\boldsymbol{x}_{\mathcal{M}}(k)\|_2^2\right\} = 2E\left\{\varepsilon_a(k)\frac{e(k)}{f(\boldsymbol{x}(k))}\right\} \quad \text{as } k \to \infty \quad (3.22)$$

which is also an exact expression and does not depend on any assumption on the signal statistics.

In steady-state analysis the main objective is to compute either the steady-state MSE, $E\{e^2(k)\}$, or steady-state EMSE, $E\{e_a^2(k)\}$, as $k \to \infty$. We will next see how this is accomplished for partial-update LMS and NLMS algorithms by using the steady-state variance relation in (3.22).

3.3.1 Partial-update LMS algorithms

For partial-update LMS algorithms the scalar function $f(\cdot)$ in (3.5c) is $f(\boldsymbol{x}(k)) = 1$. The coefficient selection matrix $\boldsymbol{I}_M(k)$ determines which particular partial-update scheme is implemented. We will consider periodic, sequential, stochastic and M-max partial updates in the following discussion.

Substituting (3.7c) into (3.22), where $f(\boldsymbol{x}(k)) = 1$, we obtain:

$$\mu E\{(e_a^2(k) + 2e_a(k)v(k) + v^2(k))\|\boldsymbol{x}_{\mathcal{M}}(k)\|_2^2\} = 2E\{\varepsilon_a(k)e_a(k) + \varepsilon_a(k)v(k)\} \quad \text{as } k \to \infty. \tag{3.23}$$

Since $v(k)$ is assumed to be statistically independent of both $e_a(k)$ and $\boldsymbol{x}(k)$, the above expression simplifies to:

$$\mu E\{e_a^2(k)\|\boldsymbol{x}_{\mathcal{M}}(k)\|_2^2\} + \mu\sigma_v^2 \mathrm{tr}\boldsymbol{R}_M = 2E\{\varepsilon_a(k)e_a(k)\} \quad \text{as } k \to \infty \tag{3.24}$$

where:

$$\begin{aligned} \mathrm{tr}\boldsymbol{R}_M &= \mathrm{tr}E\{\boldsymbol{I}_M(k)\boldsymbol{x}(k)\boldsymbol{x}^T(k)\} \\ &= E\{\boldsymbol{x}^T(k)\boldsymbol{I}_M(k)\boldsymbol{x}(k)\} \\ &= E\{\|\boldsymbol{x}_{\mathcal{M}}(k)\|_2^2\} \end{aligned} \tag{3.25}$$

Here tr denotes trace of a matrix.

Let:

$$\begin{aligned} E\{\varepsilon_a(k)e_a(k)\} &= \beta_M E\{e_a^2(k)\} \\ &= \beta_M \,\mathrm{EMSE} \end{aligned} \tag{3.26}$$

where $0 < \beta_M \leq 1$ is a constant that indicates how much the *a priori* estimation error is reduced as a result of partial coefficient updates. Substituting (3.26) into (3.24) yields:

$$\mathrm{EMSE} = \frac{\mu}{2\beta_M}(E\{e_a^2(k)\|\boldsymbol{x}_{\mathcal{M}}(k)\|_2^2\} + \sigma_v^2 \mathrm{tr}\boldsymbol{R}_M), \quad k \to \infty \tag{3.27}$$

If the step-size parameter μ is small, at steady state we expect to have:

$$E\{e_a^2(k)\|\boldsymbol{x}_{\mathcal{M}}(k)\|_2^2\} \ll \sigma_v^2 \mathrm{tr}\boldsymbol{R}_M \tag{3.28}$$

Using this in (3.27) we get:

$$\mathrm{EMSE} = \frac{\mu\sigma_v^2 \mathrm{tr}\boldsymbol{R}_M}{2\beta_M} \tag{3.29}$$

which is referred to as the small step-size approximation of steady-state EMSE.

Another approach to derive a closed-form expression for the steady-state EMSE is based on Slutsky's theorem (Wong and Polak, 1967; Ferguson, 1996; Le Cadre and Jauffret, 1999):

Theorem 3.1. *Let x_k and y_k be sequences of random variables. If x_k converges in distribution to X (i.e., $\mathrm{dlim}_{k\to\infty} x_k \sim X$) and y_k converges in probability to a constant c (i.e., $\mathrm{plim}_{k\to\infty} y_k = c$), then $\mathrm{dlim}_{k\to\infty}(x_k + y_k) \sim X + c$, $\mathrm{dlim}_{k\to\infty}(x_k y_k) \sim cX$, and $\mathrm{dlim}_{k\to\infty}(x_k/y_k) \sim X/c$ if $c \neq 0$.*

This result can also be extended to matrix random variables.

Supposing that:

$$\operatorname*{plim}_{k\to\infty} e_a^2(k) = c$$

$$\operatorname*{dlim}_{k\to\infty} \|\boldsymbol{x}_{\mathcal{M}}(k)\|_2^2 \sim X$$

we have:

$$\operatorname*{dlim}_{k\to\infty} e_a^2(k)\|\boldsymbol{x}_{\mathcal{M}}(k)\|_2^2 \sim cX \tag{3.30}$$

and:

$$E\{e_a^2(k)\|\boldsymbol{x}_{\mathcal{M}}(k)\|_2^2\} = E\{e_a^2(k)\}\mathrm{tr}\boldsymbol{R}_M, \quad k \to \infty \tag{3.31}$$

This asymptotic result is also referred to as the separation principle (Sayed, 2003) and leads to the following steady-state EMSE expression:

$$\mathrm{EMSE} = \frac{\mu\sigma_v^2 \mathrm{tr}\boldsymbol{R}_M}{2\beta_M - \mu\mathrm{tr}\boldsymbol{R}_M} \tag{3.32}$$

Compared with (3.29) the only difference is the additional term $-\mu\mathrm{tr}\boldsymbol{R}_M$ in the denominator.

If $M = N$ (i.e. all coefficients are updated) then $\beta_M = 1$, $\boldsymbol{R}_M = \boldsymbol{R}$, and (3.29) and (3.32) reduce to:

$$\mathrm{EMSE} = \frac{\mu\sigma_v^2 \mathrm{tr}\boldsymbol{R}}{2} \quad \text{small step-size approximation} \tag{3.33a}$$

$$\mathrm{EMSE} = \frac{\mu\sigma_v^2 \mathrm{tr}\boldsymbol{R}}{2 - \mu\mathrm{tr}\boldsymbol{R}} \quad \text{asymptotic approximation.} \tag{3.33b}$$

These expressions give the steady-state EMSE for the full-update LMS algorithm.

In what follows we analyse (3.29) and (3.32) for particular partial-update techniques applicable to the LMS algorithm.

Periodic partial updates

In periodic partial updates all coefficients of the adaptive filter are updated (i.e. $\boldsymbol{I}_M(k) = \boldsymbol{I}$) albeit every Sth iteration. This implies that we can use the same steady-state EMSE expression as the full-update LMS algorithm after changing the input signal correlation to $\tilde{\boldsymbol{R}} = E\{\boldsymbol{x}(kS)\boldsymbol{x}^T(kS)\}$. For the periodic-partial-update LMS algorithm we, therefore, have

$$\text{EMSE} = \frac{\mu\sigma_v^2 \text{tr}\tilde{\boldsymbol{R}}}{2} \quad \text{small step-size approximation} \tag{3.34a}$$

$$\text{EMSE} = \frac{\mu\sigma_v^2 \text{tr}\tilde{\boldsymbol{R}}}{2 - \mu \text{tr}\tilde{\boldsymbol{R}}} \quad \text{asymptotic approximation} \tag{3.34b}$$

If the input signal is stationary, then we have $\tilde{\boldsymbol{R}} = \boldsymbol{R}$ and the steady-state EMSE of the periodic-partial-update LMS algorithm is identical to that of the full-update LMS algorithm in (3.33).

Sequential partial updates

For stationary input signals the input correlation matrix for sequential partial updates is $\boldsymbol{R}_M = (M/N)\boldsymbol{R}$. Assume that $x(k)$ is a zero-mean i.i.d. (independent and identically distributed) signal and that $\boldsymbol{w}(k)$ is independent of $\boldsymbol{x}(k)$ at steady-state. Under this assumption we can write:

$$\underbrace{\boldsymbol{x}^T(k)\boldsymbol{I}_M(k)(\boldsymbol{w}_o - \boldsymbol{w}(k))}_{\varepsilon_a(k)} \underbrace{\boldsymbol{x}^T(k)(\boldsymbol{w}_o - \boldsymbol{w}(k))}_{e_a(k)}$$

$$= \sum_{i \in \mathcal{J}_M(k)} z_i(k) \sum_{i=1}^{N} z_i(k) \tag{3.35a}$$

$$E\{\varepsilon_a(k)e_a(k)\} = \sum_{i \in \mathcal{J}_M(k)} E\{z_i^2(k)\} \tag{3.35b}$$

$$E\{\varepsilon_a(k)e_a(k)\} = \frac{M}{N} E\{e_a^2(k)\} \tag{3.35c}$$

where $\mathcal{J}_M(k)$ is the set of regressor entries selected by $\boldsymbol{I}_M(k)$ and:

$$\begin{bmatrix} z_1(k) \\ z_2(k) \\ \vdots \\ z_N(k) \end{bmatrix} = \boldsymbol{x}(k) \odot (\boldsymbol{w}_o - \boldsymbol{w}(k)) \tag{3.36}$$

Here $\odot$ denotes the Hadamard (element-wise) matrix product. In arriving at (3.35c), we also implicitly assumed that the entries of $\boldsymbol{w}_o - \boldsymbol{w}(k)$ have identical variances at steady-state. We thus conclude that the reduction in the *a priori* estimation error due to sequential partial coefficient updates is:

$$\beta_M = \frac{M}{N} \tag{3.37}$$

Even though the assumptions made to obtain β_M are not realistic for tapped-delay-line adaptive filters with overlapping input regressor vectors, they are not as restrictive as the independence assumptions traditionally used in the convergence analysis and produce results that are widely applicable. Substituting $\boldsymbol{R}_M = (M/N)\boldsymbol{R}$ and $\beta_M = M/N$ into (3.29) and (3.32), we obtain:

$$\text{EMSE} = \frac{\mu\sigma_v^2 \frac{M}{N} \text{tr}\boldsymbol{R}}{2\frac{M}{N}} = \frac{\mu\sigma_v^2 \text{tr}\boldsymbol{R}}{2} \quad \text{small step-size approximation} \tag{3.38a}$$

$$\text{EMSE} = \frac{\mu\sigma_v^2 \frac{M}{N} \text{tr}\boldsymbol{R}}{2\frac{M}{N} - \mu\frac{M}{N}\text{tr}\boldsymbol{R}} = \frac{\mu\sigma_v^2 \text{tr}\boldsymbol{R}}{2 - \mu\text{tr}\boldsymbol{R}} \quad \text{asymptotic approximation} \tag{3.38b}$$

This confirms that the sequential-partial-update LMS and full-update LMS algorithms have the same steady-state EMSE irrespective of M.

Stochastic partial updates

In stochastic partial updates the input correlation matrix is $\boldsymbol{R}_M = (M/N)\boldsymbol{R}$ whether or not the input signal is stationary. As in sequential partial updates, under the assumption of zero-mean i.i.d. $x(k)$ and independence between $\boldsymbol{w}(k)$ and $\boldsymbol{x}(k)$ at steady-state, we obtain $\beta_M = M/N$. This leads to the conclusion that the steady-state EMSE of the stochastic-partial-update LMS algorithm is identical to that of the full-update LMS algorithm.

M-max partial updates

The method of M-max updates selects the coefficients to be updated according to the magnitude of regressor entries. This data-dependent coefficient selection often results in a nontrivial relationship between $\boldsymbol{R}_M$ and $\boldsymbol{R}$. This is also the case for β_M. However, the assumption of zero-mean i.i.d. $x(k)$ and independence between $\boldsymbol{w}(k)$ and $\boldsymbol{x}(k)$ at steady-state allows us to write:

$$\begin{aligned}
\beta_M &= \frac{E\{\varepsilon_a(k)e_a(k)\}}{E\{e_a^2(k)\}} \\
&= \frac{\sum_{i\in\mathcal{J}_M(k)} E\{z_i^2(k)\}}{\sum_{i=1}^{N} E\{z_i^2(k)\}} \\
&= \frac{E\{\boldsymbol{x}^T(k)\boldsymbol{I}_M(k)\boldsymbol{x}(k)\}}{E\{\boldsymbol{x}^T(k)\boldsymbol{x}(k)\}} \\
&= \frac{\text{tr}\boldsymbol{R}_M}{\text{tr}\boldsymbol{R}}
\end{aligned} \tag{3.39}$$

Using this general definition of β_M, we obtain:

$$\text{EMSE} = \frac{\mu\sigma_v^2\beta_M \text{tr}\boldsymbol{R}}{2\beta_M} = \frac{\mu\sigma_v^2 \text{tr}\boldsymbol{R}}{2} \quad \text{small step-size approximation} \tag{3.40a}$$

$$\text{EMSE} = \frac{\mu\sigma_v^2\beta_M \text{tr}\boldsymbol{R}}{2\beta_M - \mu\beta_M \text{tr}\boldsymbol{R}} = \frac{\mu\sigma_v^2 \text{tr}\boldsymbol{R}}{2 - \mu \text{tr}\boldsymbol{R}} \quad \text{asymptotic approximation} \tag{3.40b}$$

This shows that, in common with all three partial-update techniques investigated above, the M-max LMS algorithm has the same steady-state EMSE as the full-update LMS algorithm regardless of M.

3.3.2 Partial-update NLMS algorithms

The scalar function $f(\cdot)$ in (3.5c) assumes the form of $f(\boldsymbol{x}(k)) = \epsilon + \|\boldsymbol{x}(k)\|_2^2$ and $f(\boldsymbol{x}(k)) = \epsilon + \|\boldsymbol{x}_{\mathcal{M}}(k)\|_2^2$ for the M-max NLMS algorithm and the selective-partial-update NLMS algorithm, respectively. The coefficient selection matrix $\boldsymbol{I}_M(k)$ for these algorithms is defined the same way and selects the M coefficients out of N with the largest Euclidean norm regressor entries at each iteration.

Substituting (3.7c) into (3.22) and assuming independent $e_a(k)$ and $\boldsymbol{x}(k)$, we obtain:

$$\mu E\left\{\frac{e_a^2(k)}{f^2(\boldsymbol{x}(k))}\|\boldsymbol{x}_{\mathcal{M}}(k)\|_2^2\right\} + \mu\sigma_v^2 E\left\{\frac{\|\boldsymbol{x}_{\mathcal{M}}(k)\|_2^2}{f^2(\boldsymbol{x}(k))}\right\} = 2E\left\{\varepsilon_a(k)\frac{e_a(k)}{f(\boldsymbol{x}(k))}\right\} \quad \text{as } k \to \infty \tag{3.41}$$

Consider first the selective-partial-update NLMS algorithm. Assuming $f(\boldsymbol{x}(k)) = \epsilon + \|\boldsymbol{x}_{\mathcal{M}}(k)\|_2^2$ and $\epsilon \approx 0$ we have:

$$\mu E\left\{\frac{e_a^2(k)}{\|\boldsymbol{x}_{\mathcal{M}}(k)\|_2^2}\right\} + \mu\sigma_v^2 E\left\{\frac{1}{\|\boldsymbol{x}_{\mathcal{M}}(k)\|_2^2}\right\} = 2E\left\{\varepsilon_a(k)\frac{e_a(k)}{\|\boldsymbol{x}_{\mathcal{M}}(k)\|_2^2}\right\} \quad \text{as } k \to \infty. \tag{3.42}$$

Application of Theorem 3.1 to the above expression yields:

$$\mu E\{e_a^2(k)\}E\left\{\frac{1}{\|\boldsymbol{x}_{\mathcal{M}}(k)\|_2^2}\right\} + \mu\sigma_v^2 E\left\{\frac{1}{\|\boldsymbol{x}_{\mathcal{M}}(k)\|_2^2}\right\} = 2E\{\varepsilon_a(k)e_a(k)\}E\left\{\frac{1}{\|\boldsymbol{x}_{\mathcal{M}}(k)\|_2^2}\right\} \quad \text{as } k \to \infty \tag{3.43a}$$

$$\mu E\{e_a^2(k)\} + \mu\sigma_v^2 = 2E\{\varepsilon_a(k)e_a(k)\} \quad \text{as } k \to \infty \tag{3.43b}$$

where, by Theorem 3.1:

$$\operatorname*{dlim}_{k\to\infty} \frac{e_a^2(k)}{\|\boldsymbol{x}_{\mathcal{M}}(k)\|_2^2} = \operatorname*{plim}_{k\to\infty} e_a^2(k) \operatorname*{dlim}_{k\to\infty} \frac{1}{\|\boldsymbol{x}_{\mathcal{M}}(k)\|_2^2} \tag{3.44a}$$

$$\operatorname*{dlim}_{k\to\infty} \frac{\varepsilon_a(k)e_a(k)}{\|\boldsymbol{x}_{\mathcal{M}}(k)\|_2^2} = \operatorname*{plim}_{k\to\infty} \varepsilon_a(k)e_a(k) \operatorname*{dlim}_{k\to\infty} \frac{1}{\|\boldsymbol{x}_{\mathcal{M}}(k)\|_2^2} \tag{3.44b}$$

Here it is assumed that $e_a^2(k)$ and $\varepsilon_a(k)e_a(k)$ converge in probability and $1/\|\boldsymbol{x}_{\mathcal{M}}(k)\|_2^2$ converges in distribution. Letting $E\{\varepsilon_a(k)e_a(k)\} = \beta_M E\{e_a^2(k)\}$ as in (3.26), from (3.43) we obtain:

$$\text{EMSE} = \frac{\mu\sigma_v^2}{2\beta_M - \mu} \tag{3.45}$$

which is independent of the input signal statistics.

Another approach to deriving an EMSE expression for the selective-partial-update NLMS is based on a different interpretation of Slutsky's theorem. A special case of Slutsky's theorem is that if $\operatorname{plim}_{k\to\infty} x_k = c_1$ and $\operatorname{plim}_{k\to\infty} y_k = c_2$ where c_1 and c_2 are constants, then $\operatorname{plim}_{k\to\infty} x_k/y_k = c_1/c_2$ if $c_2 \neq 0$. Applying this to $e_a^2(k)/\|\boldsymbol{x}_{\mathcal{M}}(k)\|_2^2$ and $\varepsilon_a(k)e_a(k)/\|\boldsymbol{x}_{\mathcal{M}}(k)\|_2^2$, we get:

$$\operatorname*{plim}_{N\to\infty} \frac{e_a^2(k)/N}{\|\boldsymbol{x}_{\mathcal{M}}(k)\|_2^2/N} = \frac{\operatorname{plim}_{N\to\infty} e_a^2(k)/N}{\operatorname{plim}_{N\to\infty} \|\boldsymbol{x}_{\mathcal{M}}(k)\|_2^2/N} \tag{3.46a}$$

$$\operatorname*{plim}_{N\to\infty} \frac{\varepsilon_a(k)e_a(k)/N}{\|\boldsymbol{x}_{\mathcal{M}}(k)\|_2^2/N} = \frac{\operatorname{plim}_{N\to\infty} \varepsilon_a(k)e_a(k)/N}{\operatorname{plim}_{N\to\infty} \|\boldsymbol{x}_{\mathcal{M}}(k)\|_2^2/N} \tag{3.46b}$$

For sufficiently large N, the above expressions hold approximately, leading to:

$$E\left\{\frac{e_a^2(k)}{\|\boldsymbol{x}_{\mathcal{M}}(k)\|_2^2}\right\} \approx \frac{E\{e_a^2(k)\}}{E\{\|\boldsymbol{x}_{\mathcal{M}}(k)\|_2^2\}} = \frac{E\{e_a^2(k)\}}{\operatorname{tr}\boldsymbol{R}_M} \tag{3.47a}$$

$$E\left\{\frac{\varepsilon_a(k)e_a(k)}{\|\boldsymbol{x}_{\mathcal{M}}(k)\|_2^2}\right\} \approx \frac{E\{\varepsilon_a(k)e_a(k)\}}{E\{\|\boldsymbol{x}_{\mathcal{M}}(k)\|_2^2\}} = \frac{E\{\varepsilon_a(k)e_a(k)\}}{\operatorname{tr}\boldsymbol{R}_M} \tag{3.47b}$$

Substituting this into (3.42) we get:

$$\mu\frac{E\{e_a^2(k)\}}{\operatorname{tr}\boldsymbol{R}_M} + \mu\sigma_v^2 E\left\{\frac{1}{\|\boldsymbol{x}_{\mathcal{M}}(k)\|_2^2}\right\} = 2\frac{E\{\varepsilon_a(k)e_a(k)\}}{\operatorname{tr}\boldsymbol{R}_M} \quad \text{as } k \to \infty \tag{3.48}$$

whence:

$$\text{EMSE} = \frac{\mu\sigma_v^2}{2\beta_M - \mu}\text{tr}\boldsymbol{R}_M E\left\{\frac{1}{\|\boldsymbol{x}_{\mathcal{M}}(k)\|_2^2}\right\} \tag{3.49}$$

To sum up, we have derived two closed-form EMSE expressions for the selective-partial-update NLMS algorithm:

$$\text{EMSE} = \frac{\mu\sigma_v^2}{2\beta_M - \mu} \quad \text{asymptotic approximation 1} \tag{3.50a}$$

$$\text{EMSE} = \frac{\mu\sigma_v^2}{2\beta_M - \mu}\text{tr}\boldsymbol{R}_M E\left\{\frac{1}{\|\boldsymbol{x}_{\mathcal{M}}(k)\|_2^2}\right\} \quad \text{asymptotic approximation 2} \tag{3.50b}$$

Under the assumption of zero-mean i.i.d. $x(k)$ and independence between $\boldsymbol{w}(k)$ and $\boldsymbol{x}(k)$ at steady-state, β_M can be defined as [see (3.39)]

$$\beta_M = \frac{\text{tr}\boldsymbol{R}_M}{\text{tr}\boldsymbol{R}} \tag{3.51}$$

Substituting this into the first asymptotic approximation in (3.50a) gives:

$$\begin{aligned}\text{EMSE} &= \frac{\mu\sigma_v^2\text{tr}\boldsymbol{R}}{2\text{tr}\boldsymbol{R}_M - \mu\text{tr}\boldsymbol{R}} \\ &= \frac{N\mu\sigma_v^2\sigma_x^2}{2\text{tr}\boldsymbol{R}_M - N\mu\sigma_x^2}\end{aligned} \tag{3.52}$$

which is identical to the EMSE expression derived in (Werner *et al.*, 2004) for a particular input signal model proposed in (Slock, 1993). Note that the equality $\text{tr}\boldsymbol{R} = N\sigma_x^2$ assumes a stationary input signal with variance σ_x^2.

We now consider the M-max NLMS algorithm. For $f(\boldsymbol{x}(k)) = \epsilon + \|\boldsymbol{x}(k)\|_2^2$ and $\epsilon \approx 0$, (3.41) becomes:

$$\begin{aligned}&\mu E\left\{e_a^2(k)\frac{\|\boldsymbol{x}_{\mathcal{M}}(k)\|_2^2}{\|\boldsymbol{x}(k)\|_2^4}\right\} + \mu\sigma_v^2 E\left\{\frac{\|\boldsymbol{x}_{\mathcal{M}}(k)\|_2^2}{\|\boldsymbol{x}(k)\|_2^4}\right\} \\ &\quad = 2E\left\{\frac{\varepsilon_a(k)e_a(k)}{\|\boldsymbol{x}(k)\|_2^2}\right\} \quad \text{as } k \to \infty\end{aligned} \tag{3.53}$$

Using the first asymptotic approximation we get:

$$\begin{aligned}
\mu E\{e_a^2(k)\} E\left\{\frac{\|\boldsymbol{x}_{\mathcal{M}}(k)\|_2^2}{\|\boldsymbol{x}(k)\|_2^4}\right\} + \mu\sigma_v^2 E\left\{\frac{\|\boldsymbol{x}_{\mathcal{M}}(k)\|_2^2}{\|\boldsymbol{x}(k)\|_2^4}\right\} \\
= 2E\{\varepsilon_a(k)e_a(k)\} E\left\{\frac{1}{\|\boldsymbol{x}(k)\|_2^2}\right\} \quad \text{as } k \to \infty
\end{aligned} \tag{3.54}$$

and:

$$\text{EMSE} = \frac{\mu\sigma_v^2}{2\alpha_M - \mu} \tag{3.55}$$

where:

$$\alpha_M = \beta_M E\left\{\frac{1}{\|\boldsymbol{x}(k)\|_2^2}\right\} \Big/ E\left\{\frac{\|\boldsymbol{x}_{\mathcal{M}}(k)\|_2^2}{\|\boldsymbol{x}(k)\|_2^4}\right\} \tag{3.56}$$

The second asymptotic approximation yields:

$$\begin{aligned}
\mu E\{e_a^2(k)\} \frac{E\{\|\boldsymbol{x}_{\mathcal{M}}(k)\|_2^2\}}{E^2\{\|\boldsymbol{x}(k)\|_2^2\}} + \mu\sigma_v^2 E\left\{\frac{\|\boldsymbol{x}_{\mathcal{M}}(k)\|_2^2}{\|\boldsymbol{x}(k)\|_2^4}\right\} \\
= 2\frac{E\{\varepsilon_a(k)e_a(k)\}}{E\{\|\boldsymbol{x}(k)\|_2^2\}} \quad \text{as } k \to \infty
\end{aligned} \tag{3.57}$$

and:

$$\begin{aligned}
\text{EMSE} &= \frac{\mu\sigma_v^2 \text{tr}\boldsymbol{R}}{2\beta_M - \mu \text{tr}\boldsymbol{R}_M/\text{tr}\boldsymbol{R}} E\left\{\frac{\|\boldsymbol{x}_{\mathcal{M}}(k)\|_2^2}{\|\boldsymbol{x}(k)\|_2^4}\right\} \\
&= \frac{\mu\sigma_v^2}{2-\mu} \frac{\text{tr}\boldsymbol{R}}{\beta_M} E\left\{\frac{\|\boldsymbol{x}_{\mathcal{M}}(k)\|_2^2}{\|\boldsymbol{x}(k)\|_2^4}\right\}
\end{aligned} \tag{3.58}$$

Using the approximations that follow from Slutsky's theorem:

$$E\left\{\frac{\|\boldsymbol{x}_{\mathcal{M}}(k)\|_2^2}{\|\boldsymbol{x}(k)\|_2^4}\right\} \approx \frac{\text{tr}\boldsymbol{R}_M}{(\text{tr}\boldsymbol{R})^2} \quad \text{and} \quad E\left\{\frac{1}{\|\boldsymbol{x}(k)\|_2^2}\right\} \approx \frac{1}{\text{tr}\boldsymbol{R}} \tag{3.59}$$

both (3.55) and (3.58) reduce to:

$$\text{EMSE} = \frac{\mu\sigma_v^2}{2-\mu} \tag{3.60}$$

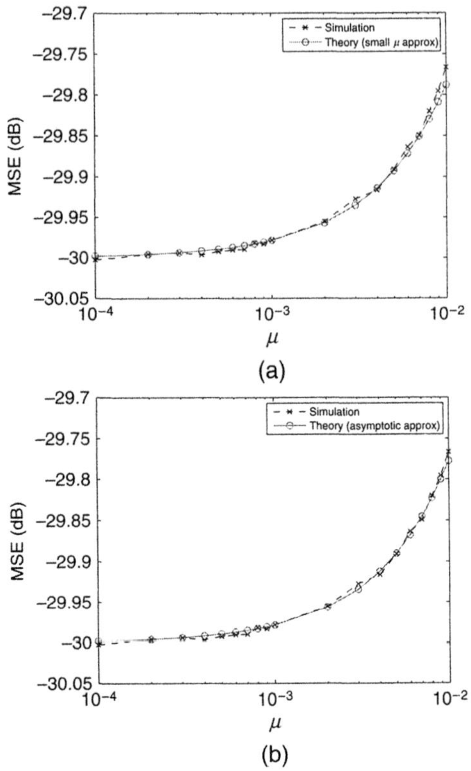

Figure 3.1 Theoretical and simulated MSE of periodic-partial-update LMS ($S = 5$) for i.i.d. Gaussian input. Theoretical MSE is computed using: (a) small μ approximation (3.34a); and (b) asymptotic approximation (3.34b).

which is the steady-state EMSE of the full-update NLMS algorithm. Thus the M-max NLMS algorithm and the full-update NLMS algorithm have identical steady-state EMSE for the same step-size.

The three closed-form EMSE expressions for the M-max NLMS algorithm are summarized below:

$$\text{EMSE} = \frac{\mu\sigma_v^2}{2\alpha_M - \mu} \quad \text{asymptotic approximation 1} \tag{3.61a}$$

$$\text{EMSE} = \frac{\mu\sigma_v^2}{2-\mu}\frac{\text{tr}\boldsymbol{R}}{\beta_M}E\left\{\frac{\|\boldsymbol{x}_M(k)\|_2^2}{\|\boldsymbol{x}(k)\|_2^4}\right\} \quad \text{asymptotic approximation 2} \tag{3.61b}$$

$$\text{EMSE} = \frac{\mu\sigma_v^2}{2-\mu} \quad \text{asymptotic approximation 3} \tag{3.61c}$$

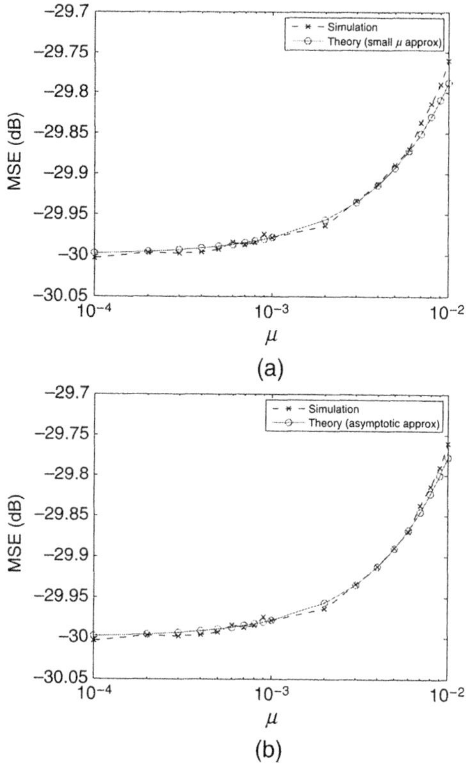

Figure 3.2 Theoretical and simulated MSE of periodic-partial-update LMS ($S = 5$) for AR(1) Gaussian input with eigenvalue spread 135.5. Theoretical MSE is computed using: (a) small μ approximation (3.34a); and (b) asymptotic approximation (3.34b).

3.3.3 Simulation examples for steady-state analysis

In this section we assess the accuracy of the steady-state analysis results by way of simulation examples. We consider a system identification problem where the unknown system is FIR of order $N - 1$ and the additive noise at the system output $n(k)$ is statistically independent of the input $x(k)$. Thus, in the signal model used in the steady-state analysis we have $v(k) = n(k)$ and $\boldsymbol{w}_o$ is the FIR system impulse response. We observe that the steady-state EMSE is not influenced by the unknown system characteristics and is only determined by the input signal statistics and the method of partial coefficient updates employed. For this reason the choice of the FIR system is immaterial as long as its length is less than the adaptive filter length N. The impulse response of the unknown FIR system to be identified has been set to:

$$\boldsymbol{h} = [1, -1.2, 0.8, 0.3, -0.4, 0.2, -0.1, -0.3, 0.1, 0.05]^T \tag{3.62}$$

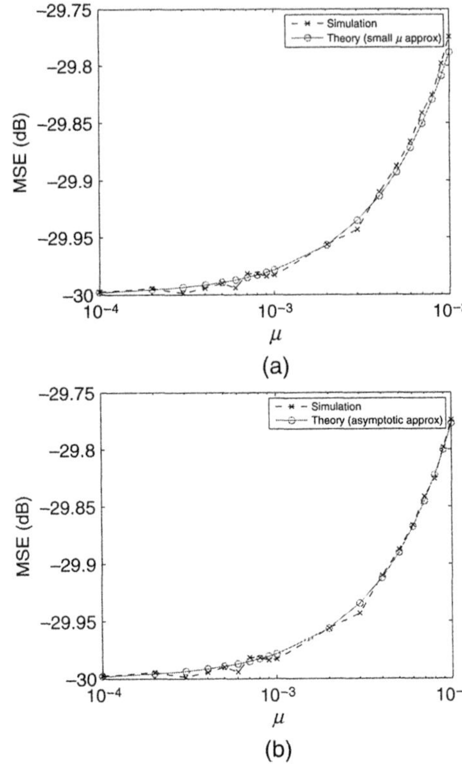

Figure 3.3 Theoretical and simulated MSE of sequential-partial-update LMS ($M = 1$) for i.i.d. Gaussian input. Theoretical MSE is computed using: (a) small μ approximation (3.38a); and (b) asymptotic approximation (3.38b).

The adaptive filter length is $N = 10$ which matches the unknown FIR system order. The additive output noise $v(k)$ is an i.i.d. zero-mean Gaussian noise with variance $\sigma_v^2 = 0.001$.

The steady-state MSE values are estimated by ensemble averaging adaptive filter output errors over 400 independent trials and taking time average of 5000 ensemble-averaged values after convergence. Recall that the steady-state MSE is given by the sum of the steady-state EMSE and the variance of $v(k)$. To speed up convergence, adaptive filter coefficients are initialized to the true system impulse response (i.e. $\boldsymbol{w}(0) = \boldsymbol{h}$). Two input signals are considered, viz. an i.i.d. Gaussian input signal with zero mean and unity variance, and a correlated AR(1) Gaussian input signal with zero mean and unity variance:

$$x(k) = -ax(k-1) + \nu(k) \tag{3.63}$$

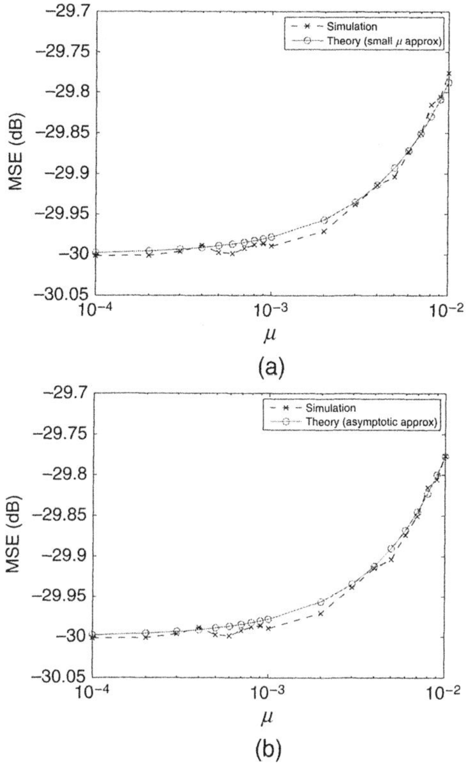

Figure 3.4 Theoretical and simulated MSE of sequential-partial-update LMS ($M = 1$) for AR(1) Gaussian input with eigenvalue spread 135.5. Theoretical MSE is computed using: (a) small μ approximation (3.38a); and (b) asymptotic approximation (3.38b).

where $a = -0.9$ and $\nu(k)$ is zero-mean i.i.d. Gaussian with variance $\sqrt{1-a^2}$ so that $\sigma_x^2 = 1$. The eigenvalue spread of this correlated Gaussian signal is $\chi(\boldsymbol{R}) = 135.5$ which indicates a strong correlation.

In Figures 3.1 and 3.2 the theoretical and simulated MSE values of the periodic-partial-update LMS algorithm are depicted for the i.i.d. and correlated Gaussian inputs, respectively. The period of coefficient updates is set to $S = 5$. The theoretical MSE is computed using the small step-size approximation (3.34a) and the asymptotic approximation (3.34b). For both input signals the asymptotic approximation appears to be more accurate than the small step-size approximation for large μ. This is intuitively expected since the small step-size approximation assumes a small μ.

The theoretical and simulated MSE values of the sequential-partial-update LMS algorithm are shown in Figures 3.3 and 3.4 for the i.i.d. and correlated Gaussian inputs, respectively. The number of coefficients to be updated is $M = 1$ (i.e. one

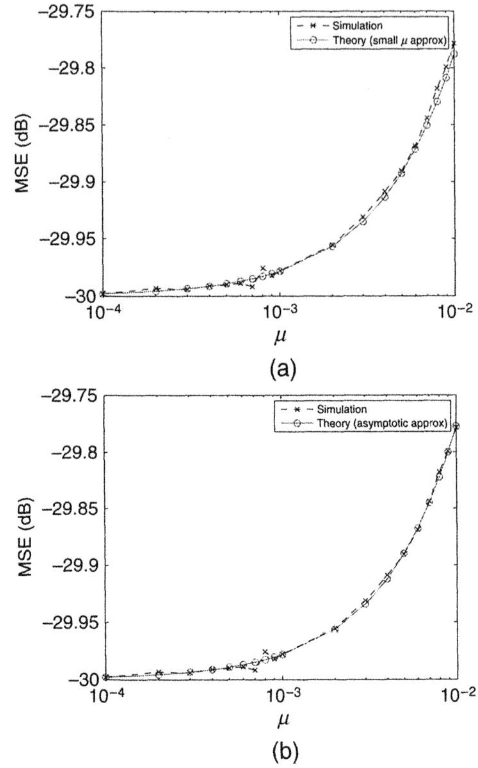

Figure 3.5 Theoretical and simulated MSE of M-max LMS ($M = 4$) for i.i.d. Gaussian input. Theoretical MSE is computed using: (a) small μ approximation (3.40a); and (b) asymptotic approximation (3.40b).

out of ten coefficients is updated at each LMS iteration). The theoretical MSE is computed using the small step-size approximation (3.38a) and the asymptotic approximation (3.38b), which are identical to those used for periodic partial updates. The asymptotic approximation yields slightly more accurate results than the small step-size approximation for large μ.

The MSE values of the M-max LMS algorithm are shown in Figures 3.5 and 3.6 for the i.i.d. and correlated Gaussian inputs, respectively. The number of coefficients to be updated is set to $M = 4$. The theoretical MSE is computed using the small step-size approximation (3.40a) and the asymptotic approximation (3.40b), which are identical to those used for periodic partial updates. For data-dependent partial coefficient updates, the input signal correlation reduces the accuracy of the MSE approximations. One of the reasons for this is that the definitions of β_M in (3.26) and (3.39) are not identical for correlated inputs.

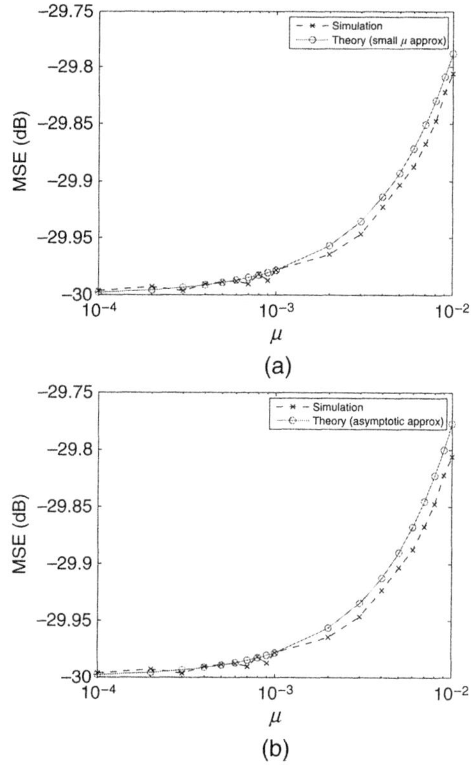

Figure 3.6 Theoretical and simulated MSE of M-max LMS ($M = 4$) for AR(1) Gaussian input with eigenvalue spread 135.5. Theoretical MSE is computed using: (a) small μ approximation (3.40a); and (b) asymptotic approximation (3.40b).

For the selective-partial-update NLMS algorithm the theoretical and simulated MSE values are shown in Figures 3.7 and 3.8. The number of coefficients to be updated is set to $M = 1$. The theoretical MSE is computed using the two asymptotic approximations in (3.50a) and (3.50b). For i.i.d. input, (3.50b) produces a better approximation of the steady-state MSE. On the other hand, strong input correlation and small M/N ratio tend to diminish the accuracy of the approximations. This is common to all data-dependent partial coefficient update methods.

Figures 3.9 and 3.10 show the theoretical and simulated MSE values of the M-max NLMS algorithm. The number of coefficients to be updated is $M = 1$. The theoretical MSE is computed using the asymptotic approximations given in (3.61). For i.i.d. input, the second asymptotic approximation in (3.61b) produces the best approximation of the steady-state MSE. The third approximation in (3.61c) gives the best MSE

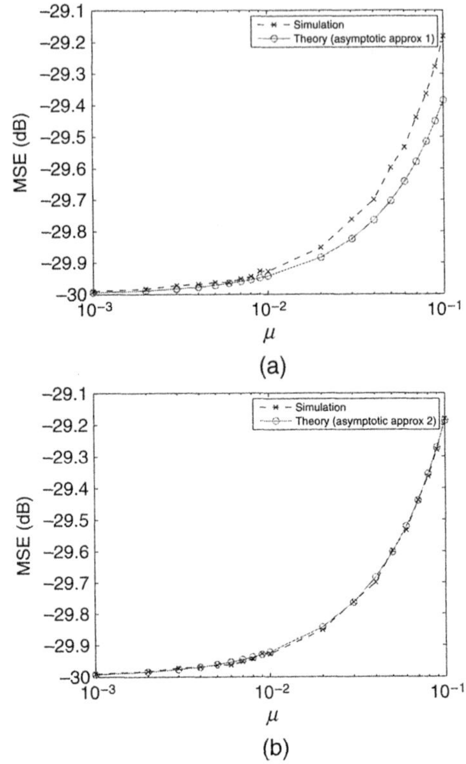

Figure 3.7 Theoretical and simulated MSE of selective-partial-update NLMS ($M = 1$) for i.i.d. Gaussian input. Theoretical MSE is computed using: (a) asymptotic approximation 1 (3.50a); and (b) asymptotic approximation 2 (3.50b).

approximation for the strongly correlated AR(1) input signal. We again observe that strong input correlation adversely impacts the accuracy of the approximations.

3.4 CONVERGENCE ANALYSIS

How fast a given partial-update adaptive filter converges and the range of step-size parameter that ensures convergence are two important questions addressed in this section. The approach adopted here is based on the energy conservation approach (Sayed, 2003) and employs an appropriately weighted version of the error norm relation in (3.17). The weighting used will allow us to express the time evolution of MSD and EMSE in terms of readily available input signal characteristics.

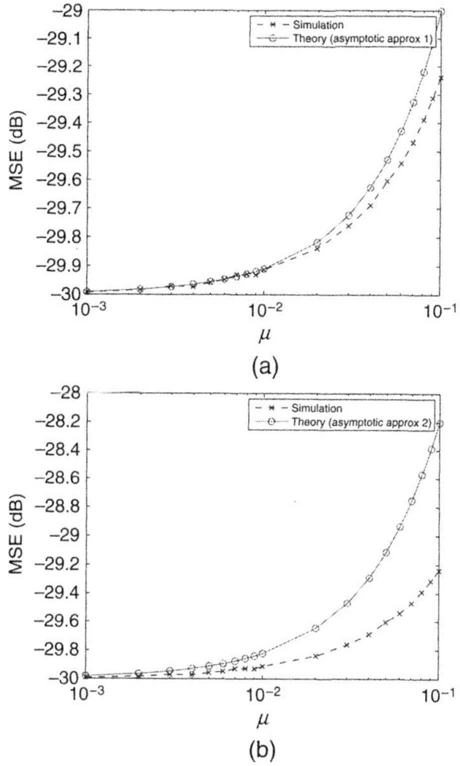

Figure 3.8 Theoretical and simulated MSE of selective-partial-update NLMS ($M = 1$) for AR(1) Gaussian input with eigenvalue spread 135.5. Theoretical MSE is computed using: (a) asymptotic approximation 1 (3.50a); and (b) asymptotic approximation 2 (3.50b).

Consider the partial-update stochastic gradient algorithm in (3.5a), which is reproduced here for convenience:

$$\boldsymbol{w}(k+1) = \boldsymbol{w}(k) + \frac{\mu}{f(\boldsymbol{x}(k))} e(k) \boldsymbol{I}_M(k) \boldsymbol{x}(k) \tag{3.64}$$

Subtracting both sides of the above equation from the optimum solution $\boldsymbol{w}_o$ gives:

$$\Delta \boldsymbol{w}(k+1) = \Delta \boldsymbol{w}(k) - \frac{\mu}{f(\boldsymbol{x}(k))} e(k) \boldsymbol{I}_M(k) \boldsymbol{x}(k) \tag{3.65}$$

where:

$$\Delta \boldsymbol{w}(k) = \boldsymbol{w}_o - \boldsymbol{w}(k) \tag{3.66}$$

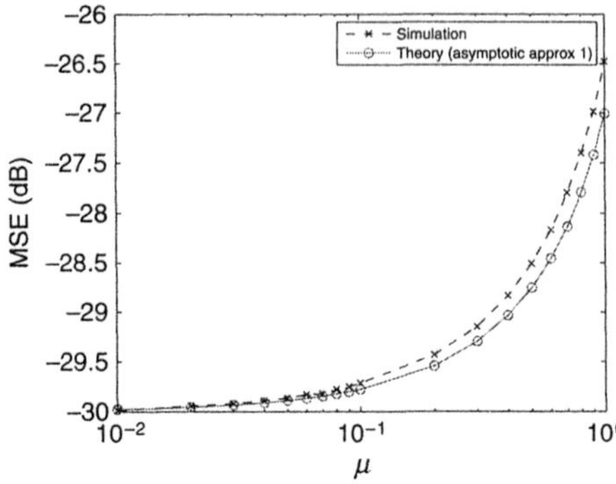

(a)

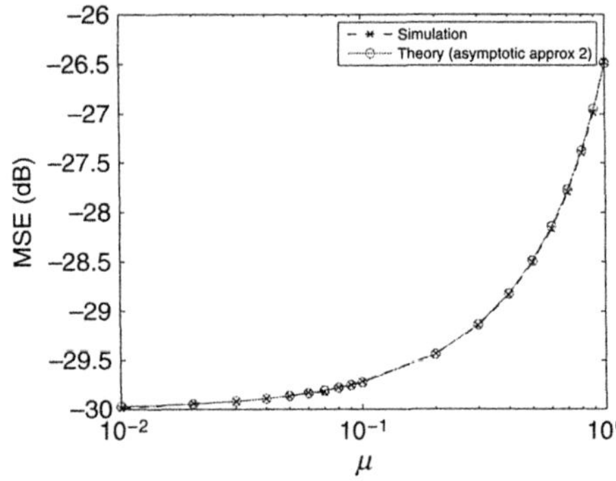

(b)

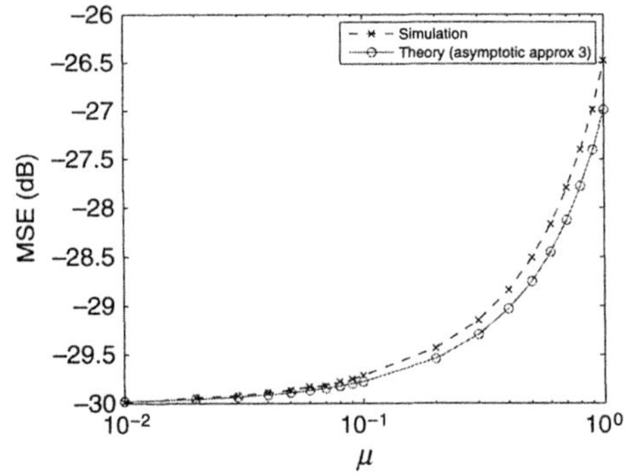

(c)

Figure 3.9 Theoretical and simulated MSE of M-max NLMS ($M = 1$) for i.i.d. Gaussian input. Theoretical MSE is computed using asymptotic approximations: (a) (3.61a); (b) (3.61b); and (c) (3.61c).

Note that in (3.65) partial coefficient updating controlled by $\boldsymbol{I}_M(k)$ leaves non-updated subsets of $\Delta\boldsymbol{w}(k)$ and $\Delta\boldsymbol{w}(k+1)$ identical.

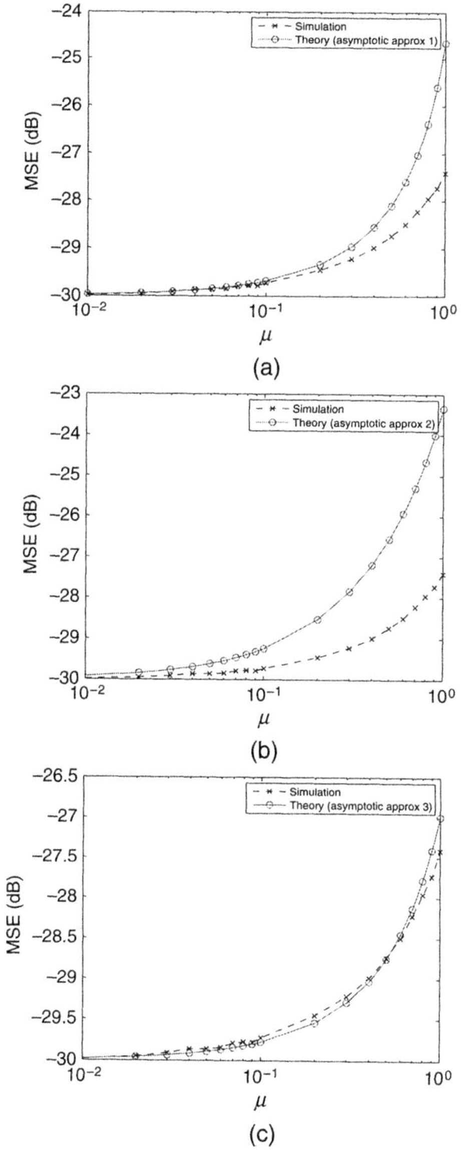

Figure 3.10 Theoretical and simulated MSE of M-max NLMS ($M = 1$) for AR(1) Gaussian input with eigenvalue spread 135.5. Theoretical MSE is computed using asymptotic approximations: (a) (3.61a); (b) (3.61b); and (c) (3.61c).

Taking the $\mathbf{\Sigma}$-weighted squared Euclidean norm of both sides, where $\mathbf{\Sigma}$ is a symmetric positive-definite weighting matrix, leads to:

$$\|\Delta \boldsymbol{w}(k+1)\|_{\mathbf{\Sigma}}^2 = \|\Delta \boldsymbol{w}(k)\|_{\mathbf{\Sigma}}^2 + \frac{\mu^2}{f^2(\boldsymbol{x}(k))} e^2(k)\boldsymbol{x}^T(k)\boldsymbol{I}_M(k)\mathbf{\Sigma}\boldsymbol{I}_M(k)\boldsymbol{x}(k)$$
$$- \frac{2\mu}{f(\boldsymbol{x}(k))} e(k)\Delta\boldsymbol{w}^T(k)\mathbf{\Sigma}\boldsymbol{I}_M(k)\boldsymbol{x}(k) \tag{3.67}$$

where:

$$\|\boldsymbol{x}\|_{\mathbf{\Sigma}}^2 = \boldsymbol{x}^T \, \mathbf{\Sigma} \, \boldsymbol{x}. \tag{3.68}$$

Substituting $e(k) = e_a(k) + v(k)$ [see (3.7c)] into (3.67) gives:

$$\|\Delta \boldsymbol{w}(k+1)\|_{\mathbf{\Sigma}}^2 = \|\Delta \boldsymbol{w}(k)\|_{\mathbf{\Sigma}}^2 + \frac{\mu^2}{f^2(\boldsymbol{x}(k))}(e_a^2(k) + v^2(k)$$
$$+ 2e_a(k)v(k))\boldsymbol{x}^T(k)\boldsymbol{I}_M(k)\mathbf{\Sigma}\boldsymbol{I}_M(k)\boldsymbol{x}(k)$$
$$- \frac{2\mu}{f(\boldsymbol{x}(k))}(e_a(k) + v(k))\Delta\boldsymbol{w}^T(k)\mathbf{\Sigma}\boldsymbol{I}_M(k)\boldsymbol{x}(k) \tag{3.69}$$

Taking the expectation of both sides we have:

$$E\{\|\Delta \boldsymbol{w}(k+1)\|_{\mathbf{\Sigma}}^2\}$$
$$= E\{\|\Delta \boldsymbol{w}(k)\|_{\mathbf{\Sigma}}^2\} + \mu^2 E\left\{\frac{e_a^2(k)\boldsymbol{x}^T(k)\boldsymbol{I}_M(k)\mathbf{\Sigma}\boldsymbol{I}_M(k)\boldsymbol{x}(k)}{f^2(\boldsymbol{x}(k))}\right\}$$
$$+ \mu^2\sigma_v^2 E\left\{\frac{\boldsymbol{x}^T(k)\boldsymbol{I}_M(k)\mathbf{\Sigma}\boldsymbol{I}_M(k)\boldsymbol{x}(k)}{f^2(\boldsymbol{x}(k))}\right\}$$
$$- 2\mu E\left\{\frac{e_a(k)\Delta\boldsymbol{w}^T(k)\mathbf{\Sigma}\boldsymbol{I}_M(k)\boldsymbol{x}(k)}{f(\boldsymbol{x}(k))}\right\} \tag{3.70}$$

where we have used the assumption that $v(k)$ is zero-mean and statistically independent of both $e_a(k)$ and $\boldsymbol{x}(k)$. Recall that the same assumption was also used in the steady-state analysis. After substituting $e_a(k) = \Delta\boldsymbol{w}^T(k)\boldsymbol{x}(k)$ [see (3.8)] we obtain:

$$E\{\|\Delta \boldsymbol{w}(k+1)\|_{\mathbf{\Sigma}}^2\} = E\{\|\Delta \boldsymbol{w}(k)\|_{\mathbf{\Sigma}}^2\}$$
$$+ \mu^2 E\left\{\frac{\Delta\boldsymbol{w}^T(k)\boldsymbol{x}(k)\boldsymbol{x}^T(k)\boldsymbol{I}_M(k)\mathbf{\Sigma}\boldsymbol{I}_M(k)\boldsymbol{x}(k)\boldsymbol{x}^T(k)\Delta\boldsymbol{w}(k)}{f^2(\boldsymbol{x}(k))}\right\}$$
$$+ \mu^2\sigma_v^2 E\left\{\frac{\boldsymbol{x}^T(k)\boldsymbol{I}_M(k)\mathbf{\Sigma}\boldsymbol{I}_M(k)\boldsymbol{x}(k)}{f^2(\boldsymbol{x}(k))}\right\}$$
$$- 2\mu E\left\{\frac{\Delta\boldsymbol{w}^T(k)\mathbf{\Sigma}\boldsymbol{I}_M(k)\boldsymbol{x}(k)\boldsymbol{x}^T(k)\Delta\boldsymbol{w}(k)}{f(\boldsymbol{x}(k))}\right\} \tag{3.71}$$

Rearranging the terms in the right-hand side of (3.71) yields:

$$
\begin{aligned}
E\{\|\Delta \boldsymbol{w}(k+1)\|_{\boldsymbol{\Sigma}}^2\}
&= E\left\{\Delta \boldsymbol{w}^T(k)\left(\boldsymbol{\Sigma} + \mu^2 \frac{\boldsymbol{x}(k)\boldsymbol{x}^T(k)\boldsymbol{I}_M(k)\boldsymbol{\Sigma}\boldsymbol{I}_M(k)\boldsymbol{x}(k)\boldsymbol{x}^T(k)}{f^2(\boldsymbol{x}(k))}\right.\right. \\
&\quad \left.\left. - 2\mu \frac{\boldsymbol{\Sigma}\boldsymbol{I}_M(k)\boldsymbol{x}(k)\boldsymbol{x}^T(k)}{f(\boldsymbol{x}(k))}\right)\Delta \boldsymbol{w}(k)\right\} \\
&\quad + \mu^2\sigma_v^2 E\left\{\frac{\boldsymbol{x}^T(k)\boldsymbol{I}_M(k)\boldsymbol{\Sigma}\boldsymbol{I}_M(k)\boldsymbol{x}(k)}{f^2(\boldsymbol{x}(k))}\right\}
\end{aligned} \tag{3.72}
$$

or:

$$
\begin{aligned}
E\{\|\Delta \boldsymbol{w}(k+1)\|_{\boldsymbol{\Sigma}}^2\} &= E\{\|\Delta \boldsymbol{w}(k)\|_{\boldsymbol{\Sigma}'}^2\} \\
&\quad + \mu^2\sigma_v^2 E\left\{\frac{\boldsymbol{x}^T(k)\boldsymbol{I}_M(k)\boldsymbol{\Sigma}\boldsymbol{I}_M(k)\boldsymbol{x}(k)}{f^2(\boldsymbol{x}(k))}\right\}
\end{aligned} \tag{3.73}
$$

where

$$
\begin{aligned}
\boldsymbol{\Sigma}' &= \boldsymbol{\Sigma} + \mu^2 \frac{\boldsymbol{x}(k)\boldsymbol{x}^T(k)\boldsymbol{I}_M(k)\boldsymbol{\Sigma}\boldsymbol{I}_M(k)\boldsymbol{x}(k)\boldsymbol{x}^T(k)}{f^2(\boldsymbol{x}(k))} \\
&\quad - 2\mu \frac{\boldsymbol{\Sigma}\boldsymbol{I}_M(k)\boldsymbol{x}(k)\boldsymbol{x}^T(k)}{f(\boldsymbol{x}(k))}
\end{aligned} \tag{3.74}
$$

Equations (3.73) and (3.74) are fundamental to the convergence analysis of not only partial-update but also conventional full-update adaptive filters.

The computation of $E\{\|\Delta \boldsymbol{w}(k)\|_{\boldsymbol{\Sigma}'}^2\}$ is not trivial mainly because of the statistical dependence between the vectors involved. To make the task at hand easier we make a simplifying assumption; viz. the input regressor vector $\boldsymbol{x}(k)$ is i.i.d. This implies that $\boldsymbol{w}(k)$ and $\Delta \boldsymbol{w}(k)$ are both statistically independent of $\boldsymbol{x}(k)$ since they are determined by $\boldsymbol{x}(k-1), \boldsymbol{x}(k-2), \ldots$ Recall that we used a similar assumption in the steady-state analysis of partial-update adaptive filters. Strictly speaking, this independence assumption is not true for tapped-delay-line adaptive filters and correlated input signals. Despite this, we will see that the theoretical convergence results based on this simplifying assumption produce good agreement with experimental results even for strongly correlated input signals.

For i.i.d. $\boldsymbol{x}(k)$ we can write:

$$
\begin{aligned}
E\{\|\Delta \boldsymbol{w}(k)\|_{\boldsymbol{\Sigma}'}^2\} &= E\{\Delta \boldsymbol{w}^T(k)\boldsymbol{\Sigma}'\Delta \boldsymbol{w}(k)\} && \text{(3.75a)} \\
&= E\{E\{\Delta \boldsymbol{w}^T(k)\boldsymbol{\Sigma}'\Delta \boldsymbol{w}(k) \mid \Delta \boldsymbol{w}(k)\}\} && \text{(3.75b)} \\
&= E\{\Delta \boldsymbol{w}^T(k)E\{\boldsymbol{\Sigma}'\}\Delta \boldsymbol{w}(k)\} && \text{(3.75c)} \\
&= E\{\|\Delta \boldsymbol{w}(k)\|_{\boldsymbol{\Gamma}}^2\} && \text{(3.75d)}
\end{aligned}
$$

where $\boldsymbol{\Gamma} = E\{\boldsymbol{\Sigma}'\}$. Here we have used the fact that $\Delta\boldsymbol{w}(k)$ and $\boldsymbol{\Sigma}'$ are statistically independent as a consequence of the i.i.d. input regressor assumption. Note also that for data-dependent partial-update techniques, $\boldsymbol{I}_M(k)$ is determined by $\boldsymbol{x}(k)$ only, thereby not affecting the i.i.d. assumption. Using (3.75) in (3.73), the fundamental weighted variance recursion in (3.73) and (3.74) is now written as:

$$E\{\|\Delta\boldsymbol{w}(k+1)\|^2_{\boldsymbol{\Sigma}}\} = E\{\|\Delta\boldsymbol{w}(k)\|^2_{\boldsymbol{\Gamma}}\} + \mu^2\sigma_v^2 E\left\{\frac{\boldsymbol{x}^T(k)\boldsymbol{I}_M(k)\boldsymbol{\Sigma}\boldsymbol{I}_M(k)\boldsymbol{x}(k)}{f^2(\boldsymbol{x}(k))}\right\} \tag{3.76a}$$

$$\boldsymbol{\Gamma} = \boldsymbol{\Sigma} + \mu^2 E\left\{\frac{\boldsymbol{x}(k)\boldsymbol{x}^T(k)\boldsymbol{I}_M(k)\boldsymbol{\Sigma}\boldsymbol{I}_M(k)\boldsymbol{x}(k)\boldsymbol{x}^T(k)}{f^2(\boldsymbol{x}(k))}\right\} - 2\mu\boldsymbol{\Sigma}E\left\{\frac{\boldsymbol{I}_M(k)\boldsymbol{x}(k)\boldsymbol{x}^T(k)}{f(\boldsymbol{x}(k))}\right\} \tag{3.76b}$$

This recursion requires computation of the following expectations which are functions of the input signal only:

$$E\left\{\frac{\boldsymbol{x}^T(k)\boldsymbol{I}_M(k)\boldsymbol{\Sigma}\boldsymbol{I}_M(k)\boldsymbol{x}(k)}{f^2(\boldsymbol{x}(k))}\right\}, \quad E\left\{\frac{\boldsymbol{x}(k)\boldsymbol{x}^T(k)\boldsymbol{I}_M(k)\boldsymbol{\Sigma}\boldsymbol{I}_M(k)\boldsymbol{x}(k)\boldsymbol{x}^T(k)}{f^2(\boldsymbol{x}(k))}\right\}$$
$$\text{and} \quad E\left\{\frac{\boldsymbol{I}_M(k)\boldsymbol{x}(k)\boldsymbol{x}^T(k)}{f(\boldsymbol{x}(k))}\right\} \tag{3.77}$$

Being independent of $\Delta\boldsymbol{w}(k)$, the above expectations can be computed numerically by sample averaging. In the following sections we will investigate how to derive simple recursions for the time evolution of EMSE and MSD based on (3.76).

We now briefly digress to focus on the mean of the coefficient error vector. With $e(k) = \boldsymbol{x}^T(k)\Delta\boldsymbol{w}(k) + v(k)$ substituted, the coefficient error recursion in (3.65) becomes

$$\Delta\boldsymbol{w}(k+1) = \left(\boldsymbol{I} - \frac{\mu}{f(\boldsymbol{x}(k))}\boldsymbol{I}_M(k)\boldsymbol{x}(k)\boldsymbol{x}^T(k)\right)\Delta\boldsymbol{w}(k) - \frac{\mu}{f(\boldsymbol{x}(k))}v(k)\boldsymbol{I}_M(k)\boldsymbol{x}(k) \tag{3.78}$$

Applying the i.i.d. assumption for $\boldsymbol{x}(k)$ and taking expectation of both sides, the above recursion is written as:

$$E\{\Delta\boldsymbol{w}(k+1)\} = \left(\boldsymbol{I} - \mu E\left\{\frac{1}{f(\boldsymbol{x}(k))}\boldsymbol{I}_M(k)\boldsymbol{x}(k)\boldsymbol{x}^T(k)\right\}\right)E\{\Delta\boldsymbol{w}(k)\} \tag{3.79}$$

We will make use of this expression when studying convergence in the mean.

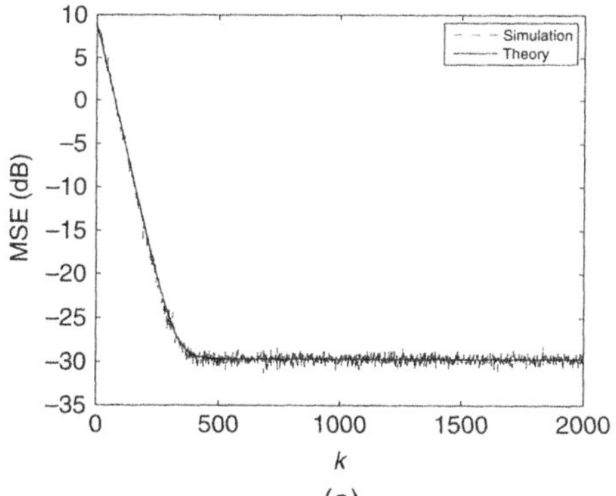

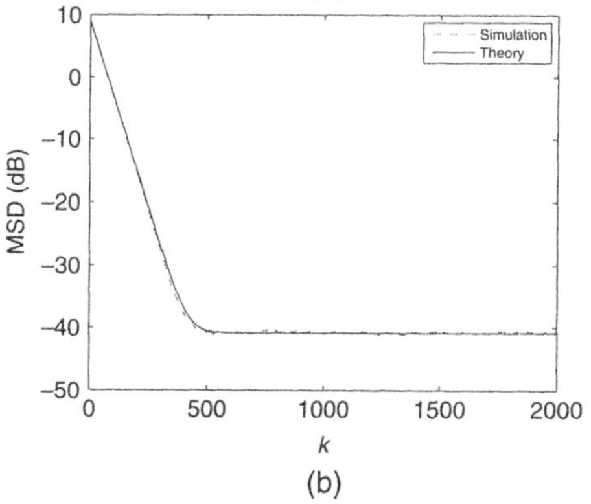

Figure 3.11 Theoretical and simulated MSE and MSD of full-update LMS algorithm for i.i.d. Gaussian input.

3.4.1 Partial-update LMS algorithms

Vectorization

We will now analyse the weighted variance relation in (3.76) for the special case of partial-update LMS algorithms for which $f(\boldsymbol{x}(k)) = 1$. Setting $f(\boldsymbol{x}(k)) = 1$ simplifies (3.76) to:

$$E\{\|\Delta \boldsymbol{w}(k+1)\|_{\boldsymbol{\Sigma}}^2\} = E\{\|\Delta \boldsymbol{w}(k)\|_{\boldsymbol{\Gamma}}^2\} + \mu^2 \sigma_v^2 E\{\|\boldsymbol{I}_M(k)\boldsymbol{x}(k)\|_{\boldsymbol{\Sigma}}^2\} \quad (3.80a)$$

$$\boldsymbol{\Gamma} = \boldsymbol{\Sigma} + \mu^2 E\{\|\boldsymbol{I}_M(k)\boldsymbol{x}(k)\|_{\boldsymbol{\Sigma}}^2 \boldsymbol{x}(k)\boldsymbol{x}^T(k)\} - 2\mu \boldsymbol{\Sigma} \boldsymbol{R}_M \quad (3.80b)$$

where $\boldsymbol{R}_M = E\{\boldsymbol{I}_M(k)\boldsymbol{x}(k)\boldsymbol{x}^T(k)\}$. To reveal the relationship between the weighting matrices $\boldsymbol{\Sigma}$ and $\boldsymbol{\Gamma}$ we will vectorize the latter. The vectorization operation $\text{vec}(\boldsymbol{A})$ stacks each column of the matrix $\boldsymbol{A}$ to create a column vector of dimension equal to

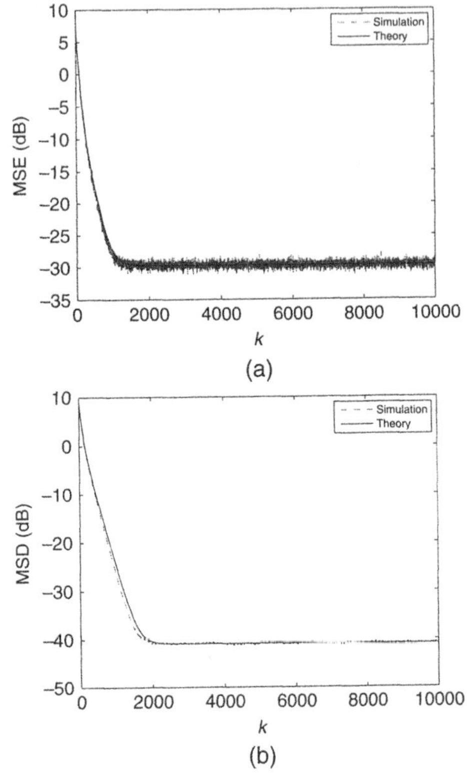

Figure 3.12 Theoretical and simulated MSE and MSD of full-update LMS algorithm for AR(1) Gaussian input with eigenvalue spread 23.5.

the number of entries of $\boldsymbol{A}$. Vectorizing $\boldsymbol{\Gamma}$ yields:

$$\begin{aligned}\mathrm{vec}(\boldsymbol{\Gamma}) &= \mathrm{vec}(\boldsymbol{\Sigma}) + \mu^2 E\{\mathrm{vec}(\|\boldsymbol{I}_M(k)\boldsymbol{x}(k)\|^2_{\boldsymbol{\Sigma}}\boldsymbol{x}(k)\boldsymbol{x}^T(k))\} \\ &\quad - 2\mu \mathrm{vec}(\boldsymbol{\Sigma}\boldsymbol{R}_M) \qquad (3.81a)\\ &= \Big(\boldsymbol{I}\otimes\boldsymbol{I} + \mu^2 E\{(\boldsymbol{x}(k)\boldsymbol{x}^T(k)\boldsymbol{I}_M(k))\otimes(\boldsymbol{x}(k)\boldsymbol{x}^T(k)\boldsymbol{I}_M(k))\} \\ &\quad - 2\mu(\boldsymbol{R}_M\otimes\boldsymbol{I})\Big)\boldsymbol{\sigma} \qquad (3.81b)\\ &= \boldsymbol{F}\boldsymbol{\sigma} \qquad (3.81c)\end{aligned}$$

where $\boldsymbol{\sigma} = \mathrm{vec}(\boldsymbol{\Sigma})$, $\otimes$ denotes the Kronecker product, and $\boldsymbol{R}_M$ is assumed to be a symmetric matrix. We have used the following identity in (3.81):

$$\mathrm{vec}(\boldsymbol{A}\boldsymbol{\Sigma}\boldsymbol{B}) = (\boldsymbol{B}^T \otimes \boldsymbol{A})\mathrm{vec}(\boldsymbol{\Sigma}) \qquad (3.82)$$

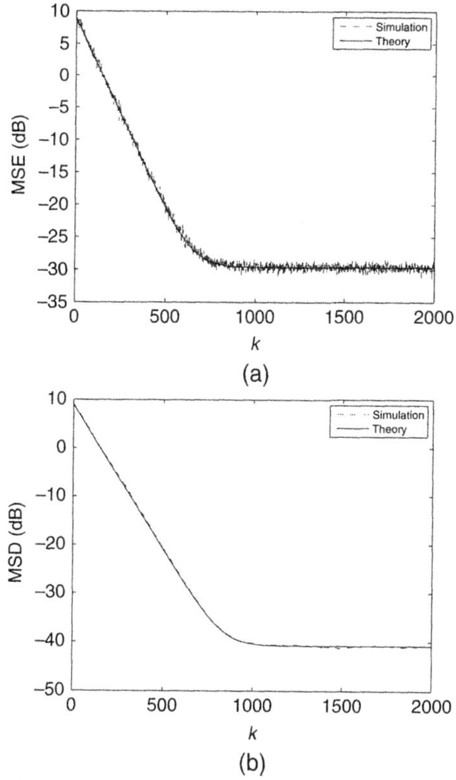

Figure 3.13 Theoretical and simulated MSE and MSD of periodic-partial-update LMS algorithm with $S = 2$ for i.i.d. Gaussian input.

As we shall see later, the matrix $\boldsymbol{F}$ plays a significant role in determining the convergence and stability properties of partial-update adaptive filters.

After vectorization the weighted variance relation in (3.80) becomes:

$$E\{\|\Delta \boldsymbol{w}(k+1)\|^2_{\boldsymbol{\sigma}}\} = E\{\|\Delta \boldsymbol{w}(k)\|^2_{\boldsymbol{F}\boldsymbol{\sigma}}\} + \mu^2 \sigma_v^2 \boldsymbol{r}_M^T \boldsymbol{\sigma} \tag{3.83a}$$

$$\begin{aligned}\boldsymbol{F} = \boldsymbol{I}_{N^2} &+ \mu^2 E\{(\boldsymbol{x}(k)\boldsymbol{x}^T(k)\boldsymbol{I}_M(k)) \otimes (\boldsymbol{x}(k)\boldsymbol{x}^T(k)\boldsymbol{I}_M(k))\} \\ &- 2\mu(\boldsymbol{R}_M \otimes \boldsymbol{I})\end{aligned} \tag{3.83b}$$

where $\boldsymbol{I}_{N^2}$ is the identity matrix of size N^2

$$\boldsymbol{r}_M = \text{vec}(E\{\boldsymbol{I}_M(k)\boldsymbol{x}(k)\boldsymbol{x}^T(k)\boldsymbol{I}_M(k)\}) \tag{3.84}$$

and $\|\boldsymbol{x}\|^2_{\boldsymbol{\sigma}} = \boldsymbol{x}^T \text{vec}^{-1}(\boldsymbol{\sigma})\boldsymbol{x}$, i.e., a vector weight $\boldsymbol{\sigma}$ implies the use of the weighting matrix $\boldsymbol{\Sigma}$ such that $\boldsymbol{\sigma} = \text{vec}(\boldsymbol{\Sigma})$. The vector $\boldsymbol{r}_M$ is derived from:

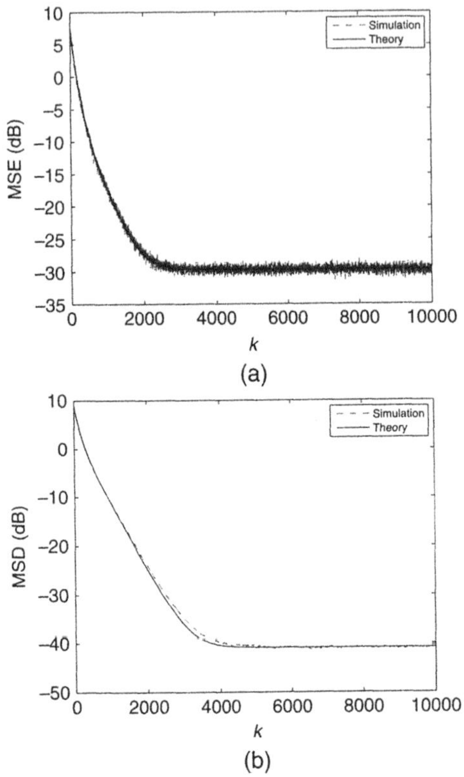

Figure 3.14 Theoretical and simulated MSE and MSD of periodic-partial-update LMS algorithm with $S = 2$ for AR(1) Gaussian input with eigenvalue spread 23.5.

$$\begin{aligned} E\{\|\boldsymbol{I}_M(k)\boldsymbol{x}(k)\|^2_{\boldsymbol{\Sigma}}\} &= E\{\boldsymbol{x}^T(k)\boldsymbol{I}_M(k)\boldsymbol{\Sigma}\boldsymbol{I}_M(k)\boldsymbol{x}(k)\} && (3.85a)\\ &= \mathrm{tr}E\{\boldsymbol{I}_M(k)\boldsymbol{x}(k)\boldsymbol{x}^T(k)\boldsymbol{I}_M(k)\boldsymbol{\Sigma}\} && (3.85b)\\ &= \mathrm{tr}(E\{\boldsymbol{I}_M(k)\boldsymbol{x}(k)\boldsymbol{x}^T(k)\boldsymbol{I}_M(k)\}\boldsymbol{\Sigma}) && (3.85c)\\ &= \mathrm{vec}^T(E\{\boldsymbol{I}_M(k)\boldsymbol{x}(k)\boldsymbol{x}^T(k)\boldsymbol{I}_M(k)\})\boldsymbol{\sigma} && (3.85d)\\ &= \boldsymbol{r}_M^T\boldsymbol{\sigma} && (3.85e) \end{aligned}$$

Iterating (3.83) back to the initialization $\boldsymbol{w}(0) = \boldsymbol{0}$ yields:

$$\begin{aligned} E\{\|\Delta\boldsymbol{w}(k+1)\|^2_{\boldsymbol{\sigma}}\} &= E\{\|\Delta\boldsymbol{w}(k)\|^2_{\boldsymbol{F}\boldsymbol{\sigma}}\} + \mu^2\sigma_v^2\boldsymbol{r}_M^T\boldsymbol{\sigma} && (3.86a)\\ E\{\|\Delta\boldsymbol{w}(k)\|^2_{\boldsymbol{F}\boldsymbol{\sigma}}\} &= E\{\|\Delta\boldsymbol{w}(k-1)\|^2_{\boldsymbol{F}^2\boldsymbol{\sigma}}\} + \mu^2\sigma_v^2\boldsymbol{r}_M^T\boldsymbol{F}\boldsymbol{\sigma} && (3.86b)\\ &\;\;\vdots && (3.86c)\\ E\{\|\Delta\boldsymbol{w}(1)\|^2_{\boldsymbol{F}^k\boldsymbol{\sigma}}\} &= \|\boldsymbol{w}_o\|^2_{\boldsymbol{F}^{k+1}\boldsymbol{\sigma}} + \mu^2\sigma_v^2\boldsymbol{r}_M^T\boldsymbol{F}^k\boldsymbol{\sigma} && (3.86d) \end{aligned}$$

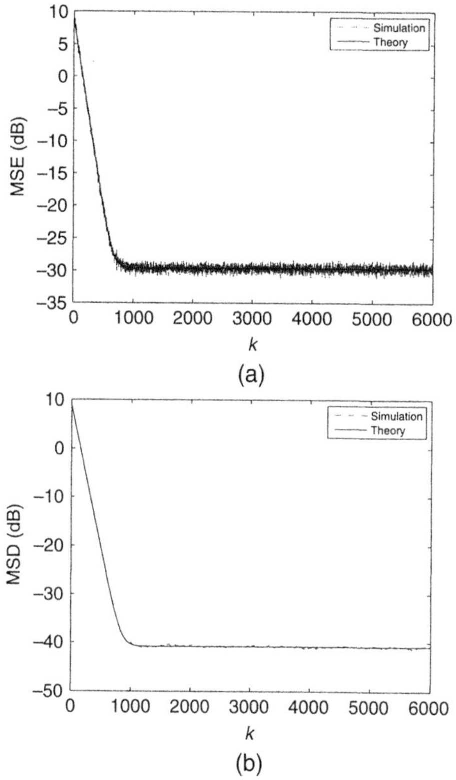

Figure 3.15 Theoretical and simulated MSE and MSD of sequential-partial-update LMS algorithm with $M = 5$ for i.i.d. Gaussian input.

The above recursion is more compactly written as:

$$E\{\|\Delta \boldsymbol{w}(k+1)\|^2_{\boldsymbol{\sigma}}\} = \|\boldsymbol{w}_o\|^2_{\boldsymbol{F}^{k+1}\boldsymbol{\sigma}} + \mu^2\sigma_v^2 \sum_{i=0}^{k} \boldsymbol{r}_M^T \boldsymbol{F}^i \boldsymbol{\sigma} \tag{3.87}$$

Subtracting $E\{\|\Delta \boldsymbol{w}(k)\|^2_{\boldsymbol{\sigma}}\}$ from $E\{\|\Delta \boldsymbol{w}(k+1)\|^2_{\boldsymbol{\sigma}}\}$ we obtain:

$$\begin{aligned} E\{\|\Delta \boldsymbol{w}(k+1)\|^2_{\boldsymbol{\sigma}}\} &= E\{\|\Delta \boldsymbol{w}(k)\|^2_{\boldsymbol{\sigma}}\} + \|\boldsymbol{w}_o\|^2_{\boldsymbol{F}^{k+1}\boldsymbol{\sigma}} \\ &\quad - \|\boldsymbol{w}_o\|^2_{\boldsymbol{F}^{k}\boldsymbol{\sigma}} + \mu^2\sigma_v^2 \boldsymbol{r}_M^T \boldsymbol{F}^k \boldsymbol{\sigma} \\ &= E\{\|\Delta \boldsymbol{w}(k)\|^2_{\boldsymbol{\sigma}}\} + \|\boldsymbol{w}_o\|^2_{\boldsymbol{F}^{k}(\boldsymbol{F}-\boldsymbol{I}_{N^2})\boldsymbol{\sigma}} \\ &\quad + \mu^2\sigma_v^2 \boldsymbol{r}_M^T \boldsymbol{F}^k \boldsymbol{\sigma} \end{aligned} \tag{3.88}$$

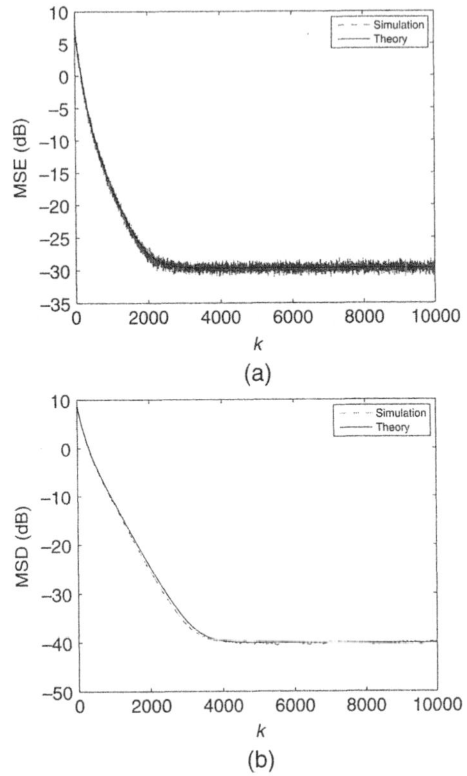

Figure 3.16 Theoretical and simulated MSE and MSD of sequential-partial-update LMS algorithm with $M = 5$ for AR(1) Gaussian input with eigenvalue spread 23.5.

Equations (3.87) and (3.88) give the time evolution of MSD if $\boldsymbol{\sigma} = \text{vec}(\boldsymbol{I})$ (i.e., $\boldsymbol{\Sigma} = \boldsymbol{I}$).

The independence assumption between $\boldsymbol{w}(k)$ and $\boldsymbol{x}(k)$ leads to:

$$E\{e_a^2(k)\} = E\{\Delta \boldsymbol{w}^T(k)\boldsymbol{x}(k)\boldsymbol{x}^T(k)\Delta \boldsymbol{w}(k)\} \tag{3.89a}$$

$$= E\{E\{\Delta \boldsymbol{w}^T(k)\boldsymbol{x}(k)\boldsymbol{x}^T(k)\Delta \boldsymbol{w}(k) \mid \Delta \boldsymbol{w}(k)\}\} \tag{3.89b}$$

$$= E\{\Delta \boldsymbol{w}^T(k)\boldsymbol{R}\Delta \boldsymbol{w}(k)\} \tag{3.89c}$$

$$= E\{\|\Delta \boldsymbol{w}(k)\|_{\boldsymbol{R}}^2\} \tag{3.89d}$$

Thus, setting $\boldsymbol{\sigma} = \text{vec}(\boldsymbol{R})$ in (3.87) and (3.88), where $\boldsymbol{R} = E\{\boldsymbol{x}(k)\boldsymbol{x}^T(k)\}$, gives the time evolution of EMSE.

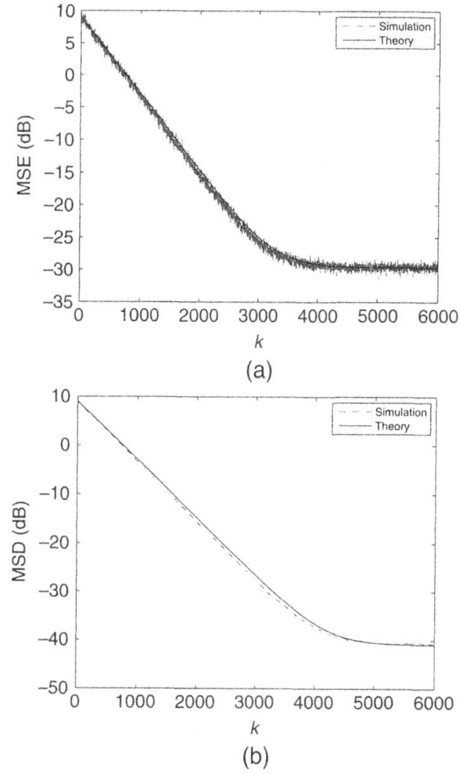

Figure 3.17 Theoretical and simulated MSE and MSD of sequential-partial-update LMS algorithm with $M = 1$ for i.i.d. Gaussian input.

Steady-state analysis

Assuming mean-squared convergence for the partial-update LMS algorithm under consideration, as $k \to \infty$ (3.83) becomes:

$$E\{\|\Delta \boldsymbol{w}(\infty)\|^2_{\boldsymbol{\sigma}}\} - E\{\|\Delta \boldsymbol{w}(\infty)\|^2_{\boldsymbol{F\sigma}}\} = \mu^2 \sigma_v^2 \boldsymbol{r}_M^T \boldsymbol{\sigma} \tag{3.90}$$

which can be rewritten as:

$$E\left\{\Delta \boldsymbol{w}^T(\infty)\text{vec}^{-1}(\boldsymbol{\sigma})\Delta \boldsymbol{w}(\infty) - \Delta \boldsymbol{w}^T(\infty)\text{vec}^{-1}(\boldsymbol{F\sigma})\Delta \boldsymbol{w}(\infty)\right\} = \mu^2 \sigma_v^2 \boldsymbol{r}_M^T \boldsymbol{\sigma} \tag{3.91a}$$

$$E\left\{\Delta \boldsymbol{w}^T(\infty)\left(\text{vec}^{-1}(\boldsymbol{\sigma}) - \text{vec}^{-1}(\boldsymbol{F\sigma})\right)\Delta \boldsymbol{w}(\infty)\right\} = \mu^2 \sigma_v^2 \boldsymbol{r}_M^T \boldsymbol{\sigma} \tag{3.91b}$$

$$E\{\|\Delta \boldsymbol{w}^T(\infty)\|^2_{(\boldsymbol{I}_{N^2} - \boldsymbol{F})\boldsymbol{\sigma}}\} = \mu^2 \sigma_v^2 \boldsymbol{r}_M^T \boldsymbol{\sigma} \tag{3.91c}$$

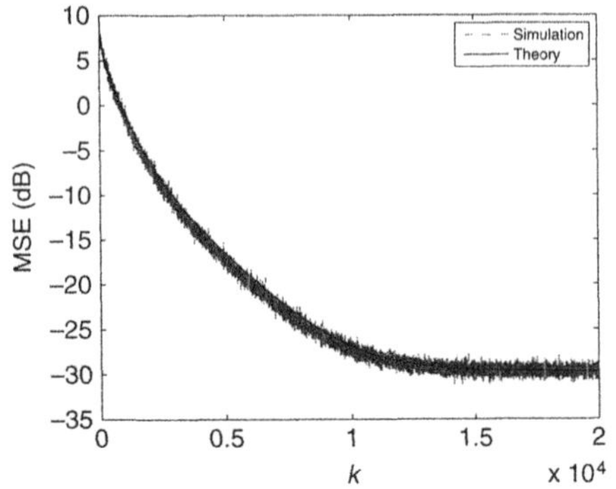

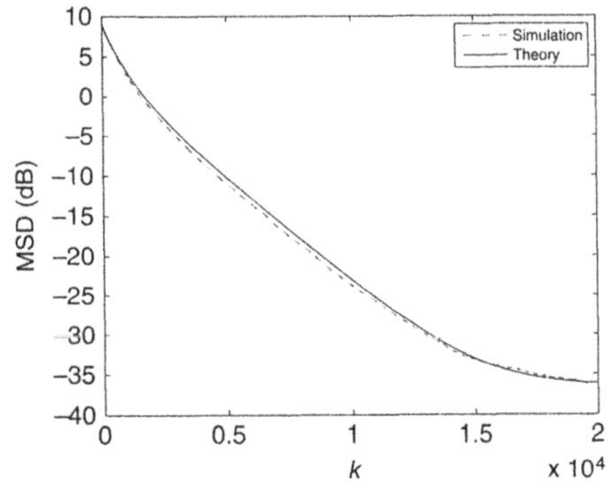

Figure 3.18 Theoretical and simulated MSE and MSD of sequential-partial-update LMS algorithm with $M = 1$ for AR(1) Gaussian input with eigenvalue spread 23.5.

The steady-state MSD $E\{\|\Delta \boldsymbol{w}^T(\infty)\|^2_{\text{vec}(\boldsymbol{I})}\}$ is obtained by setting the weighting vector to $\boldsymbol{\sigma} = \left(\boldsymbol{I}_{N^2} - \boldsymbol{F}\right)^{-1} \text{vec}(\boldsymbol{I})$:

$$\text{Steady-state MSD} = \mu^2 \sigma_v^2 \boldsymbol{r}_M^T \left(\boldsymbol{I}_{N^2} - \boldsymbol{F}\right)^{-1} \text{vec}(\boldsymbol{I}) \tag{3.92}$$

The steady-state EMSE $E\{\|\Delta \boldsymbol{w}^T(\infty)\|^2_{\text{vec}(\boldsymbol{R})}\}$ is given by setting $\boldsymbol{\sigma} = \left(\boldsymbol{I}_{N^2} - \boldsymbol{F}\right)^{-1} \text{vec}(\boldsymbol{R})$:

$$\text{Steady-state EMSE} = \mu^2 \sigma_v^2 \boldsymbol{r}_M^T \left(\boldsymbol{I}_{N^2} - \boldsymbol{F}\right)^{-1} \text{vec}(\boldsymbol{R}) \tag{3.93}$$

Suppose that the step-size parameter μ is sufficiently small so that (3.83b) can be approximated by:

$$\boldsymbol{F} \approx \boldsymbol{I}_{N^2} - 2\mu(\boldsymbol{R}_M \otimes \boldsymbol{I}) \tag{3.94}$$

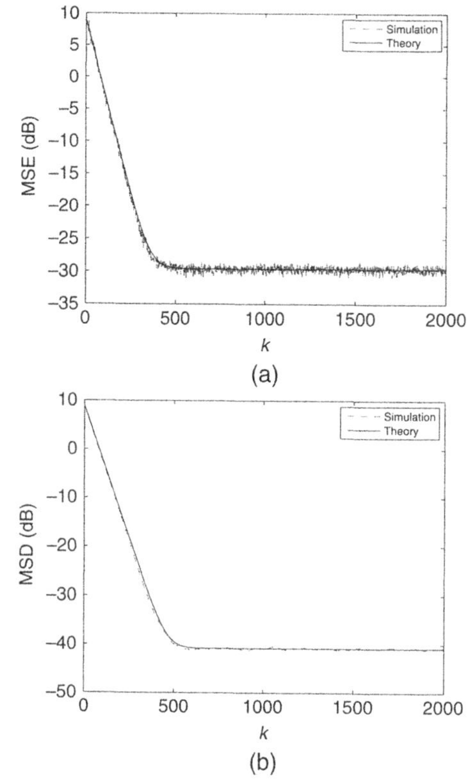

Figure 3.19 Theoretical and simulated MSE and MSD of M-max LMS algorithm with $M = 5$ for i.i.d. Gaussian input.

Substituting (3.94) into (3.93) yields:

$$\begin{aligned}\mu^2\sigma_v^2 \boldsymbol{r}_M^T \left(\boldsymbol{I}_{N^2} - \boldsymbol{F}\right)^{-1} \mathrm{vec}(\boldsymbol{R}) &\approx \frac{\mu\sigma_v^2}{2}\boldsymbol{r}_M^T(\boldsymbol{R}_M \otimes \boldsymbol{I})^{-1}\mathrm{vec}(\boldsymbol{R}) \\ &\approx \frac{\mu\sigma_v^2}{2}\boldsymbol{r}_M^T(\boldsymbol{R}_M^{-1} \otimes \boldsymbol{I})\mathrm{vec}(\boldsymbol{R}) \\ &\approx \frac{\mu\sigma_v^2}{2}\boldsymbol{r}_M^T\mathrm{vec}(\boldsymbol{R}\boldsymbol{R}_M^{-1}) \\ &\approx \frac{\mu\sigma_v^2}{2}\mathrm{tr}(E\{\boldsymbol{I}_M(k)\boldsymbol{x}(k)\boldsymbol{x}^T(k)\boldsymbol{I}_M(k)\} \\ &\quad \times \boldsymbol{R}\boldsymbol{R}_M^{-1})\end{aligned} \tag{3.95}$$

If $x(k)$ is an i.i.d. signal, then:

$$E\{\boldsymbol{I}_M(k)\boldsymbol{x}(k)\boldsymbol{x}^T(k)\boldsymbol{I}_M(k)\} = \boldsymbol{R}_M \tag{3.96}$$

$$\mathrm{tr}(E\{\boldsymbol{I}_M(k)\boldsymbol{x}(k)\boldsymbol{x}^T(k)\boldsymbol{I}_M(k)\}\boldsymbol{R}\boldsymbol{R}_M^{-1}) = \mathrm{tr}\boldsymbol{R} \tag{3.97}$$

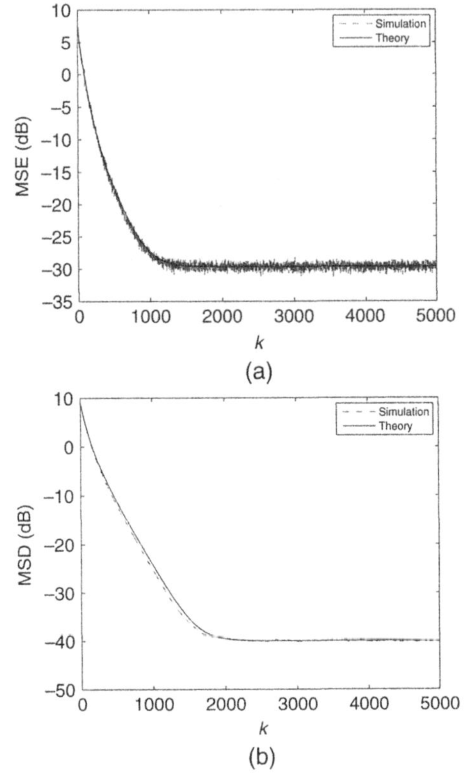

Figure 3.20 Theoretical and simulated MSE and MSD of M-max LMS algorithm with $M = 5$ for AR(1) Gaussian input with eigenvalue spread 23.5.

and we have:

$$\text{Steady-state EMSE} \approx \frac{\mu \sigma_v^2}{2} \mathrm{tr} \boldsymbol{R} \tag{3.98}$$

This result is identical to the small step-size approximation for steady-state EMSE derived in Section 3.3.1.

Mean-square stability

In this section we study the stability bounds for the step-size parameter. Our objective is to determine the maximum value of the positive step-size parameter μ that ensures convergence for the weighted variance recursion in (3.83a). In the following analysis the state-space formulation developed in (Sayed, 2003) is employed in order to derive the stability conditions for the step-size parameter. Stacking (3.83a) for weighting vectors $\boldsymbol{\sigma}$, $\boldsymbol{F}\boldsymbol{\sigma}, \ldots, \boldsymbol{F}^{N^2-1}\boldsymbol{\sigma}$ gives a state-space equation with N^2 states:

$$\underbrace{\begin{bmatrix} E\{\|\Delta\boldsymbol{w}(k+1)\|^2_{\boldsymbol{\sigma}}\} \\ E\{\|\Delta\boldsymbol{w}(k+1)\|^2_{\boldsymbol{F}\boldsymbol{\sigma}}\} \\ E\{\|\Delta\boldsymbol{w}(k+1)\|^2_{\boldsymbol{F}^2\boldsymbol{\sigma}}\} \\ \vdots \\ E\{\|\Delta\boldsymbol{w}(k+1)\|^2_{\boldsymbol{F}^{N^2-1}\boldsymbol{\sigma}}\} \end{bmatrix}}_{\boldsymbol{\xi}(k+1)} = \underbrace{\begin{bmatrix} 0 & 1 & & & \boldsymbol{0} \\ & 0 & 1 & & \\ \vdots & & 0 & 1 & \\ & & & \ddots & \ddots \\ 0 & & \cdots & 0 & 1 \\ -p_0 & -p_1 & & \cdots & -p_{N^2-1} \end{bmatrix}}_{\boldsymbol{\mathcal{F}}} \underbrace{\begin{bmatrix} E\{\|\Delta\boldsymbol{w}(k)\|^2_{\boldsymbol{\sigma}}\} \\ E\{\|\Delta\boldsymbol{w}(k)\|^2_{\boldsymbol{F}\boldsymbol{\sigma}}\} \\ E\{\|\Delta\boldsymbol{w}(k)\|^2_{\boldsymbol{F}^2\boldsymbol{\sigma}}\} \\ \vdots \\ E\{\|\Delta\boldsymbol{w}(k)\|^2_{\boldsymbol{F}^{N^2-1}\boldsymbol{\sigma}}\} \end{bmatrix}}_{\boldsymbol{\xi}(k)} + \mu^2\sigma_v^2 \underbrace{\begin{bmatrix} \boldsymbol{r}_M^T\boldsymbol{\sigma} \\ \boldsymbol{r}_M^T\boldsymbol{F}\boldsymbol{\sigma} \\ \boldsymbol{r}_M^T\boldsymbol{F}^2\boldsymbol{\sigma} \\ \vdots \\ \boldsymbol{r}_M^T\boldsymbol{F}^{N^2-1}\boldsymbol{\sigma} \end{bmatrix}}_{\boldsymbol{\rho}} \tag{3.99}$$

which is more compactly written as:

$$\boldsymbol{\xi}(k+1) = \boldsymbol{\mathcal{F}}\,\boldsymbol{\xi}(k) + \mu^2\sigma_v^2\boldsymbol{\rho} \tag{3.100}$$

In (3.99) the p_i are the coefficients of the characteristic polynomial of $\boldsymbol{F}$:

$$p(x) = \det(x\boldsymbol{I}_{N^2} - \boldsymbol{F}) = x^{N^2} + p_{N^2-1}x^{N^2-1} + \cdots + p_1 x + p_0 \tag{3.101}$$

Since $p(\boldsymbol{F}) = \boldsymbol{0}$, we have:

$$\boldsymbol{F}^{N^2} = -p_0\boldsymbol{I}_{N^2} - p_1\boldsymbol{F} - \cdots - p_{N^2-1}\boldsymbol{F}^{N^2-1} \tag{3.102}$$

and:

$$E\{\|\Delta\boldsymbol{w}(k)\|^2_{\boldsymbol{F}^{N^2}\boldsymbol{\sigma}}\} = -p_0 E\{\|\Delta\boldsymbol{w}(k)\|^2_{\boldsymbol{\sigma}}\} - p_1 E\{\|\Delta\boldsymbol{w}(k)\|^2_{\boldsymbol{F}\boldsymbol{\sigma}}\} - \cdots - p_{N^2-1}E\{\|\Delta\boldsymbol{w}(k)\|^2_{\boldsymbol{F}^{N^2-1}\boldsymbol{\sigma}}\} \tag{3.103}$$

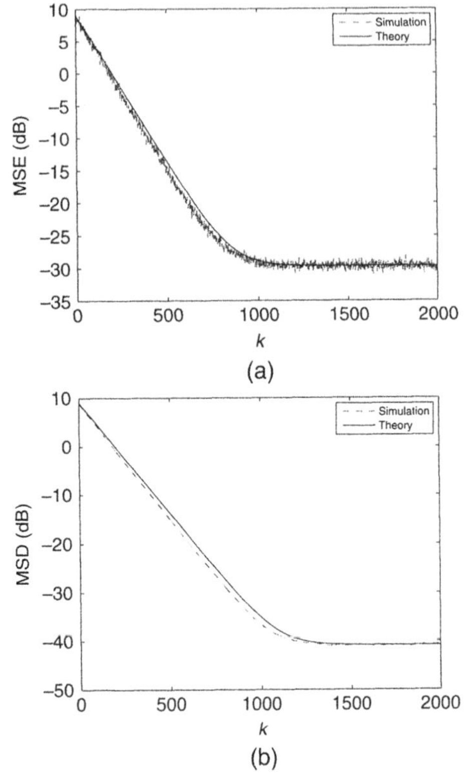

Figure 3.21 Theoretical and simulated MSE and MSD of M-max LMS algorithm with $M = 1$ for i.i.d. Gaussian input.

From (3.83a) we have:

$$E\{\|\Delta \boldsymbol{w}(k+1)\|^2_{\boldsymbol{F}^{N^2-1}\boldsymbol{\sigma}}\} = E\{\|\Delta \boldsymbol{w}(k)\|^2_{\boldsymbol{F}^{N^2}\boldsymbol{\sigma}}\} + \mu^2 \sigma_v^2 \boldsymbol{r}_M^T \boldsymbol{F}^{N^2-1} \boldsymbol{\sigma} \quad (3.104)$$

Substituting (3.103) into the above expression we obtain:

$$E\{\|\Delta \boldsymbol{w}(k+1)\|^2_{\boldsymbol{F}^{N^2-1}\boldsymbol{\sigma}}\} = -\sum_{i=0}^{N^2-1} p_i E\{\|\Delta \boldsymbol{w}(k)\|^2_{\boldsymbol{F}^i \boldsymbol{\sigma}}\} + \mu^2 \sigma_v^2 \boldsymbol{r}_M^T \boldsymbol{F}^{N^2-1} \boldsymbol{\sigma} \quad (3.105)$$

which gives the last row of the state-space equation in (3.99).

The $N^2 \times N^2$ matrix $\mathcal{F}$ is a companion matrix whose eigenvalues are given by the roots of the characteristic polynomial of $\boldsymbol{F}$ or equivalently the eigenvalues of $\boldsymbol{F}$. The mean-squared stability of the partial-update adaptive filter is guaranteed if and

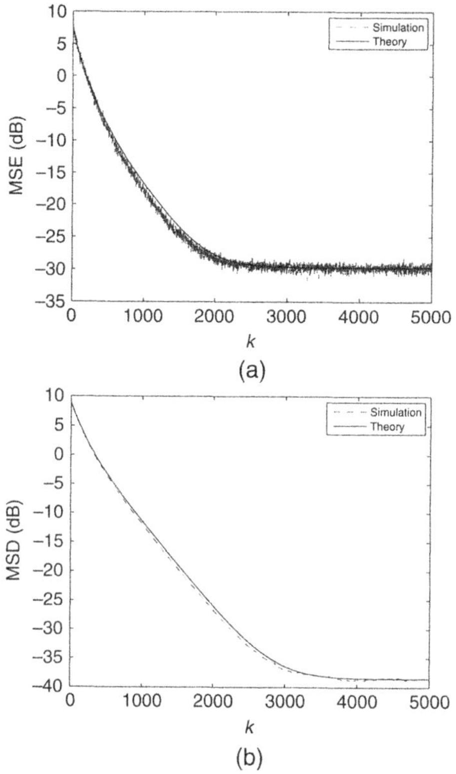

Figure 3.22 Theoretical and simulated MSE and MSD of M-max LMS algorithm with $M = 1$ for AR(1) Gaussian input with eigenvalue spread 23.5.

only if the state-space equation (3.99) is convergent (i.e. the eigenvalues of $\mathcal{F}$ are on the open interval $(-1, 1)$). This condition is satisfied if the eigenvalues of $\boldsymbol{F}$, $\lambda_i(\boldsymbol{F})$, $i = 0, 1, \ldots, N^2 - 1$, obey the inequality:

$$|\lambda_i(\boldsymbol{F})| < 1, \quad i = 0, 1, \ldots, N^2 - 1 \tag{3.106}$$

Let us write (3.83b) as:

$$\boldsymbol{F} = \boldsymbol{I}_{N^2} - \mu \boldsymbol{U} + \mu^2 \boldsymbol{V} \tag{3.107}$$

where:

$$\boldsymbol{U} = 2(\boldsymbol{R}_M \otimes \boldsymbol{I}), \quad \boldsymbol{V} = E\{(\boldsymbol{x}(k)\boldsymbol{x}^T(k)\boldsymbol{I}_M(k)) \otimes (\boldsymbol{x}(k)\boldsymbol{x}^T(k)\boldsymbol{I}_M(k))\} \tag{3.108}$$

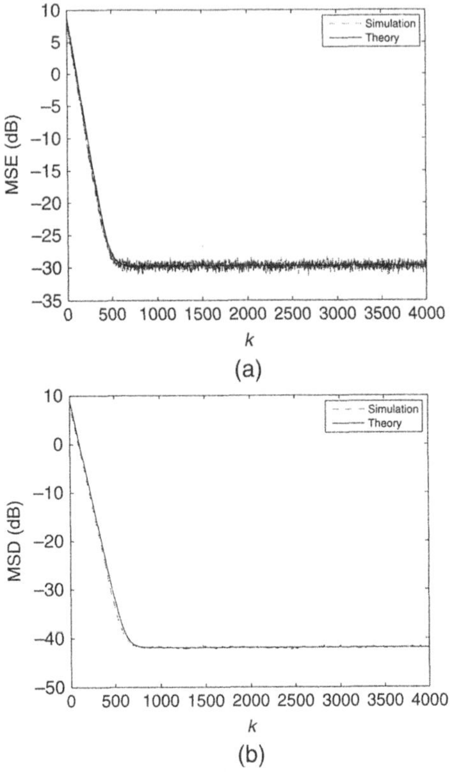

Figure 3.23 Theoretical and simulated MSE and MSD of full-update NLMS algorithm for i.i.d. Gaussian input.

The stability condition in (3.106) is met if:

$$-1 < \boldsymbol{I}_{N^2} - \mu \boldsymbol{U} + \mu^2 \boldsymbol{V} < 1 \tag{3.109}$$

or equivalently:

$$\boldsymbol{U} - \mu \boldsymbol{V} > 0 \tag{3.110a}$$

and:

$$\underbrace{2\boldsymbol{I}_{N^2} - \mu \boldsymbol{U} + \mu^2 \boldsymbol{V}}_{\boldsymbol{G}(\mu)} > 0 \tag{3.110b}$$

where the inequalities apply to the eigenvalues of matrices. The above inequalities satisfy the stability conditions $\boldsymbol{F} < 1$ and $\boldsymbol{F} > -1$, respectively. The first inequality is satisfied if:

$$0 < \mu < \frac{1}{\lambda_{\max}(\boldsymbol{U}^{-1}\boldsymbol{V})} \tag{3.111}$$

Since the eigenvalues of $\boldsymbol{G}(\mu)$ change continuously with $\mu > 0$, the maximum step-size parameter satisfying (3.110b) is obtained from the roots of $\boldsymbol{G}(\mu)$:

$$\det \boldsymbol{G}(\mu) = 0 \tag{3.112}$$

Using a block matrix determinant identity, we can write (Sayed, 2003)

$$\det \boldsymbol{G}(\mu) = \det \begin{bmatrix} 2\boldsymbol{I}_{N^2} - \mu \boldsymbol{U} & \mu \boldsymbol{V} \\ -\mu \boldsymbol{I}_{N^2} & \boldsymbol{I}_{N^2} \end{bmatrix} \tag{3.113a}$$

$$= \det \left(\begin{bmatrix} 2\boldsymbol{I}_{N^2} & \boldsymbol{0} \\ \boldsymbol{0} & \boldsymbol{I}_{N^2} \end{bmatrix} \left(\begin{bmatrix} \boldsymbol{I}_{N^2} & \boldsymbol{0} \\ \boldsymbol{0} & \boldsymbol{I}_{N^2} \end{bmatrix} - \mu \underbrace{\begin{bmatrix} \frac{1}{2}\boldsymbol{U} & -\frac{1}{2}\boldsymbol{V} \\ \boldsymbol{I}_{N^2} & \boldsymbol{0} \end{bmatrix}}_{\boldsymbol{H}} \right) \right) \tag{3.113b}$$

whence it is clear that μ must satisfy the following inequality in order to guarantee $\boldsymbol{F} > -1$:

$$0 < \mu < \frac{1}{\lambda_{\max}(\boldsymbol{H})} \tag{3.114}$$

Note that only the real eigenvalues of $\boldsymbol{H}$ are considered. Thus the mean-square stability bound for μ is:

$$0 < \mu < \min \left\{ \frac{1}{\lambda_{\max}(\boldsymbol{U}^{-1}\boldsymbol{V})}, \frac{1}{\lambda_{\max}(\boldsymbol{H})} \right\} \tag{3.115}$$

For purposes of illustration consider the method of sequential partial updates for which we have $\boldsymbol{R}_M = (M/N)\boldsymbol{R}$ and:

$$\boldsymbol{U} = \frac{2M}{N}(\boldsymbol{R} \otimes \boldsymbol{I}) \tag{3.116}$$

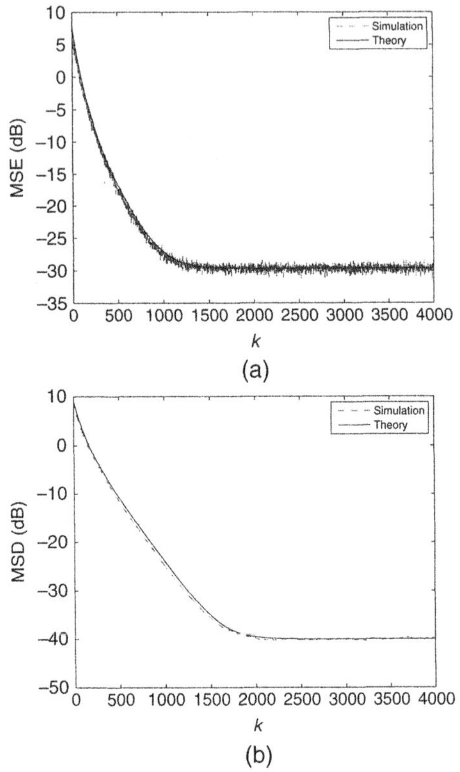

Figure 3.24 Theoretical and simulated MSE and MSD of full-update NLMS algorithm for AR(1) Gaussian input with eigenvalue spread 23.5.

Suppose that $x(k)$ is an i.i.d. signal. Setting $\boldsymbol{\Sigma} = \boldsymbol{R}$, as we did in the time evolution of MSE, leads to:

$$\begin{aligned}
\text{vec}^{-1}(\boldsymbol{V}\boldsymbol{\sigma}) &= \text{vec}^{-1}(E\{(\boldsymbol{x}(k)\boldsymbol{x}^T(k)\boldsymbol{I}_M(k)) \otimes (\boldsymbol{x}(k)\boldsymbol{x}^T(k)\boldsymbol{I}_M(k))\}\text{vec}(\boldsymbol{R})) \\
&= E\{(\boldsymbol{x}^T(k)\boldsymbol{I}_M(k)\boldsymbol{R}\boldsymbol{I}_M(k)\boldsymbol{x}(k))\,\boldsymbol{x}(k)\boldsymbol{x}^T(k)\} \\
&= E\{\boldsymbol{x}(k)\boldsymbol{x}^T(k)\boldsymbol{R}\boldsymbol{I}_M(k)\boldsymbol{x}(k)\boldsymbol{x}^T(k)\} \qquad (3.117)
\end{aligned}$$

since $\boldsymbol{R} = \sigma_x^2\boldsymbol{I}$ due to the i.i.d. $x(k)$ assumption. Vectorizing the last expression above leads to:

$$\begin{aligned}
\boldsymbol{V} &= E\{(\boldsymbol{x}(k)\boldsymbol{x}^T(k)\boldsymbol{I}_M(k)) \otimes (\boldsymbol{x}(k)\boldsymbol{x}^T(k))\} \\
&= \frac{M}{N}E\{(\boldsymbol{x}(k)\boldsymbol{x}^T(k)) \otimes (\boldsymbol{x}(k)\boldsymbol{x}^T(k))\}. \qquad (3.118)
\end{aligned}$$

Hence we have:

$$U^{-1}V = \frac{1}{2}(R \otimes I)^{-1} E\{(x(k)x^T(k)) \otimes (x(k)x^T(k))\} \qquad (3.119)$$

which corresponds to the full-update LMS algorithm. Thus for i.i.d. $x(k)$ the sequential-partial-update LMS and the full-update LMS algorithms have the same mean-square stability bound. Similar conclusions can be extended to other partial update approaches.

Mean stability

For partial-update LMS algorithms the mean coefficient error recursion (3.79) becomes:

$$E\{\Delta w(k+1)\} = \left(I - \mu E\{I_M(k)x(k)x^T(k)\}\right) E\{\Delta w(k)\} \qquad (3.120)$$

This recursion is convergent if and only if:

$$-1 < I - \mu R_M < 1 \qquad (3.121)$$

Thus the mean stability is achieved if:

$$0 < \mu < \frac{2}{\lambda_{\max}(R_M)} \qquad (3.122)$$

For data-independent partial-update techniques we have $R_M = (M/N)R$, which simplifies the above inequality to:

$$0 < \mu < \frac{2N}{M\lambda_{\max}(R)} \qquad (3.123)$$

Combining the mean stability condition (3.122) with the mean-square stability condition (3.115) gives:

$$0 < \mu < \min\left\{\frac{2}{\lambda_{\max}(R_M)}, \frac{1}{\lambda_{\max}(U^{-1}V)}, \frac{1}{\lambda_{\max}(H)}\right\} \qquad (3.124)$$

3.4.2 Partial-update NLMS algorithms

For the partial-update NLMS algorithms we use the weighted variance relation in (3.76) with $f(x(k)) = \epsilon + \|x(k)\|_2^2$ and $f(x(k)) = \epsilon + x^T(k)I_M(k)x(k)$ for M-max and selective partial updates, respectively.

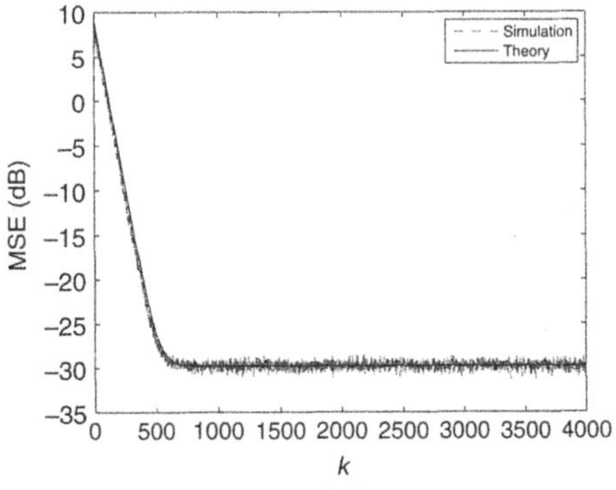

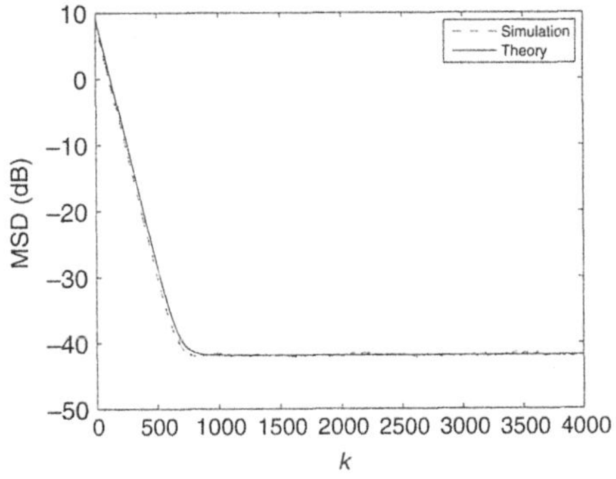

Figure 3.25 Theoretical and simulated MSE and MSD of M-max NLMS algorithm with $M = 5$ for i.i.d. Gaussian input.

The corresponding vectorized weighted variance relation is:

$$E\{\|\Delta \boldsymbol{w}(k+1)\|^2_{\boldsymbol{\sigma}}\} = E\{\|\Delta \boldsymbol{w}(k)\|^2_{\boldsymbol{F}\boldsymbol{\sigma}}\} + \mu^2 \sigma_v^2 \boldsymbol{r}_M^T \boldsymbol{\sigma} \tag{3.125a}$$

$$\boldsymbol{F} = \boldsymbol{I}_{N^2} + \mu^2 E\left\{\frac{(\boldsymbol{x}(k)\boldsymbol{x}^T(k)\boldsymbol{I}_M(k)) \otimes (\boldsymbol{x}(k)\boldsymbol{x}^T(k)\boldsymbol{I}_M(k))}{f^2(\boldsymbol{x}(k))}\right\} - 2\mu\left(E\left\{\frac{\boldsymbol{x}(k)\boldsymbol{x}^T(k)\boldsymbol{I}_M(k)}{f(\boldsymbol{x}(k))}\right\} \otimes \boldsymbol{I}\right) \tag{3.125b}$$

where:

$$\boldsymbol{r}_M = \text{vec}\left(E\left\{\frac{\boldsymbol{I}_M(k)\boldsymbol{x}(k)\boldsymbol{x}^T(k)\boldsymbol{I}_M(k)}{f^2(\boldsymbol{x}(k))}\right\}\right) \tag{3.126}$$

The time-invariant $N^2 \times N^2$ matrix $\boldsymbol{F}$ can be computed numerically using the input signal moments in (3.77).

The time evolution of MSD and EMSE is obtained from (3.87):

$$E\{\|\Delta \boldsymbol{w}(k+1)\|_{\boldsymbol{\sigma}}^2\} = \|\boldsymbol{w}_o\|^2_{\boldsymbol{F}^{k+1}\boldsymbol{\sigma}} + \mu^2\sigma_v^2 \sum_{i=0}^{k} \boldsymbol{r}_M^T \boldsymbol{F}^i \boldsymbol{\sigma} \tag{3.127}$$

where $\boldsymbol{r}_M$ and $\boldsymbol{F}$ are defined in (3.126) and (3.125b), respectively. Here setting $\boldsymbol{\sigma} = \text{vec}(\boldsymbol{I})$ gives the MSD curve whereas $\boldsymbol{\sigma} = \text{vec}(\boldsymbol{R})$ results in the EMSE curve.

The steady-state MSD and EMSE follows from (3.92) and (3.93) as:

$$\text{Steady-state MSD} = \mu^2\sigma_v^2 \boldsymbol{r}_M^T \left(\boldsymbol{I}_{N^2} - \boldsymbol{F}\right)^{-1} \text{vec}(\boldsymbol{I}) \tag{3.128}$$

$$\text{Steady-state EMSE} = \mu^2\sigma_v^2 \boldsymbol{r}_M^T \left(\boldsymbol{I}_{N^2} - \boldsymbol{F}\right)^{-1} \text{vec}(\boldsymbol{R}) \tag{3.129}$$

where $\boldsymbol{r}_M$ and $\boldsymbol{F}$ are given by (3.126) and (3.125b), respectively. For a sufficiently small μ and i.i.d. $x(k)$, the steady-state EMSE is approximately given by:

$$\mu^2\sigma_v^2 \boldsymbol{r}_M^T \left(\boldsymbol{I}_{N^2} - \boldsymbol{F}\right)^{-1} \text{vec}(\boldsymbol{R})$$

$$\approx \frac{\mu}{2}\sigma_v^2 \boldsymbol{r}_M^T \text{vec}\left(\boldsymbol{R}\, E^{-1}\left\{\frac{\boldsymbol{x}(k)\boldsymbol{x}^T(k)\boldsymbol{I}_M(k)}{f(\boldsymbol{x}(k))}\right\}\right) \tag{3.130a}$$

$$\approx \frac{\mu}{2}\sigma_v^2 \text{tr}\left(E\left\{\frac{\boldsymbol{I}_M(k)\boldsymbol{x}(k)\boldsymbol{x}^T(k)\boldsymbol{I}_M(k)}{f^2(\boldsymbol{x}(k))}\right\}\right.$$

$$\left.\times\ \boldsymbol{R}\, E^{-1}\left\{\frac{\boldsymbol{x}(k)\boldsymbol{x}^T(k)\boldsymbol{I}_M(k)}{f(\boldsymbol{x}(k))}\right\}\right) \tag{3.130b}$$

$$\approx \frac{\mu}{2}\sigma_v^2 E\left\{\frac{1}{f(\boldsymbol{x}(k))}\right\} \text{tr}\boldsymbol{R} \tag{3.130c}$$

which approximates (3.50b) and (3.60) for small μ and ϵ.

Using the state space formulation (3.100):

$$\boldsymbol{\xi}(k+1) = \boldsymbol{\mathcal{F}}\,\boldsymbol{\xi}(k) + \mu^2\sigma_v^2\boldsymbol{\rho} \tag{3.131}$$

the mean-square stability of partial-update NLMS algorithms is determined from the eigenvalues of $\boldsymbol{F}$ in (3.125b). The mean-square stability is guaranteed if:

$$-1 < \boldsymbol{F} < 1 \tag{3.132}$$

Define:

$$\begin{aligned}\boldsymbol{U} &= 2\left(E\left\{\frac{\boldsymbol{x}(k)\boldsymbol{x}^T(k)\boldsymbol{I}_M(k)}{f(\boldsymbol{x}(k))}\right\} \otimes \boldsymbol{I}\right)\\ \boldsymbol{V} &= E\left\{\left(\frac{\boldsymbol{x}(k)\boldsymbol{x}^T(k)\boldsymbol{I}_M(k)}{f(\boldsymbol{x}(k))}\right) \otimes \left(\frac{\boldsymbol{x}(k)\boldsymbol{x}^T(k)\boldsymbol{I}_M(k)}{f(\boldsymbol{x}(k))}\right)\right\}\end{aligned} \tag{3.133}$$

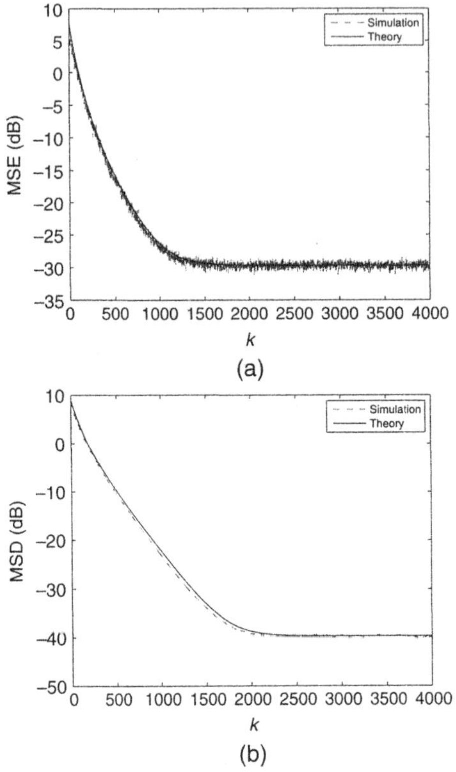

Figure 3.26 Theoretical and simulated MSE and MSD of M-max NLMS algorithm with $M = 5$ for AR(1) Gaussian input with eigenvalue spread 23.5.

In terms of $\boldsymbol{U}$ and $\boldsymbol{V}$ the stability bound for μ is given by (3.115):

$$0 < \mu < \min\left\{\frac{1}{\lambda_{\max}(\boldsymbol{U}^{-1}\boldsymbol{V})}, \frac{1}{\lambda_{\max}(\boldsymbol{H})}\right\} \tag{3.134}$$

The mean stability condition is given by (cf. (3.122)):

$$0 < \mu < \frac{2}{\lambda_{\max}\left(E\left\{\frac{\boldsymbol{I}_M(k)\boldsymbol{x}(k)\boldsymbol{x}^T(k)}{f(\boldsymbol{x}(k))}\right\}\right)} \tag{3.135}$$

Consider first the M-max NLMS algorithm for which $f(\boldsymbol{x}(k)) = \epsilon + \|\boldsymbol{x}(k)\|_2^2$. Suppose that $x(k)$ is i.i.d. so that, following the same arguments leading to (3.118), $\boldsymbol{V}$

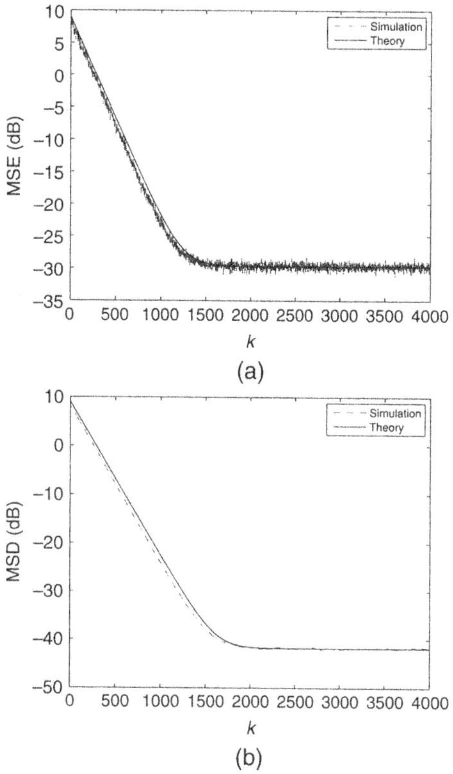

Figure 3.27 Theoretical and simulated MSE and MSD of M-max NLMS algorithm with $M = 1$ for i.i.d. Gaussian input.

can be replaced by:

$$V = E\left\{\left(\frac{\boldsymbol{x}(k)\boldsymbol{x}^T(k)\boldsymbol{I}_M(k)}{f(\boldsymbol{x}(k))}\right)\otimes\left(\frac{\boldsymbol{x}(k)\boldsymbol{x}^T(k)}{f(\boldsymbol{x}(k))}\right)\right\} \tag{3.136}$$

It can be shown that (Sayed, 2003):

$$E\left\{\left(\frac{\boldsymbol{x}(k)\boldsymbol{x}^T(k)\boldsymbol{I}_M(k)}{\epsilon+\|\boldsymbol{x}(k)\|_2^2}\right)\otimes\left(\frac{\boldsymbol{x}(k)\boldsymbol{x}^T(k)}{\epsilon+\|\boldsymbol{x}(k)\|_2^2}\right)\right\} \leq E\left\{\left(\frac{\boldsymbol{x}(k)\boldsymbol{x}^T(k)\boldsymbol{I}_M(k)}{\epsilon+\|\boldsymbol{x}(k)\|_2^2}\right)\otimes \boldsymbol{I}\right\} \tag{3.137}$$

which implies:

$$\boldsymbol{U} - 2\boldsymbol{V} \geq 0 \tag{3.138}$$

where $\boldsymbol{V}$ is a non-negative-definite matrix, i.e., $\boldsymbol{V} \geq 0$. Therefore $\boldsymbol{F} < 1$ is satisfied if:

$$0 < \mu < 2 \tag{3.139}$$

In (Sayed, 2003) it was shown that this bound also satisfies $\boldsymbol{F} > -1$ and mean stability. Thus, for i.i.d. $x(k)$ the mean-square stability bound for the step-size of the M-max NLMS algorithm is the same as that of the full-update NLMS algorithm.

Under the i.i.d. $x(k)$ assumption, for the selective-partial-update NLMS algorithm the inequality in (3.137) takes the following form:

$$\begin{aligned} E \Bigg\{ \frac{\epsilon + \boldsymbol{x}^T(k)\boldsymbol{I}_M(k)\boldsymbol{x}(k)}{\epsilon + \|\boldsymbol{x}(k)\|_2^2} \left(\frac{\boldsymbol{x}(k)\boldsymbol{x}^T(k)\boldsymbol{I}_M(k)}{\epsilon + \boldsymbol{x}^T(k)\boldsymbol{I}_M(k)\boldsymbol{x}(k)} \right) \\ \otimes \left(\frac{\boldsymbol{x}(k)\boldsymbol{x}^T(k)}{\epsilon + \boldsymbol{x}^T(k)\boldsymbol{I}_M(k)\boldsymbol{x}(k)} \right) \Bigg\} \\ \leq E \left\{ \left(\frac{\boldsymbol{x}(k)\boldsymbol{x}^T(k)\boldsymbol{I}_M(k)}{\epsilon + \boldsymbol{x}^T(k)\boldsymbol{I}_M(k)\boldsymbol{x}(k)} \right) \otimes \boldsymbol{I} \right\} \end{aligned} \tag{3.140}$$

where the factor introduced to the left side of the inequality guarantees that:

$$\frac{\epsilon + \boldsymbol{x}^T(k)\boldsymbol{I}_M(k)\boldsymbol{x}(k)}{\epsilon + \|\boldsymbol{x}(k)\|_2^2} \left(\frac{\boldsymbol{x}(k)\boldsymbol{x}^T(k)}{\epsilon + \boldsymbol{x}^T(k)\boldsymbol{I}_M(k)\boldsymbol{x}(k)} \right) \leq 1 \tag{3.141}$$

For sufficiently large N, (3.140) may be approximated by:

$$\begin{aligned} \beta_M E \left\{ \left(\frac{\boldsymbol{x}(k)\boldsymbol{x}^T(k)\boldsymbol{I}_M(k)}{\epsilon + \boldsymbol{x}^T(k)\boldsymbol{I}_M(k)\boldsymbol{x}(k)} \right) \otimes \left(\frac{\boldsymbol{x}(k)\boldsymbol{x}^T(k)}{\epsilon + \boldsymbol{x}^T(k)\boldsymbol{I}_M(k)\boldsymbol{x}(k)} \right) \right\} \\ \leq E \left\{ \left(\frac{\boldsymbol{x}(k)\boldsymbol{x}^T(k)\boldsymbol{I}_M(k)}{\epsilon + \boldsymbol{x}^T(k)\boldsymbol{I}_M(k)\boldsymbol{x}(k)} \right) \otimes \boldsymbol{I} \right\} \end{aligned} \tag{3.142}$$

where [see (3.39)]:

$$\beta_M = \frac{E\{\boldsymbol{x}^T(k)\boldsymbol{I}_M(k)\boldsymbol{x}(k)\}}{E\{\boldsymbol{x}^T(k)\boldsymbol{x}(k)\}} = \frac{\mathrm{tr}\boldsymbol{R}_M}{\mathrm{tr}\boldsymbol{R}} \tag{3.143}$$

In terms of $\boldsymbol{U}$ and $\boldsymbol{V}$ (3.142) becomes:

$$\boldsymbol{U} - 2\beta_M \boldsymbol{V} \geq 0 \tag{3.144}$$

Thus, for i.i.d. $x(k)$ the condition $\boldsymbol{F} < 1$ is satisfied if:

$$0 < \mu < 2\beta_M \tag{3.145}$$

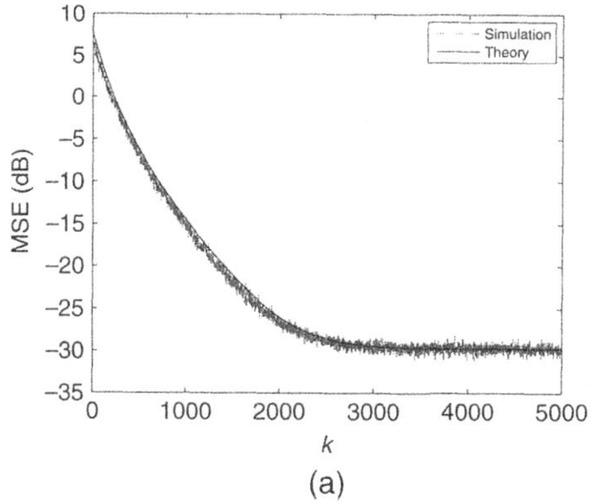

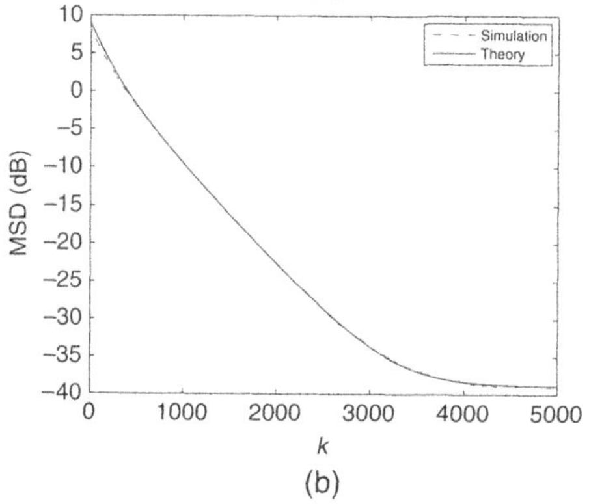

Figure 3.28 Theoretical and simulated MSE and MSD of M-max NLMS algorithm with $M = 1$ for AR(1) Gaussian input with eigenvalue spread 23.5.

It can be shown that the above inequality also satisfies the mean stability condition and $\boldsymbol{F} > -1$. The stability bound in (3.145) was derived in (Werner *et al.*, 2004) under a different set of assumptions.

3.4.3 Simulation examples for convergence analysis

In the simulation examples presented here we consider the identification of an unknown FIR system of order $N - 1$ by partial-update LMS and NLMS algorithms. The unknown system characteristics have no effect on the convergence behaviour. Thus an arbitrary FIR system has been selected as the unknown system:

$$\boldsymbol{h} = [-1.5, -1.8, 1.2, 0.8, -0.2, -0.5, -0.3, 0.2, 0.1, -0.03]^T \qquad (3.146)$$

The additive noise at the system output $n(k)$ is assumed to be statistically independent of the input $x(k)$ in order to comply with the signal model adopted in the convergence analysis. We therefore have $v(k) = n(k)$ and the optimum estimate is $\boldsymbol{w}_o = \boldsymbol{h}$. The

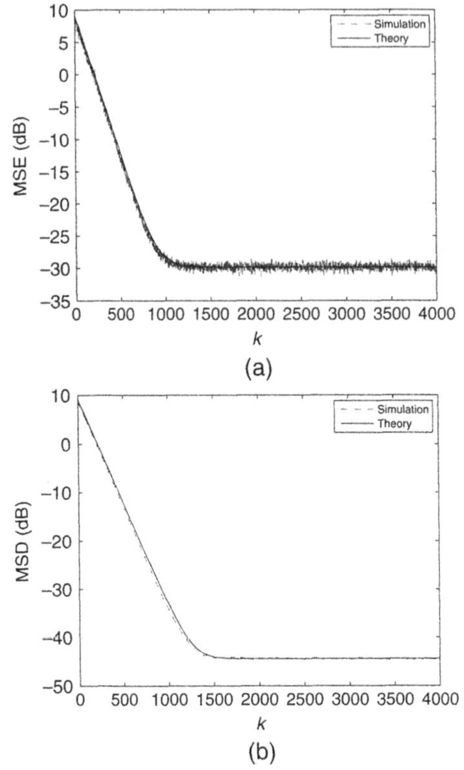

Figure 3.29 Theoretical and simulated MSE and MSD of selective-partial-update NLMS algorithm with $M = 5$ for i.i.d. Gaussian input.

adaptive filter length is $N = 10$ and the additive output noise $v(k)$ is an i.i.d. zero-mean Gaussian noise with variance $\sigma_v^2 = 0.001$. The simulated adaptive filters have tapped-delay-line structures. Thus input regressor vectors are not i.i.d. This violates the i.i.d. $\boldsymbol{x}(k)$ assumption made in the convergence analysis. As we shall see, this does not cause theoretical convergence curves to significantly deviate from simulation results.

The time evolution of MSD and MSE is estimated by ensemble averaging adaptive filter coefficient and output errors over 200 independent trials. Adaptive filter coefficients are initialized to $\boldsymbol{w}(0) = \boldsymbol{0}$. In the simulations two input signals are considered, viz. an i.i.d. Gaussian input signal with zero mean and unity variance, and a correlated AR(1) Gaussian input signal with zero mean, unity variance and eigenvalue spread 23.5 ($a = -0.7$) (see Section 3.3.3). In LMS simulations the step-size parameter is $\mu = 0.015$. It is set to $\mu = 0.1$ and $\mu = 0.05$ for M-max NLMS and selective-partial-update NLMS, respectively.

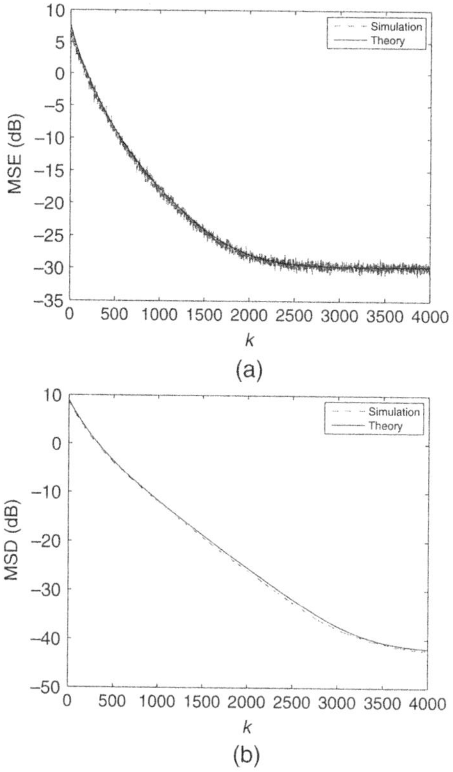

Figure 3.30 Theoretical and simulated MSE and MSD of selective-partial-update NLMS algorithm with $M = 5$ for AR(1) Gaussian input with eigenvalue spread 23.5.

The theoretical and simulated MSE and MSD curves of the full-update LMS algorithm are depicted in Figures 3.11 and 3.12 for the i.i.d. and correlated Gaussian inputs, respectively. The theoretical MSE and MSD curves were computed using (3.87). A good agreement between the theoretical and simulated results is observed even for the correlated input signal, which violates the assumptions underlying the theoretical results.

Figures 3.13 and 3.14 show the theoretical and simulated MSE and MSD curves of the periodic-partial-update LMS algorithm for $S = 2$ which halves the update complexity of LMS. The theoretical MSE and MSD curves are obtained from (3.87) simply by setting $M = N$ and applying time scaling S to the iteration index. As a consequence, we observe a slower convergence for the MSE and MSD curves by a factor of two compared with the full-update LMS algorithm.

Figures 3.15 and 3.16 show the theoretical and simulated MSE and MSD curves of the sequential-partial-update LMS algorithm for $M = 5$ which also halves the update

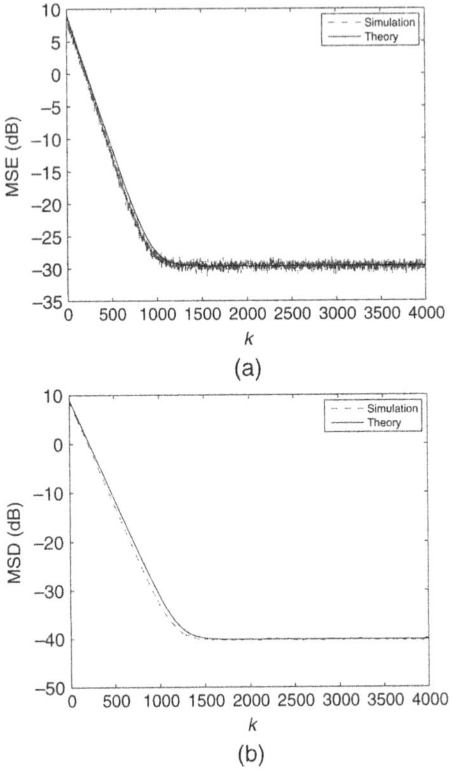

Figure 3.31 Theoretical and simulated MSE and MSD of selective-partial-update NLMS algorithm with $M = 1$ for i.i.d. Gaussian input.

complexity of LMS. The theoretical MSE and MSD curves obtained from (3.87) are seen to approximate the simulated curves well. A slow-down in convergence by a factor of two is observed in line with expectations.

In Figures 3.17 and 3.18 the theoretical and simulated MSE and MSD curves of the sequential-partial-update LMS algorithm are shown this time for $M = 1$ which provides the lowest update complexity. Again the predicted and simulated learning curves are in good agreement despite the strong correlation of the AR(1) Gaussian input. The convergence rate is slowed down by a factor of 10.

In Figures 3.19 and 3.20 we illustrate the theoretical and simulated MSE and MSD curves of the M-max LMS algorithm for $M = 5$. The method of M-max updates provides a significantly faster convergence than the data-independent partial update methods. This is seen by comparing the convergence curves of M-max LMS with the previous convergence curves.

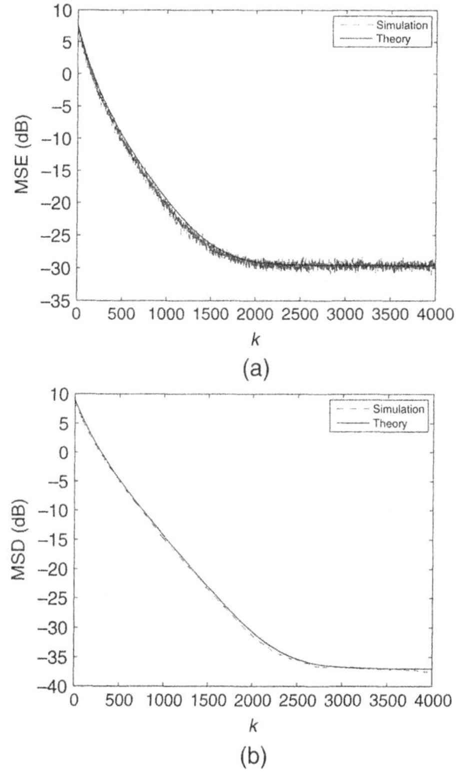

Figure 3.32 Theoretical and simulated MSE and MSD of selective-partial-update NLMS algorithm with $M = 1$ for AR(1) Gaussian input with eigenvalue spread 23.5.

Table 3.1 Theoretical step-size bounds for i.i.d. Gaussian input

		Mean stability	$F < 1$	$F > -1$	Mean-square stability
Full LMS		2.0000	0.1650	1.0000	0.1650
Sequential LMS	$M = 1$	20.0000	0.1691	10.0000	0.1691
	$M = 5$	4.0000	0.1646	2.0000	0.1646
M-max LMS	$M = 1$	5.1932	0.1665	2.5939	0.1665
	$M = 5$	2.1324	0.1660	1.0673	0.1660
Full NLMS		19.7470	2.0000	9.8783	2.0000
M-max NLMS	$M = 1$	51.6437	2.0000	25.8083	2.0000
	$M = 5$	21.7277	2.0000	10.8710	2.0000
SPU NLMS	$M = 1$	19.5789	0.6992	9.7896	0.6992
	$M = 5$	19.6126	1.7978	9.8106	1.7978

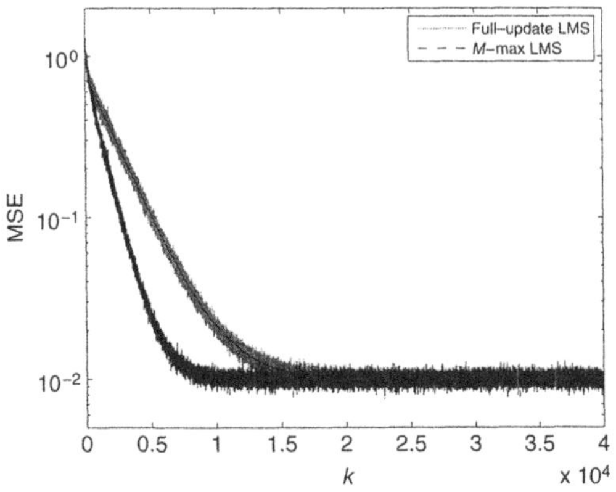

Figure 3.33 Theoretical (smooth curves) and ensemble averaged MSE of full-update LMS and M-max LMS for AR(1) Gaussian input with $a = -0.9$. Note that M-max LMS outperforms full-update LMS.

Table 3.2 Theoretical step-size bounds for AR(1) Gaussian input

		Mean stability	$F < 1$	$F > -1$	Mean-square stability
Full LMS		0.4707	0.1255	0.2405	0.1255
Sequential LMS	$M = 1$	4.7075	0.1738	2.4140	0.1738
	$M = 5$	0.9415	0.1376	0.4811	0.1376
M-max LMS	$M = 1$	2.0204	0.2059	1.0475	0.2059
	$M = 5$	0.5951	0.1391	0.3070	0.1391
Full NLMS		6.0453	2.0000	3.1127	2.0000
M-max NLMS	$M = 1$	23.4223	2.2789	12.3406	2.2789
	$M = 5$	7.2311	2.0132	3.7536	2.0132
SPU NLMS	$M = 1$	7.3899	0.7411	3.8041	0.7411
	$M = 5$	6.3020	1.7760	3.2690	1.7760

Figures 3.21 and 3.22 depict the theoretical and simulated MSE and MSD curves of the M-max LMS algorithm for $M = 1$. Despite the strong input signal correlation the theoretical and simulated MSE and MSD curves demonstrate a close match. The convergence rate is still considerably faster than the sequential-partial-update LMS algorithm with $M = 1$ (see Figures 3.17 and 3.18).

We next demonstrate the convergence performance of the NLMS algorithm and its partial-update variants. The theoretical and simulated MSE and MSD curves of the full-update NLMS algorithm are depicted in Figures 3.23 and 3.24 for the i.i.d. and correlated Gaussian inputs, respectively. The theoretical MSE and MSD curves were

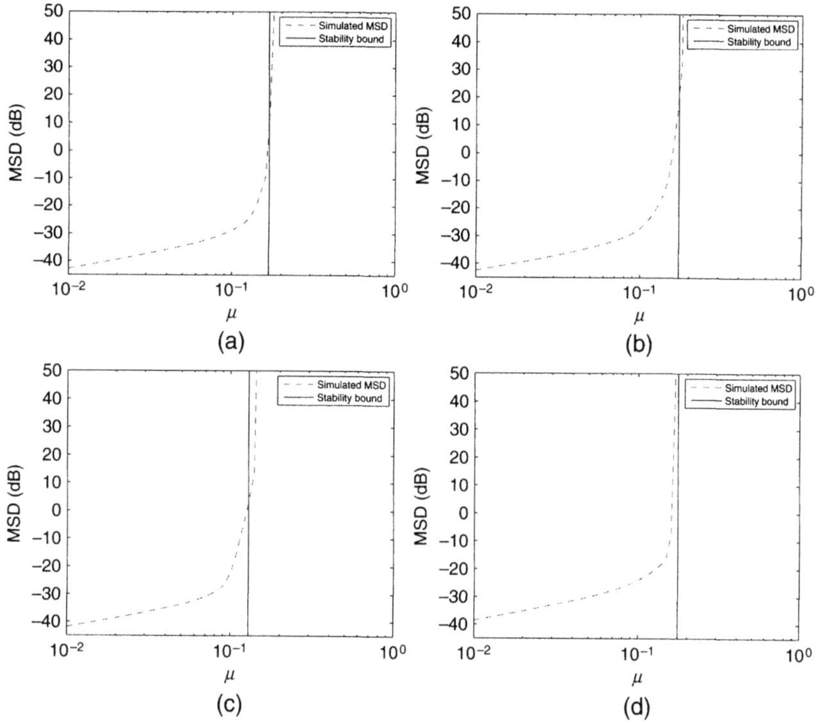

Figure 3.34 Simulated MSD versus step-size and theoretical step-size bound for sequential-partial-update LMS algorithm: (a) $M = 5$, i.i.d. input; (b) $M = 1$, i.i.d. input; (c) $M = 5$, AR(1) input; (d) $M = 1$, AR(1) input.

computed using (3.127). The theoretical and simulated results are in good agreement even for the strongly correlated input signal.

Figures 3.25 and 3.26 depict the theoretical and simulated MSE and MSD curves of the M-max NLMS algorithm for $M = 5$. Compared with the full-update NLMS algorithm the M-max NLMS algorithm does not incur significant performance loss as a result of updating half of the adaptive filter coefficients. This is mainly due to the use of M-max partial updates.

Figures 3.27 and 3.28 show the theoretical and simulated MSE and MSD curves of the M-max NLMS algorithm for $M = 1$. The convergence rate of the M-max NLMS algorithm is only slowed down by a factor of two despite $1/10$th of the coefficients being updated at each iteration. This again demonstrates the performance advantage of M-max updates at a small complexity overhead compared with sequential partial updates.

We now turn our attention to the selective-partial-update NLMS algorithm. The theoretical and simulated MSE and MSD curves of the selective-partial-update NLMS

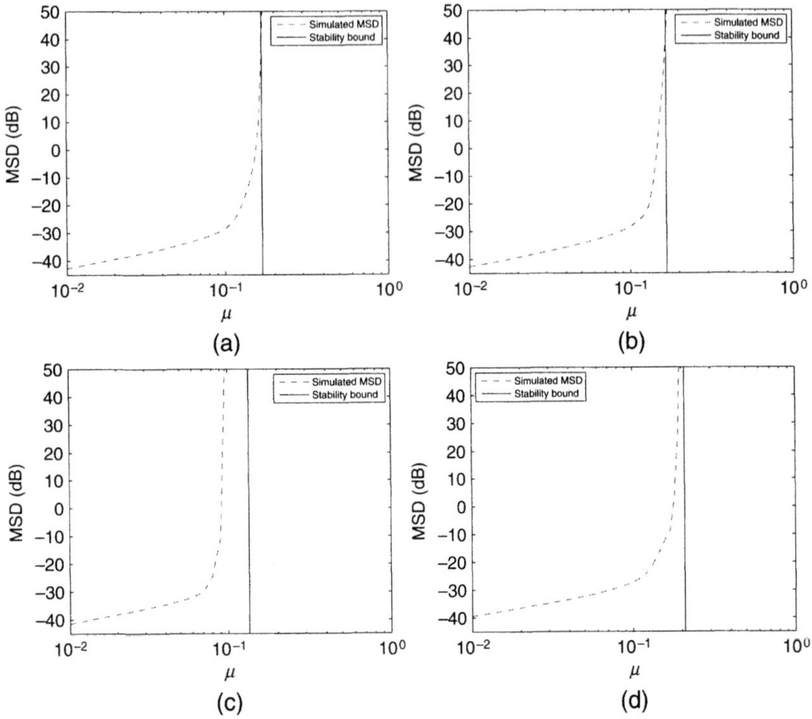

Figure 3.35 Simulated MSD versus step-size and theoretical step-size bound for M-max LMS algorithm: (a) $M = 5$, i.i.d. input; (b) $M = 1$, i.i.d. input; (c) $M = 5$, AR(1) input; (d) $M = 1$, AR(1) input.

algorithm are shown in Figures 3.29 and 3.30 for the i.i.d. and correlated Gaussian inputs, respectively. These figures demonstrate an excellent match between the theoretical and simulated convergence results.

Figures 3.31 and 3.32 finally depict for the theoretical and simulated MSE and MSD curves of the selective-partial-update NLMS algorithm for $M = 1$.

We now revisit the convergence simulations for a correlated input presented in Chapter 1. The simulation parameters were $\boldsymbol{h} = [1, -0.7]^T$, $N = 2$, $\mu = 4 \times 10^{-4}$, $\sigma_v^2 = 0.01$ and $a = -0.9$. In the simulated system identification example we observed a significant convergence improvement by the M-max LMS algorithm over the full-update LMS algorithm. In Chapter 1 we justified this by analysing the reduced eigenvalue spread of M-max updates. The theoretical and ensemble averaged MSE curves for $M = 1$ are depicted in Figure 3.33. To compute ensemble average of MSE, 500 independent realizations were used. Figure 3.33 demonstrates an excellent agreement between theoretical and simulated learning curves. This figure

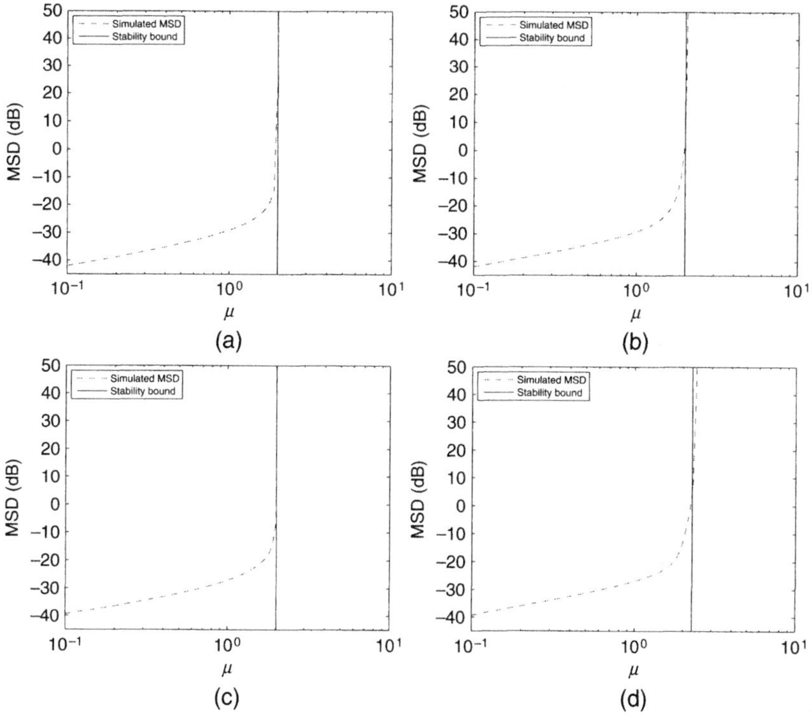

Figure 3.36 Simulated MSD versus step-size and theoretical step-size bound for M-max NLMS algorithm: (a) $M = 5$, i.i.d. input; (b) $M = 1$, i.i.d. input; (c) $M = 5$, AR(1) input; (d) $M = 1$, AR(1) input.

also confirms that the convergence improvement of the M-max LMS algorithm can be accurately predicted by using the theoretical MSE results in (3.87).

The next set of simulations is concerned with the stability bound for the step-size parameter. The theoretical convergence analysis has led to the conclusion that convergence in the mean-square sense is guaranteed if the eigenvalues of the matrix $\boldsymbol{F}$ are on the interval $(-1, 1)$. This condition is satisfied by the stability bounds in (3.115) and (3.134) for partial-update LMS and NLMS algorithms, respectively. These two bounds differ only in the way $\boldsymbol{U}$ and $\boldsymbol{V}$ are defined. We will illustrate the validity of these bounds through numerical simulations. In the simulations the input signal to the adaptive filter is assumed to be either i.i.d. Gaussian with zero mean and unit variance or an AR(1) Gaussian signal with $a = -0.7$ and unit variance.

The theoretical bounds on the step-size parameter μ for several implementations of partial-update LMS and NLMS algorithms are listed in Tables 3.1 and 3.2 for i.i.d. Gaussian input and AR(1) Gaussian input with unit variance and eigenvalue spread 23.5 ($a = -0.7$), respectively. The adaptive filter length is assumed to be

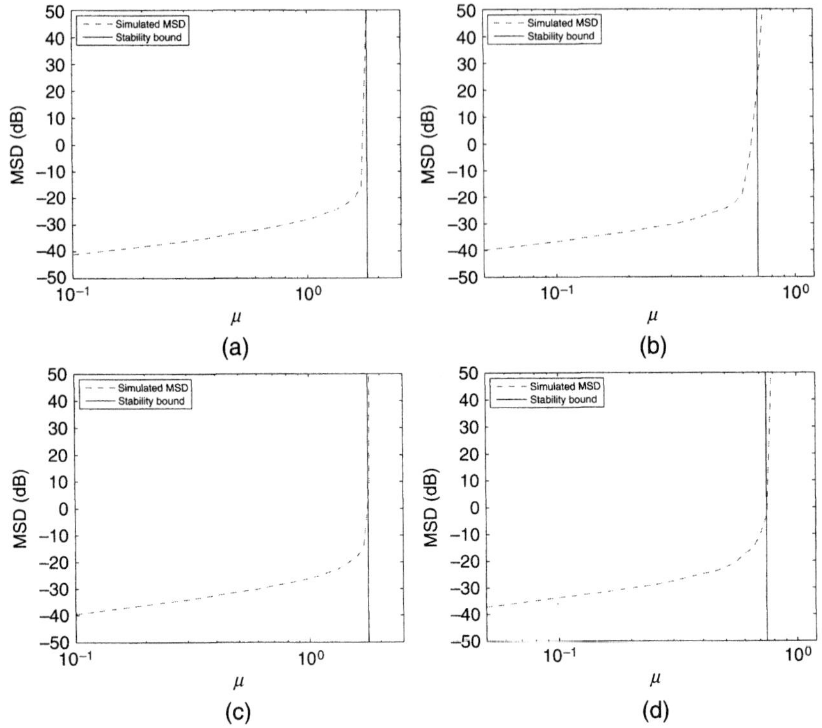

Figure 3.37 Simulated MSD versus step-size and theoretical step-size bound for selective-partial-update NLMS algorithm: (a) $M = 5$, i.i.d. input; (b) $M = 1$, i.i.d. input; (c) $M = 5$, AR(1) input; (d) $M = 1$, AR(1) input.

$N = 10$ and the NLMS regularization parameter is set to $\epsilon = 10^{-4}$. The tables provide bounds for mean stability, $\boldsymbol{F} < 1$ and $\boldsymbol{F} > -1$. These bounds were numerically computed using (3.115) and (3.134) with $\boldsymbol{U}$ and $\boldsymbol{V}$ estimated by sample averaging over data records of length 50000. The parameter β_M for M-max and selective partial updates was also estimated. In the case of i.i.d. Gaussian input $\beta_M = 0.3802$ for $M = 1$ and $\beta_M = 0.9089$ for $M = 5$. For the AR(1) Gaussian input $\beta_M = 0.3212$ for $M = 1$ and $\beta_M = 0.8645$ for $M = 5$.

We observe that the theoretical step-size bounds for the full-update LMS and partial-update LMS algorithms are identical if the input signal is i.i.d. Gaussian. The same is also true for the full-update NLMS and M-max NLMS algorithms. The step-size bound for the selective-partial-update NLMS algorithm appears to be closely approximated by $2\beta_M$ for the i.i.d. Gaussian input [see (3.145)]. For the correlated input signal the theoretical step-size bounds for partial-update adaptive filters seem to vary with the number of coefficients updated M and the method of partial coefficient updates being employed.

To verify the accuracy of the theoretical step-size bounds several simulations have been performed using the steady-state MSD. Figure 3.34 depicts the simulated steady-state MSD versus the step-size parameter along with the theoretical stability bounds in Tables 3.1 and 3.2 for the sequential-partial-update LMS algorithm. The stability bounds for μ are seen to agree with the simulated MSD behaviour.

The simulated steady-state MSD and the theoretical stability bounds for the M-max LMS algorithm are shown in Figure 3.35. Overall the stability bounds for the M-max LMS algorithm are in good agreement with the simulated steady-state MSD curves. In practical applications the step-size parameter is usually chosen well below the theoretical stability bounds in order to ensure a good steady-state convergence performance.

Figure 3.36 shows the simulated steady-state MSD and the theoretical stability bounds for the M-max NLMS algorithm. The theoretical stability bounds appear to predict the divergence region for MSD very accurately.

Finally, in Figure 3.37 we see the simulated steady-state MSD and the theoretical stability bounds for the selective-partial-update NLMS algorithm. Again an excellent agreement between the theoretical stability bounds and the steady-state MSD behaviour is observed.

Chapter | four

Partial-update adaptive filters

4.1 INTRODUCTION

This chapter provides a detailed description of several stochastic gradient algorithms with partial coefficient updates. Both time-domain and transform-domain adaptive filters are covered. The treatment is focused on the following adaptive filtering algorithms and their partial-update versions: least-mean-square, normalized least-mean-square, affine projection algorithm (APA), recursive least squares (RLS), transform-domain LMS, and generalized-subband-decomposition LMS. The computational complexity of each algorithm is examined in detail. The convergence performance of the partial-update algorithms is compared in a channel equalization example.

Overall the largest complexity reduction is achieved by periodic partial updates. The method of partial coefficient updates has the most dramatic complexity and convergence impact on LMS and NLMS algorithms. Interestingly, LMS and NLMS employ the simplest approximation to steepest descent algorithm and Newton's method, respectively, and therefore they present more opportunity of performance improvement. On the other hand, as we move to computationally more complicated algorithms such as the RLS algorithm, the complexity and convergence impact of partial coefficient updates is not so impressive. In fact, RLS has stability issues for partial coefficient updates arising from sparse time-averaging of the input correlation matrix except for periodic partial updates. Thus, the periodic-partial-update RLS algorithm is the only partial-update RLS algorithm that may prove to be useful in practice.

As we saw in Chapter 2, the method of selective partial updates is based on Newton's method. For more sophisticated approximations of Newton's method, such as RLS and APA, we shall see that the method of M-max updates either is not applicable or does not perform as well.

4.2 LEAST-MEAN-SQUARE ALGORITHM

Consider the mean-squared error cost function:

$$J(\boldsymbol{w}(k)) = E\{e^2(k)\} \tag{4.1}$$

where $e(k)$ is the error signal defined by:

$$e(k) = d(k) - \boldsymbol{x}^T(k)\boldsymbol{w}(k) \tag{4.2}$$

and $\boldsymbol{w}(k) = [w_1(k), \ldots, w_N(k)]^T$ is the adaptive filter coefficient vector of size $N \times 1$. The minimization of $J(\boldsymbol{w}(k))$ over $\boldsymbol{w}(k)$ gives:

$$\boldsymbol{w}_o = \arg\min_{\boldsymbol{w}(k)} J(\boldsymbol{w}(k)) \tag{4.3}$$

where the minimizing solution is obtained from:

$$\left.\frac{\partial J(\boldsymbol{w}(k))}{\partial \boldsymbol{w}(k)}\right|_{\boldsymbol{w}(k)=\boldsymbol{w}_o} = \boldsymbol{0} \tag{4.4}$$

The gradient of $J(\boldsymbol{w}(k))$ is:

$$\begin{aligned}\frac{\partial J(\boldsymbol{w}(k))}{\partial \boldsymbol{w}(k)} &= -2E\{\boldsymbol{x}(k)e(k)\} \\ &= 2E\{\boldsymbol{x}(k)\boldsymbol{x}^T(k)\}\boldsymbol{w}(k) - 2E\{\boldsymbol{x}(k)d(k)\} \\ &= 2(\boldsymbol{R}\boldsymbol{w}(k) - \boldsymbol{p})\end{aligned} \tag{4.5}$$

whence:

$$\boldsymbol{w}_o = \boldsymbol{R}^{-1}\boldsymbol{p} \tag{4.6}$$

which is the minimum MSE estimate of $\boldsymbol{w}(k)$.

With an appropriate step-size parameter μ, the following gradient search algorithm can also find the minimum MSE estimate:

$$\boldsymbol{w}(k+1) = \boldsymbol{w}(k) - \frac{\mu}{2}\frac{\partial J(\boldsymbol{w}(k))}{\partial \boldsymbol{w}(k)}, \quad k = 0, 1, \ldots \tag{4.7}$$

Substituting (4.5) into the above recursion gives:

$$\boldsymbol{w}(k+1) = \boldsymbol{w}(k) + \mu(\boldsymbol{p} - \boldsymbol{R}\boldsymbol{w}(k)), \quad k = 0, 1, \ldots \tag{4.8}$$

which is the steepest descent algorithm. Note that (4.8) avoids the matrix inversion in (4.6).

In practical applications the autocorrelation matrix $\boldsymbol{R}$ and the cross-correlation vector $\boldsymbol{p}$ are usually unknown and must be learned on-the-fly. This is where the least-mean-square (LMS) algorithm becomes an attractive alternative since it does not require prior knowledge of $\boldsymbol{R}$ and $\boldsymbol{p}$. Consider the instantaneous approximation of the MSE cost function in (4.1):

$$J_{\text{approx}}(\boldsymbol{w}(k)) = e^2(k) \tag{4.9}$$

The gradient of $J_{\text{approx}}(\boldsymbol{w}(k))$ is simply given by:

$$\frac{\partial J_{\text{approx}}(\boldsymbol{w}(k))}{\partial \boldsymbol{w}(k)} = -2\boldsymbol{x}(k)e(k) \tag{4.10}$$

Substituting this into (4.7), we have:

$$\boldsymbol{w}(k+1) = \boldsymbol{w}(k) + \mu e(k)\boldsymbol{x}(k), \quad k = 0, 1, \ldots \tag{4.11}$$

which is known as the LMS algorithm (Widrow and Stearns, 1985). The LMS algorithm is one of the simplest stochastic gradient algorithms. In addition to not requiring prior knowledge of $\boldsymbol{R}$ and $\boldsymbol{p}$, the LMS algorithm is able to track changes in $\boldsymbol{R}$ and $\boldsymbol{p}$ as a result of time variations in the unknown system. Stochastic gradient algorithms are distinguished from gradient search algorithms such as the steepest descent algorithm by their use of an approximate gradient in the update term. Approximation of cost function gradient often leads to gradient noise which is responsible for the excess mean-square error.

4.3 PARTIAL-UPDATE LMS ALGORITHMS

4.3.1 Periodic-partial-update LMS algorithm

The periodic-partial-update LMS algorithm is described by:

$$\boldsymbol{w}((k+1)S) = \boldsymbol{w}(kS) + \mu e(kS)\boldsymbol{x}(kS), \quad k = 0, 1, \ldots \tag{4.12}$$

The update period S determines how often the adaptive filter coefficients are updated. All coefficients are updated every S iterations. Sparse time updates of $\boldsymbol{w}(k)$ lead to a reduction in average computational complexity per iteration.

4.3.2 Sequential-partial-update LMS algorithm

The sequential-partial-update LMS algorithm differs from the full-update LMS algorithm in (4.11) in the way the update term is defined:

$$\boldsymbol{w}(k+1) = \boldsymbol{w}(k) + \mu e(k)\boldsymbol{I}_M(k)\boldsymbol{x}(k), \quad k = 0, 1, \ldots \tag{4.13}$$

where the coefficient selection matrix $\boldsymbol{I}_M(k)$ is a diagonal matrix defined by:

$$\boldsymbol{I}_M(k) = \begin{bmatrix} i_1(k) & 0 & \cdots & 0 \\ 0 & i_2(k) & \ddots & \vdots \\ \vdots & \ddots & \ddots & 0 \\ 0 & \cdots & 0 & i_N(k) \end{bmatrix}, \quad i_j(k) = \begin{cases} 1 & \text{if } j \in \mathcal{J}_{(k \bmod \bar{B})+1} \\ 0 & \text{otherwise} \end{cases} \tag{4.14}$$

with $\bar{B} = \lceil N/M \rceil$. The number of coefficients updated at each iteration is limited by M.

The coefficient subsets $\mathcal{J}_i$ are not unique as long as they obey the following requirements:

1. Cardinality of $\mathcal{J}_i$ is between 1 and M;
2. $\bigcup_{i=1}^{\bar{B}} \mathcal{J}_i = \mathcal{S}$ where $\mathcal{S} = \{1, 2, \ldots, N\}$;
3. $\mathcal{J}_i \cap \mathcal{J}_j = \emptyset, \forall i, j \in \{1, \ldots, \bar{B}\}$ and $i \neq j$.

Note that if $\bar{B} = N/M$ (i.e. N/M is an integer) then the cardinality of each $\mathcal{J}_i$ must be M by necessity.

The following procedure yields a set of $\mathcal{J}_i$ compliant with the above requirements (cf. (2.36)):

$$\begin{aligned} \mathcal{J}_1 &= \{1, 2, \ldots, M\} \\ \mathcal{J}_2 &= \{M+1, M+2, \ldots, 2M\} \\ &\vdots \\ \mathcal{J}_{\bar{B}} &= \{(\bar{B}-1)M+1, (\bar{B}-1)M+2, \ldots, N\} \end{aligned} \tag{4.15}$$

where the cardinality of $\mathcal{J}_1, \ldots, \mathcal{J}_{\bar{B}-1}$ is M and that of $\mathcal{J}_{\bar{B}}$ is $N - (\bar{B}-1)M \leq M$.

4.3.3 Stochastic-partial-update LMS algorithm

The stochastic-partial-update LMS algorithm is given by:

$$\boldsymbol{w}(k+1) = \boldsymbol{w}(k) + \mu e(k) \boldsymbol{I}_M(k) \boldsymbol{x}(k), \quad k = 0, 1, \ldots \tag{4.16}$$

where the coefficient selection matrix $\boldsymbol{I}_M(k)$ is a diagonal matrix:

$$\boldsymbol{I}_M(k) = \begin{bmatrix} i_1(k) & 0 & \cdots & 0 \\ 0 & i_2(k) & \ddots & \vdots \\ \vdots & \ddots & \ddots & 0 \\ 0 & \cdots & 0 & i_N(k) \end{bmatrix}, \quad i_j(k) = \begin{cases} 1 & \text{if } j \in \mathcal{J}_{m(k)} \\ 0 & \text{otherwise} \end{cases} \tag{4.17}$$

Here the coefficient subsets $\mathcal{J}_i$ are defined identically to the sequential-partial-update LMS algorithm, and $m(k)$ is an independent discrete random variable with uniform distribution:

$$\Pr\{m(k) = i\} = \frac{1}{\tilde{B}}, \quad 1 \leq i \leq \tilde{B} \tag{4.18}$$

4.3.4 M-max LMS algorithm

The M-max LMS algorithm is the data-dependent partial-update version of the LMS algorithm and is described by:

$$\boldsymbol{w}(k+1) = \boldsymbol{w}(k) + \mu e(k)\boldsymbol{I}_M(k)\boldsymbol{x}(k), \quad k = 0, 1, \ldots \tag{4.19}$$

where the coefficient selection matrix $\boldsymbol{I}_M(k)$ is dependent on how the regressor magnitudes are ranked:

$$\boldsymbol{I}_M(k) = \begin{bmatrix} i_1(k) & 0 & \cdots & 0 \\ 0 & i_2(k) & \ddots & \vdots \\ \vdots & \ddots & \ddots & 0 \\ 0 & \cdots & 0 & i_N(k) \end{bmatrix},$$

$$i_j(k) = \begin{cases} 1 & \text{if } |x_j(k)| \in \max_{1 \leq l \leq N}(|x_l(k)|, M) \\ 0 & \text{otherwise} \end{cases} \tag{4.20}$$

Here $\max_j(w_j, M)$ denotes the set of M maxima of the w_j, and:

$$\boldsymbol{x}(k) = \begin{bmatrix} x_1(k) \\ x_2(k) \\ \vdots \\ x_N(k) \end{bmatrix} \tag{4.21}$$

In the M-max LMS algorithm M adaptive filter coefficients with the largest Euclidean norm regressor entries are selected for update at each iteration. This partial-update method is optimal in the sense of maintaining the smallest distance between M dimensional regressor subvector and the original regressor vector.

4.3.5 Computational complexity

The full-update LMS algorithm has $O(N)$ complexity. The full complexity analysis is provided below:

- At each k, the LMS recursion in (4.11) requires one multiplication to compute $\mu e(k)$ and N multiplications to compute $\mu e(k)\boldsymbol{x}(k)$.

- The computation of $e(k)$ requires N multiplications and $N - 1$ additions to compute the filter output $y(k)$ and one addition to obtain $e(k) = d(k) - y(k)$. Here we assume that the filtering is done by time-domain convolution (i.e. by computing the inner product $\boldsymbol{x}^T(k)\boldsymbol{w}(k)$). Computationally efficient FFT based methods are also available for the convolution operation.
- The update of $\boldsymbol{w}(k)$ takes N additions.

The total computational complexity of the LMS algorithm is $2N + 1$ multiplications and $2N$ additions at each iteration.

The periodic-partial-update LMS algorithm is a time-decimated version of the LMS algorithm with all computations spread over S iterations except for the filtering operation. Thus no complexity reduction in computing $e(kS)$ is achieved by periodic partial updates. For an update period of S the computational steps of the periodic-partial-update LMS algorithm are given by:

- N multiplications and N additions at each iteration to compute the filter output $y(k)$ and the decimated error $e(kS)$.
- One multiplication to compute $\mu e(kS)$ and N multiplications to compute $\mu e(kS)\boldsymbol{x}(kS)$ over S iterations.
- N additions to update $\boldsymbol{w}(kS)$ over S iterations.

The total computational complexity of the periodic-partial-update LMS algorithm is $N + (N + 1)/S$ multiplications and $N + N/S$ additions at each iteration.

The computational steps of the sequential-partial-update LMS algorithm at each iteration are:

- N multiplications and N additions to compute $e(k) = d(k) - y(k)$.
- One multiplication to compute $\mu e(k)$ and M multiplications to compute $\mu e(k)\boldsymbol{I}_M(k)\boldsymbol{x}(k)$.
- M additions to update $\boldsymbol{w}(k)$.

For partial update of M coefficients the total computational complexity of the sequential-partial-update LMS algorithm is $N + M + 1$ multiplications and $N + M$ additions at each iteration.

The computational complexity of the stochastic-partial-update LMS algorithm is given by the computational complexity of the sequential-partial-update LMS algorithm *and* the random number generator. A particularly simple random generator, which should be adequate for the implementation of random coefficient selection, is the *linear congruential generator* which is defined by (Press *et al.*, 1993)

$$u(k + 1) = (a\,u(k) + c) \bmod b, \quad k = 0, 1, \ldots \tag{4.22}$$

where a and c are positive integers called the multiplier and increment, respectively, b is the modulus, and $u(0)$ is a positive integer used as a seed for the random number generator. With appropriate choices of a, b and c, the above recursion generates $u(1), u(2), \ldots$ taking integer values between 0 and $b-1$. The generated sequence $u(k)$ is periodic with a maximum period not exceeding b. The range of $u(k)$ can be changed using:

$$m(k) = \frac{B-1}{b-1}u(k) + 1 \tag{4.23}$$

which gives the random integer $m(k)$ that can be used in the stochastic-partial-update coefficient selection matrix $\boldsymbol{I}_M(k)$ defined in (4.17). The computational complexity associated with (4.22) and (4.23) is 2 multiplications and 2 additions. Thus, assuming the use of a linear congruential generator as described above, the total computational complexity of the stochastic-partial-update LMS algorithm is $N + M + 3$ multiplications and $N + M + 2$ additions at each iteration.

The computational complexity of the M-max LMS algorithm is given by the computational complexity of the sequential-partial-update LMS algorithm plus the complexity of coefficient selection. The M-max coefficient selection requires ranking of the regressor vector element magnitudes. Assuming the regressor vector $\boldsymbol{x}(k)$ has the shift structure, the sortline algorithm (Pitas, 1989) can be used to find the M maxima of the regressor magnitudes. The computational complexity of sortline is $2\lceil \log_2 N \rceil + 2$ comparisons (see Appendix A). Therefore, the total computational complexity of the M-max LMS algorithm is $N + M + 1$ multiplications, $N + M$ additions and $2\lceil \log_2 N \rceil + 2$ comparisons at each iteration.

Table 4.1 summarizes the computational complexity of the LMS algorithm and its partial-update implementations. The number of multiplications is an important indication of the computational complexity since the overall hardware complexity is to a large extent dominated by multiply operations. For large N the partial-update LMS algorithms are capable of reducing the full complexity approximately by a factor of two.

4.4 NORMALIZED LEAST-MEAN-SQUARE ALGORITHM

The normalized least-mean-square (NLMS) algorithm is derived from the minimum-norm solution to a constrained optimization problem and can be considered an instantaneous approximation of Newton's method. The constrained optimization problem is defined by:

$$\min_{\boldsymbol{w}(k+1)} \| \underbrace{\boldsymbol{w}(k+1) - \boldsymbol{w}(k)}_{\delta \boldsymbol{w}(k+1)} \|_2^2 \tag{4.24a}$$

Table 4.1 Computational complexity of LMS and partial-update LMS algorithms at each k

Algorithm	$\times$	$+$	$\lessgtr$
LMS	$2N+1$	$2N$	
Per-par-upd LMS	$N+(N+1)/S$	$N+N/S$	
Seq-par-upd LMS	$N+M+1$	$N+M$	
Stoch-par-upd LMS	$N+M+3$	$N+M+2$	
M-max LMS	$N+M+1$	$N+M$	$2\lceil\log_2 N\rceil+2$

subject to:

$$e_p(k) - \left(1 - \frac{\mu\|\boldsymbol{x}(k)\|_2^2}{\epsilon + \|\boldsymbol{x}(k)\|_2^2}\right) e(k) = 0 \tag{4.24b}$$

where $e(k) = d(k) - \boldsymbol{w}^T(k)\boldsymbol{x}(k)$, $e_p(k) = d(k) - \boldsymbol{w}^T(k+1)\boldsymbol{x}(k)$ and ϵ is a positive regularization parameter. Substituting the definitions of $e(k)$ and $e_p(k)$ back into (4.24b), the constrained optimization problem in (4.24) is equivalently written as:

$$\boldsymbol{x}^T(k)\,\delta\boldsymbol{w}(k+1) = \frac{\mu\|\boldsymbol{x}(k)\|_2^2}{\epsilon + \|\boldsymbol{x}(k)\|_2^2} e(k), \quad \|\delta\boldsymbol{w}(k+1)\|_2^2 \text{ is minimum} \tag{4.25}$$

In other words, we now have an underdetermined least-squares problem for which we seek the unique minimum-norm solution. The unique minimum-norm solution is given by:

$$\delta\boldsymbol{w}(k+1) = (\boldsymbol{x}^T(k))^\dagger \frac{\mu\|\boldsymbol{x}(k)\|_2^2}{\epsilon + \|\boldsymbol{x}(k)\|_2^2} e(k) \tag{4.26}$$

where $(\boldsymbol{x}^T(k))^\dagger$ is the pseudoinverse of $\boldsymbol{x}^T(k)$ and is given by:

$$(\boldsymbol{x}^T(k))^\dagger = \frac{\boldsymbol{x}(k)}{\|\boldsymbol{x}(k)\|_2^2} \tag{4.27}$$

Thus we have:

$$\delta\boldsymbol{w}(k+1) = \frac{\mu}{\epsilon + \|\boldsymbol{x}(k)\|_2^2} e(k)\boldsymbol{x}(k) \tag{4.28}$$

or:

$$\boldsymbol{w}(k+1) = \boldsymbol{w}(k) + \frac{\mu}{\epsilon + \|\boldsymbol{x}(k)\|_2^2} e(k)\boldsymbol{x}(k) \tag{4.29}$$

which is called the NLMS algorithm. The constrained optimization problem in (4.24) is referred to as the principle of minimum disturbance.

Alternatively, the NLMS algorithm can be derived as an approximation to Newton's method:

$$\boldsymbol{w}(k+1) = \boldsymbol{w}(k) + \mu(\epsilon \boldsymbol{I} + \boldsymbol{R})^{-1}(\boldsymbol{p} - \boldsymbol{R}\boldsymbol{w}(k)), \quad k = 0, 1, \ldots \tag{4.30}$$

Stripping the expectation operator off $\boldsymbol{R}$ and $\boldsymbol{p}$ we obtain:

$$\boldsymbol{w}(k+1) = \boldsymbol{w}(k) + \mu(\epsilon \boldsymbol{I} + \boldsymbol{x}(k)\boldsymbol{x}^T(k))^{-1} e(k)\boldsymbol{x}(k) \tag{4.31}$$

Applying the matrix inversion lemma to $(\epsilon \boldsymbol{I} + \boldsymbol{x}(k)\boldsymbol{x}^T(k))^{-1}$ (see Section 2.6.2) eventually results in (4.29). The resemblance between the NLMS algorithm and Newton's method is particularly important because of the accelerated convergence properties of the latter.

4.5 PARTIAL-UPDATE NLMS ALGORITHMS

4.5.1 Periodic-partial-update NLMS algorithm

The periodic-partial-update NLMS algorithm is described by:

$$\boldsymbol{w}((k+1)S) = \boldsymbol{w}(kS) + \frac{\mu}{\epsilon + \|\boldsymbol{x}(kS)\|_2^2} e(kS)\boldsymbol{x}(kS), \quad k = 0, 1, \ldots \tag{4.32}$$

The update period S determines the frequency of adaptive filter coefficient updates. All coefficients are updated every S iterations. This reduces the average per iteration complexity of all update operations by a factor of S except for the $e(kS)$ computation. The adaptive filter output $y(k)$ is computed at every k regardless of S. This means that $e(kS)$ can be obtained from $y(kS)$ using one addition only since $e(kS) = d(kS) - y(kS)$. If $S < N$ the update complexity may be further reduced by computing:

$$\begin{aligned} \epsilon + \|\boldsymbol{x}(k)\|_2^2 &= \epsilon + \sum_{i=0}^{N-1} x^2(k-i) \\ &= \epsilon + \|\boldsymbol{x}(k-1)\|_2^2 + x^2(k) - x^2(k-N) \end{aligned} \tag{4.33}$$

at every iteration. This requires one multiplication and two additions per iteration rather than N/S multiplications and N/S additions per iteration.

4.5.2 Sequential-partial-update NLMS algorithm

The sequential-partial-update NLMS algorithm is defined by:

$$\boldsymbol{w}(k+1) = \boldsymbol{w}(k) + \frac{\mu}{\epsilon + \|\boldsymbol{x}(k)\|_2^2} e(k)\boldsymbol{I}_M(k)\boldsymbol{x}(k), \quad k = 0, 1, \ldots \tag{4.34}$$

where the coefficient selection matrix $\boldsymbol{I}_M(k)$ is given by (4.14). A variant of the sequential-partial-update NLMS algorithm is

$$\boldsymbol{w}(k+1) = \boldsymbol{w}(k) + \frac{\mu}{\epsilon + \|\boldsymbol{I}_M(k)\boldsymbol{x}(k)\|_2^2} e(k)\boldsymbol{I}_M(k)\boldsymbol{x}(k), \quad k = 0, 1, \ldots \tag{4.35}$$

where the sequential partial update is also applied to step-size normalization. This version of the sequential-partial-update NLMS algorithm may prove computationally more efficient in adaptive filtering applications where $\boldsymbol{x}(k)$ does not have a shift structure, which we have exploited to reduce the complexity of step-size normalization [see (4.33)].

4.5.3 Stochastic-partial-update NLMS algorithm

The stochastic-partial-update NLMS algorithm is given by:

$$\boldsymbol{w}(k+1) = \boldsymbol{w}(k) + \frac{\mu}{\epsilon + \|\boldsymbol{x}(k)\|_2^2} e(k)\boldsymbol{I}_M(k)\boldsymbol{x}(k), \quad k = 0, 1, \ldots \tag{4.36}$$

where the coefficient selection matrix $\boldsymbol{I}_M(k)$ is the randomized version of sequential partial updates as given in (4.17).

4.5.4 M-max NLMS algorithm

The M-max NLMS algorithm is a data-dependent partial-update version of the NLMS algorithm defined by:

$$\boldsymbol{w}(k+1) = \boldsymbol{w}(k) + \frac{\mu}{\epsilon + \|\boldsymbol{x}(k)\|_2^2} e(k)\boldsymbol{I}_M(k)\boldsymbol{x}(k), \quad k = 0, 1, \ldots \tag{4.37}$$

where the coefficient selection matrix $\boldsymbol{I}_M(k)$ is given by (4.20).

4.5.5 Selective-partial-update NLMS algorithm

The selective-partial-update NLMS algorithm is derived from the solution of the minimum-norm constrained optimization problem:

$$\min_{\boldsymbol{I}_M(k)} \min_{\boldsymbol{w}_{\mathcal{M}}(k+1)} \|\boldsymbol{w}_{\mathcal{M}}(k+1) - \boldsymbol{w}_{\mathcal{M}}(k)\|_2^2 \tag{4.38a}$$

subject to:

$$e_p(k) - \left(1 - \frac{\mu\|\boldsymbol{x}_{\mathcal{M}}(k)\|_2^2}{\epsilon + \|\boldsymbol{x}_{\mathcal{M}}(k)\|_2^2}\right) e(k) = 0 \tag{4.38b}$$

In Section 2.6.1 the solution to the above optimization problem was shown to be given by:

$$\boldsymbol{w}(k+1) = \boldsymbol{w}(k) + \frac{\mu}{\epsilon + \boldsymbol{x}^T(k)\boldsymbol{I}_M(k)\boldsymbol{x}(k)} e(k)\boldsymbol{I}_M(k)\boldsymbol{x}(k), \quad k = 0, 1, \ldots \tag{4.39}$$

where the coefficient selection matrix $\boldsymbol{I}_M(k)$ is identical to the M-max LMS coefficient selection matrix in (4.20). Equation (4.39) is referred to as the selective-partial-update NLMS algorithm and it only differs from the M-max NLMS algorithm in the way the normalization term is defined. It was also shown in Section 2.6.2 that the selective-partial-update NLMS is an instantaneous approximation of the regularized Newton's method employing M-max updates for instantaneous approximations of $\boldsymbol{R}_M$ and $\boldsymbol{p}_M$.

4.5.6 Set-membership partial-update NLMS algorithm

As we saw in Section 2.7, the set-membership partial-update NLMS algorithm is obtained from the solution of:

$$\min_{\boldsymbol{I}_M(k)} \min_{\boldsymbol{w}_{\mathcal{M}}(k+1)} \|\boldsymbol{w}_{\mathcal{M}}(k+1) - \boldsymbol{w}_{\mathcal{M}}(k)\|_2^2 \tag{4.40a}$$

subject to:

$$e_p(k) - \left(1 - \alpha(k)\frac{\|\boldsymbol{x}_{\mathcal{M}}(k)\|_2^2}{\epsilon + \|\boldsymbol{x}_{\mathcal{M}}(k)\|_2^2}\right) e(k) = 0 \tag{4.40b}$$

where $\alpha(k)$ is a time-varying step-size defined by:

$$\alpha(k) = \begin{cases} \left(1 - \dfrac{\gamma}{|e(k)|}\right)\left(1 + \dfrac{\epsilon}{\|\boldsymbol{x}_{\mathcal{M}}(k)\|_2^2}\right) & \text{if } |\, e(k)\,| > \gamma \\ 0 & \text{otherwise} \end{cases} \tag{4.41}$$

Here γ is a bound on the magnitude of the *a priori* error $e(k)$. The solution to (4.40) is given by:

$$\boldsymbol{w}(k+1) = \boldsymbol{w}(k) + \frac{\alpha(k)}{\epsilon + \boldsymbol{x}^T(k)\boldsymbol{I}_M(k)\boldsymbol{x}(k)} e(k)\boldsymbol{I}_M(k)\boldsymbol{x}(k) \tag{4.42}$$

which is known as the set-membership partial-update NLMS algorithm.

4.5.7 Computational complexity

The full-update NLMS algorithm differs from the full-update LMS algorithm in two aspects; viz. computation of $\epsilon + \|\boldsymbol{x}(k)\|_2^2$ and division of μ by $\epsilon + \|\boldsymbol{x}(k)\|_2^2$. These additional operations result in some computational overheads. The computational steps of the NLMS algorithm at each iteration k are:

- N multiplications and N additions to compute $e(k)$.
- One multiplication to compute $\mu e(k)$.
- One multiplication and two additions to compute $\epsilon + \|\boldsymbol{x}(k)\|_2^2$. We assume that $\boldsymbol{x}(k)$ has a shift structure so that we can write:

$$\epsilon + \|\boldsymbol{x}(k)\|_2^2 = \epsilon + \|\boldsymbol{x}(k-1)\|_2^2 + x^2(k) - x^2(k-N) \qquad (4.43)$$

  where only the new signal $x^2(k)$ needs to be computed since $x^2(k-N)$ has already been computed in an earlier iteration.
- One division to compute $\mu e(k)/(\epsilon + \|\boldsymbol{x}(k)\|_2^2)$.
- N multiplications to compute $\mu e(k)/(\epsilon + \|\boldsymbol{x}(k)\|_2^2)\boldsymbol{x}(k)$.
- N additions to update $\boldsymbol{w}(k)$.

The total computational complexity of the NLMS algorithm is $2N+2$ multiplications, $2N+2$ additions and one division at each iteration.

For an update period of S the computational steps of the periodic-partial-update NLMS algorithm are given by:

- N multiplications and N additions at each iteration to compute the filter output $y(k)$ and the decimated error $e(kS)$.
- One multiplication to compute $\mu e(kS)$ over S iterations.
- To compute $\epsilon + \|\boldsymbol{x}(kS)\|_2^2$, one multiplication and two additions per iteration, if $S < N$, or N/S multiplications and N/S additions per iteration, otherwise.
- One division to compute $\mu e(kS)/(\epsilon + \|\boldsymbol{x}(kS)\|_2^2)$ over S iterations.
- N multiplications to compute $\mu e(kS)/(\epsilon + \|\boldsymbol{x}(kS)\|_2^2)\boldsymbol{x}(kS)$ over S iterations.
- N additions to update $\boldsymbol{w}(kS)$ over S iterations.

The total computational complexity of the periodic-partial-update NLMS algorithm is $N+1+(N+1)/S$ multiplications, $N+1+N/S$ additions and $1/S$ divisions at each iteration, if $S < N$, and $N+(2N+1)/S$ multiplications, $N+2N/S$ additions and $1/S$ divisions at each iteration, otherwise.

The computational steps of the sequential-partial-update NLMS algorithm at each iteration are:

- N multiplications and N additions to compute $e(k) = d(k) - y(k)$.
- One multiplication to compute $\mu e(k)$.
- One multiplication and two additions to compute $\epsilon + \|\boldsymbol{x}(k)\|_2^2$.
- One division to compute $\mu e(k)/(\epsilon + \|\boldsymbol{x}(k)\|_2^2)$.

- M multiplications to compute $\mu e(k)/(\epsilon + \|\boldsymbol{x}(k)\|_2^2)\boldsymbol{I}_M(k)\boldsymbol{x}(k)$.
- M additions to update $\boldsymbol{w}(k)$.

For partial update of M coefficients the total computational complexity of the sequential-partial-update NLMS algorithm is $N + M + 2$ multiplications, $N + M + 2$ additions and one division at each iteration.

The stochastic-partial-update NLMS requires a further 2 multiplications and 2 additions to implement a linear congruential generator as described in Section 4.3.5. Therefore the total computational complexity of the stochastic-partial-update NLMS algorithm is $N + M + 4$ multiplications, $N + M + 4$ additions and one division at each iteration.

The computational complexity of the M-max NLMS algorithm is given by the sum of computational complexities of the sequential-partial-update NLMS algorithm and M-max coefficient selection. Using the sortline algorithm the computational complexity of the M-max coefficient selection is $2\lceil \log_2 N \rceil + 2$ comparisons (see Appendix A). Thus the total computational complexity of the M-max NLMS algorithm is $N + M + 2$ multiplications, $N + M + 2$ additions, one division and $2\lceil \log_2 N \rceil + 2$ comparisons at each iteration.

The selective-partial-update NLMS algorithm is similar to the M-max NLMS algorithm. The only difference between the two partial-update algorithms is the way the step-size parameter is normalized. The computation of $\epsilon + \boldsymbol{x}^T(k)\boldsymbol{I}_M(k)\boldsymbol{x}(k)$ requires one multiplication to obtain $x^2(k)$ and two additions to update the normalization term if $|x(k)|$ is one of the new M maxima or $|x(k - N)|$ was one of the M maxima at iteration $k - 1$. Therefore, the worst-case complexity of the selective-partial-update NLMS algorithm is identical to the computational complexity of the M-max NLMS algorithm.

To save some computation, the time-varying step-size parameter of the set-membership partial-update NLMS algorithm in (4.41) may be replaced by:

$$\alpha(k) = \begin{cases} \left(1 - \dfrac{\gamma}{|e(k)|}\right) & \text{if } |e(k)| > \gamma \\ 0 & \text{otherwise} \end{cases} \tag{4.44}$$

This simplification is justified if the regularization parameter ϵ is chosen very small. The set-membership partial-update NLMS algorithm proceeds as follows:

- N multiplications and N additions to compute $e(k) = d(k) - y(k)$.
- $2\lceil \log_2 N \rceil + 2$ comparisons to obtain $\boldsymbol{I}_M(k)$ using the sortline algorithm.
- One multiplication and two additions to compute $\epsilon + \|\boldsymbol{I}_M(k)\boldsymbol{x}(k)\|_2^2$.
- One comparison to decide if $|e(k)| > \gamma$, i.e., update is required.
- If update is required,
  - One division and one addition to compute approximate $\alpha(k)$ in (4.44).

Table 4.2 Computational complexity of NLMS and partial-update NLMS algorithms at each k

Algorithm		$\times$	$+$	$\div$	$\lessgtr$
NLMS		$2N+2$	$2N+2$	1	
Per-par-upd NLMS	$S<N$	$N+1+(N+1)/S$	$N+1+N/S$	$1/S$	
	$S \geq N$	$N+(2N+1)/S$	$N+2N/S$	$1/S$	
Seq-par-upd NLMS		$N+M+2$	$N+M+2$	1	
Stoch-par-upd NLMS		$N+M+4$	$N+M+4$	1	
M-max NLMS		$N+M+2$	$N+M+2$	1	$2\lceil \log_2 N \rceil + 2$
Sel-par-upd NLMS		$N+M+2$	$N+M+2$	1	$2\lceil \log_2 N \rceil + 2$
Set-mem par-upd NLMS		$N+M+2$	$N+M+3$	2	$2\lceil \log_2 N \rceil + 3$

- One multiplication to compute $\alpha(k)e(k)$.
- One division to compute $\alpha(k)e(k)/(\epsilon + \|\boldsymbol{I}_M(k)\boldsymbol{x}(k)\|_2^2)$.
- M multiplications to compute $\mu e(k)/(\epsilon + \|\boldsymbol{I}_M(k)\boldsymbol{x}(k)\|_2^2)\boldsymbol{I}_M(k)\boldsymbol{x}(k)$.
- M additions to update $\boldsymbol{w}(k)$.

Note that ranking of regressor data and computation of $\epsilon + \|\boldsymbol{I}_M(k)\boldsymbol{x}(k)\|_2^2$ are carried out at every iteration regardless of update decision. This is to ensure that the shift structure of the regressor data is maintained in order to take advantage of complexity reduction resulting from it. The total computational complexity of the set-membership partial-update NLMS algorithm is $N+1$ multiplications, $N+2$ additions and $2\lceil \log_2 N \rceil + 3$ comparisons for each iteration that does not require a coefficient update, and $N+M+2$ multiplications, $N+M+3$ additions, 2 divisions and $2\lceil \log_2 N \rceil + 3$ comparisons, otherwise. Since the adaptive filter coefficients are not necessarily updated at every iteration the average computational complexity is usually smaller than that of the selective-partial-update NLMS algorithm. A significant advantage of low average computational complexity is reduced power consumption.

Table 4.2 summarizes the computational complexity of the NLMS algorithm and its partial-update variants. In the case of set-membership partial-update NLMS, the peak computational complexity is shown even though the average complexity may be significantly smaller. This is explained by the fact that the adaptive filter hardware must be able to cope with the computational demand placed on it at every iteration. This means that it must have the computational resources to handle the highest complexity per iteration.

4.6 AFFINE PROJECTION ALGORITHM

The APA is derived in a similar way to the NLMS algorithm by seeking the minimum-norm solution to a constrained optimization problem. As different from the NLMS algorithm, the constrained optimization problem involves multiple constraints:

$$\min_{\boldsymbol{w}(k+1)} \| \underbrace{\boldsymbol{w}(k+1) - \boldsymbol{w}(k)}_{\delta \boldsymbol{w}(k+1)} \|_2^2 \tag{4.45a}$$

subject to:

$$\boldsymbol{e}_p(k) - \left(\boldsymbol{I} - \mu \boldsymbol{X}(k)\boldsymbol{X}^T(k)(\epsilon \boldsymbol{I} + \boldsymbol{X}(k)\boldsymbol{X}^T(k))^{-1}\right) \boldsymbol{e}(k) = \boldsymbol{0} \tag{4.45b}$$

Here $\boldsymbol{e}_p(k)$ and $\boldsymbol{e}(k)$ are $P \times 1$ *a posteriori* and *a priori* error vectors defined by:

$$\boldsymbol{e}(k) = \boldsymbol{d}(k) - \boldsymbol{X}(k)\boldsymbol{w}(k) \quad \text{and} \quad \boldsymbol{e}_p(k) = \boldsymbol{d}(k) - \boldsymbol{X}(k)\boldsymbol{w}(k+1) \tag{4.46}$$

where $\boldsymbol{X}(k)$ is the matrix of P regressors:

$$\boldsymbol{X}(k) = \begin{bmatrix} \boldsymbol{x}^T(k) \\ \boldsymbol{x}^T(k-1) \\ \vdots \\ \boldsymbol{x}^T(k-P+1) \end{bmatrix}_{P \times N} \tag{4.47}$$

and $\boldsymbol{d}(k)$ is the $P \times 1$ vector of desired outputs:

$$\boldsymbol{d}(k) = \begin{bmatrix} d(k) \\ d(k-1) \\ \vdots \\ d(k-P+1) \end{bmatrix} \tag{4.48}$$

The number of constraints P, which is also known as the APA order, obeys the condition $P \leq N$. Using the definitions of $\boldsymbol{e}(k)$ and $\boldsymbol{e}_p(k)$ in (4.46), the constrained optimization problem in (4.45) may be written as an underdetermined least-squares problem:

$$\begin{aligned} \boldsymbol{X}(k)\, \delta \boldsymbol{w}(k+1) &= \mu \boldsymbol{X}(k)\boldsymbol{X}^T(k)(\epsilon \boldsymbol{I} + \boldsymbol{X}(k)\boldsymbol{X}^T(k))^{-1}\boldsymbol{e}(k), \\ &\|\delta \boldsymbol{w}(k+1)\|_2^2 \text{ is minimum} \end{aligned} \tag{4.49}$$

The unique minimum-norm solution is given by:

$$\begin{aligned} \delta \boldsymbol{w}(k+1) &= \mu \boldsymbol{X}^\dagger(k)\boldsymbol{X}(k)\boldsymbol{X}^T(k)(\epsilon \boldsymbol{I} + \boldsymbol{X}(k)\boldsymbol{X}^T(k))^{-1}\boldsymbol{e}(k) \\ &= \mu \boldsymbol{X}^T(k)(\epsilon \boldsymbol{I} + \boldsymbol{X}(k)\boldsymbol{X}^T(k))^{-1}\boldsymbol{e}(k) \end{aligned} \tag{4.50}$$

where $\boldsymbol{X}^\dagger(k)$ is the pseudoinverse of $\boldsymbol{X}(k)$, and we have used one of the properties of pseudoinverse that $\boldsymbol{X}^\dagger(k)\boldsymbol{X}(k)\boldsymbol{X}^T(k) = \boldsymbol{X}^T(k)$. We therefore have:

$$\boldsymbol{w}(k+1) = \boldsymbol{w}(k) + \mu \boldsymbol{X}^T(k)(\epsilon \boldsymbol{I} + \boldsymbol{X}(k)\boldsymbol{X}^T(k))^{-1}\boldsymbol{e}(k) \tag{4.51}$$

which is called APA.

APA can alternatively be derived as an improved approximation to Newton's method. Recall that in the NLMS derivation we used instantaneous approximation of $\boldsymbol{R}$ and $\boldsymbol{p}$. A better approximation is given by:

$$\begin{aligned} \boldsymbol{R} &\approx \frac{1}{P} \sum_{i=k-P+1}^{k} \boldsymbol{x}(i)\boldsymbol{x}^T(i) = \frac{1}{P}\boldsymbol{X}^T(k)\boldsymbol{X}(k) \\ \boldsymbol{p} &\approx \frac{1}{P} \sum_{i=k-P+1}^{k} \boldsymbol{x}(i)d(i) = \frac{1}{P}\boldsymbol{X}^T(k)\boldsymbol{d}(k) \end{aligned} \tag{4.52}$$

Replacing $\boldsymbol{R}$ and $\boldsymbol{p}$ with the above approximations in Newton's method we obtain:

$$\boldsymbol{w}(k+1) = \boldsymbol{w}(k) + \mu(\epsilon \boldsymbol{I} + \boldsymbol{X}^T(k)\boldsymbol{X}(k))^{-1}\boldsymbol{X}^T(k)\boldsymbol{e}(k), \quad k = 0, 1, \ldots \tag{4.53}$$

Applying the matrix inversion lemma (see Section 2.6.2), we have $(\epsilon \boldsymbol{I} + \boldsymbol{X}^T(k)\boldsymbol{X}(k))^{-1}\boldsymbol{X}^T(k) = \boldsymbol{X}^T(k)(\epsilon \boldsymbol{I} + \boldsymbol{X}(k)\boldsymbol{X}^T(k))^{-1}$ which confirms that (4.53) is identical to (4.51).

Consider the regularized least-squares problem:

$$\boldsymbol{w}(k+1) = \arg\min_{\boldsymbol{w}} \epsilon \|\boldsymbol{w} - \boldsymbol{w}(k)\|_2^2 + \|\boldsymbol{d}(k) - \boldsymbol{X}(k)\boldsymbol{w}\|_2^2 \tag{4.54}$$

The closed-form solution to this problem is (Sayed, 2003):

$$\begin{aligned} \boldsymbol{w}(k+1) &= \boldsymbol{w}(k) + (\epsilon \boldsymbol{I} + \boldsymbol{X}^T(k)\boldsymbol{X}(k))^{-1}\boldsymbol{X}^T(k)\underbrace{(\boldsymbol{d}(k) - \boldsymbol{X}(k)\boldsymbol{w}(k))}_{\boldsymbol{e}(k)} \\ &= \boldsymbol{w}(k) + \boldsymbol{X}^T(k)(\epsilon \boldsymbol{I} + \boldsymbol{X}(k)\boldsymbol{X}^T(k))^{-1}\underbrace{(\boldsymbol{d}(k) - \boldsymbol{X}(k)\boldsymbol{w}(k))}_{\boldsymbol{e}(k)} \end{aligned} \tag{4.55}$$

Comparison of the above equation with (4.51) reveals that APA solves a relaxed form of the regularized least-squares problem in (4.54) at each iteration k. The APA solution is relaxed through the use of the step-size parameter μ.

Setting $P = 1$ reduces APA to the NLMS algorithm. If $P > 1$ APA offers improved performance over the NLMS algorithm especially for highly correlated input signals. The computational complexity of APA is $O(P^3)$ as we will see in Section 4.7.8. However, fast (computationally efficient) versions of APA are also available with computational complexity comparable to the NLMS algorithm (Gay and Tavathia, 1995).

4.7 PARTIAL-UPDATE AFFINE PROJECTION ALGORITHMS

4.7.1 Periodic-partial-update APA

The periodic-partial-update APA is given by:

$$\boldsymbol{w}((k+1)S) = \boldsymbol{w}(kS) + \mu \boldsymbol{X}^T(kS)(\epsilon \boldsymbol{I} + \boldsymbol{X}(kS)\boldsymbol{X}^T(kS))^{-1}\boldsymbol{e}(kS) \quad (4.56)$$

where S is the period of coefficient updates. The adaptation complexity is reduced by spreading update complexity over S iterations at the expense of slower convergence rate. A special case of the periodic-partial-update APA which sets $S = P$ is known as the partial-rank algorithm (Kratzer and Morgan, 1985).

4.7.2 Sequential-partial-update APA

The sequential-partial-update APA is given by:

$$\boldsymbol{w}(k+1) = \boldsymbol{w}(k) + \mu \boldsymbol{I}_M(k)\boldsymbol{X}^T(k)(\epsilon \boldsymbol{I} + \boldsymbol{X}(k)\boldsymbol{X}^T(k))^{-1}\boldsymbol{e}(k) \quad (4.57)$$

where $\boldsymbol{I}_M(k)$ is the coefficient selection matrix defined in (4.14). The adaptation complexity is not as small as M/Nth that of the full-update APA as a result of the requirement to compute $(\epsilon \boldsymbol{I} + \boldsymbol{X}(k)\boldsymbol{X}^T(k))^{-1}\boldsymbol{e}(k)$. The matrix product $\boldsymbol{I}_M(k)\boldsymbol{X}^T(k)$ can be expressed as:

$$\begin{aligned}
\boldsymbol{I}_M(k)\boldsymbol{X}^T(k) &= \begin{bmatrix} i_1(k) & & & \boldsymbol{0} \\ & i_2(k) & & \\ & & \ddots & \\ \boldsymbol{0} & & & i_N(k) \end{bmatrix} \begin{bmatrix} \boldsymbol{x}(k) & \boldsymbol{x}(k-1) & \cdots & \boldsymbol{x}(k-P+1) \end{bmatrix} \\
&= \begin{bmatrix} i_1(k) & & & \boldsymbol{0} \\ & i_2(k) & & \\ & & \ddots & \\ \boldsymbol{0} & & & i_N(k) \end{bmatrix} \begin{bmatrix} \boldsymbol{x}_A^T(k) \\ \boldsymbol{x}_A^T(k-1) \\ \vdots \\ \boldsymbol{x}_A^T(k-N+1) \end{bmatrix}, \\
\boldsymbol{x}_A(k) &= \begin{bmatrix} x(k) \\ x(k-1) \\ \vdots \\ x(k-P+1) \end{bmatrix} \\
&= \begin{bmatrix} i_1(k)\boldsymbol{x}_A^T(k) \\ i_2(k)\boldsymbol{x}_A^T(k-1) \\ \vdots \\ i_N(k)\boldsymbol{x}_A^T(k-N+1) \end{bmatrix}
\end{aligned} \quad (4.58)$$

By including only non-zero rows of $I_M(k)X^T(k)$ in the coefficient updates, we achieve some reduction in the adaptation complexity.

4.7.3 Stochastic-partial-update APA

The stochastic-partial-update APA has the same update equation as the sequential-partial-update APA:

$$w(k+1) = w(k) + \mu I_M(k)X^T(k)(\epsilon I + X(k)X^T(k))^{-1}e(k) \qquad (4.59)$$

The only difference is the randomized selection of adaptive filter coefficients rather than deterministic round-robin selection. The coefficient selection matrix $I_M(k)$ for stochastic-partial-update APA is given by (4.17).

4.7.4 M-max APA

The M-max APA is described by the recursion:

$$w(k+1) = w(k) + \mu I_M(k)X^T(k)(\epsilon I + X(k)X^T(k))^{-1}e(k). \qquad (4.60)$$

The M-max APA employs a data-dependent coefficient selection matrix $I_M(k)$ that aims to maintain the shortest Euclidean distance between the partial coefficient updates and full coefficient updates. Thus, for the M-max APA $I_M(k)$ is defined by:

$$\begin{aligned}
&\max_{I_M(k)} \|I_M(k)X^T(k)(\epsilon I + X(k)X^T(k))^{-1}e(k)\|_2^2 \\
&\max_{I_M(k)} e^T(k)(\epsilon I + X(k)X^T(k))^{-1}X(k)I_M(k)X^T(k) \\
&\qquad \times(\epsilon I + X(k)X^T(k))^{-1}e(k)
\end{aligned} \qquad (4.61)$$

In order to solve the above minimization problem we must first compute the full update vector. However, doing so will rule out any computational savings and we might as well use the full-update APA. To avoid any undesirable complexity increase we need to look for a computationally inexpensive approximation of (4.61). We propose to approximate (4.61) with:

$$\max_{I_M(k)} \operatorname{tr}(X(k)I_M(k)X^T(k)) \qquad (4.62)$$

which is equivalent to:

$$\begin{aligned}
&\max_{I_M(k)} \operatorname{tr}\left(\sum_{j=1}^{N} i_j(k)x_A(k-j+1)x_A^T(k-j+1)\right) \\
&\max_{I_M(k)} \sum_{j=1}^{N} i_j(k)\|x_A(k-j+1)\|_2^2
\end{aligned} \qquad (4.63)$$

The simplified coefficient selection ranks $\|\boldsymbol{x}_A(k)\|_2^2, \|\boldsymbol{x}_A(k-1)\|_2^2, \ldots, \|\boldsymbol{x}_A(k-N+1)\|_2^2$ and sets the $i_j(k)$ corresponding to the M maxima to one:

$$\boldsymbol{I}_M(k) = \begin{bmatrix} i_1(k) & 0 & \cdots & 0 \\ 0 & i_2(k) & \ddots & \vdots \\ \vdots & \ddots & \ddots & 0 \\ 0 & \cdots & 0 & i_N(k) \end{bmatrix},$$

$$i_j(k) = \begin{cases} 1 & \text{if } \|\boldsymbol{x}_A(k-j+1)\|_2^2 \in \max_{1\le l\le N}(\|\boldsymbol{x}_A(k-l+1)\|_2^2, M) \\ 0 & \text{otherwise} \end{cases} \tag{4.64}$$

Note that, for $P = 1$, the simplified criterion in (4.64) reduces to the M-max NLMS criterion in (4.20). Thus, (4.64) is consistent with the selection criterion for the M-max NLMS algorithm. To differentiate the computationally demanding full implementation of the M-max APA in (4.61) from its simplified version that uses (4.64), we will refer to the latter as the *simplified M-max APA*. Although (4.64) is not exactly equivalent to (4.61), it usually yields satisfactory convergence performance at a significantly reduced computational complexity.

4.7.5 Selective-partial-update APA

The selective-partial-update APA is formulated as the solution to the following constrained optimization problem:

$$\min_{\boldsymbol{I}_M(k)} \min_{\boldsymbol{w}_{\mathcal{M}}(k+1)} \|\underbrace{\boldsymbol{w}_{\mathcal{M}}(k+1) - \boldsymbol{w}_{\mathcal{M}}(k)}_{\delta\boldsymbol{w}_{\mathcal{M}}(k+1)}\|_2^2 \tag{4.65a}$$

subject to:

$$\boldsymbol{e}_p(k) - \left(\boldsymbol{I} - \mu\boldsymbol{X}_{\mathcal{M}}(k)\boldsymbol{X}_{\mathcal{M}}^T(k)(\epsilon\boldsymbol{I} + \boldsymbol{X}_{\mathcal{M}}(k)\boldsymbol{X}_{\mathcal{M}}^T(k))^{-1}\right)\boldsymbol{e}(k) = \boldsymbol{0} \tag{4.65b}$$

where $\boldsymbol{X}_{\mathcal{M}}(k)$ is an $P \times M$ partition of $\boldsymbol{X}(k)$ obtained by retaining the non-zero columns of $\boldsymbol{X}(k)$ corresponding to M non-zero diagonal elements of $\boldsymbol{I}_M(k)$. The matrix $\boldsymbol{X}_{\mathcal{M}}(k)$ satisfies $\boldsymbol{X}(k)\boldsymbol{I}_M(k)\boldsymbol{X}^T(k) = \boldsymbol{X}_{\mathcal{M}}(k)\boldsymbol{X}_{\mathcal{M}}^T(k)$. The number of constraints is assumed to satisfy the condition $P \le M$ so that $\boldsymbol{X}_{\mathcal{M}}(k)\boldsymbol{X}_{\mathcal{M}}^T(k)$ can be full-rank. For fixed $\boldsymbol{I}_M(k)$ this constrained optimization problem is equivalently written as:

$$\begin{aligned} \boldsymbol{X}_{\mathcal{M}}(k)\,\delta\boldsymbol{w}_{\mathcal{M}}(k+1) &= \mu\boldsymbol{X}_{\mathcal{M}}(k)\boldsymbol{X}_{\mathcal{M}}^T(k)(\epsilon\boldsymbol{I} + \boldsymbol{X}_{\mathcal{M}}(k)\boldsymbol{X}_{\mathcal{M}}^T(k))^{-1}\boldsymbol{e}(k), \\ &\|\delta\boldsymbol{w}_{\mathcal{M}}(k+1)\|_2^2 \text{ is minimum} \end{aligned} \tag{4.66}$$

The condition $P \leq M$ ensures that a unique minimum-norm solution to (4.66) exists and is given by:

$$\begin{aligned}\delta \boldsymbol{w}_{\mathcal{M}}(k+1) &= \mu \boldsymbol{X}_{\mathcal{M}}^{\dagger}(k)\boldsymbol{X}_{\mathcal{M}}(k)\boldsymbol{X}_{\mathcal{M}}^{T}(k)(\epsilon \boldsymbol{I} + \boldsymbol{X}_{\mathcal{M}}(k)\boldsymbol{X}_{\mathcal{M}}^{T}(k))^{-1}\boldsymbol{e}(k) \\ &= \mu \boldsymbol{X}_{\mathcal{M}}^{T}(k)(\epsilon \boldsymbol{I} + \boldsymbol{X}_{\mathcal{M}}(k)\boldsymbol{X}_{\mathcal{M}}^{T}(k))^{-1}\boldsymbol{e}(k) \end{aligned} \tag{4.67}$$

where we have used a property of pseudoinverse that $\boldsymbol{X}_{\mathcal{M}}^{\dagger}(k)\boldsymbol{X}_{\mathcal{M}}(k)\boldsymbol{X}_{\mathcal{M}}^{T}(k) = \boldsymbol{X}_{\mathcal{M}}^{T}(k)$. The selective-partial-update APA recursion for fixed $\boldsymbol{I}_M(k)$ is, therefore, given by:

$$\boldsymbol{w}_{\mathcal{M}}(k+1) = \boldsymbol{w}_{\mathcal{M}}(k) + \mu \boldsymbol{X}_{\mathcal{M}}^{T}(k)(\epsilon \boldsymbol{I} + \boldsymbol{X}_{\mathcal{M}}(k)\boldsymbol{X}_{\mathcal{M}}^{T}(k))^{-1}\boldsymbol{e}(k) \tag{4.68}$$

The coefficient selection matrix $\boldsymbol{I}_M(k)$ is obtained from:

$$\begin{aligned} &\min_{\boldsymbol{I}_M(k)} \left\| \boldsymbol{X}_{\mathcal{M}}^{T}(k)(\epsilon \boldsymbol{I} + \boldsymbol{X}_{\mathcal{M}}(k)\boldsymbol{X}_{\mathcal{M}}^{T}(k))^{-1}\boldsymbol{e}(k) \right\|_2^2 \\ &\min_{\boldsymbol{I}_M(k)} \boldsymbol{e}^T(k)(\epsilon \boldsymbol{I} + \boldsymbol{X}_{\mathcal{M}}(k)\boldsymbol{X}_{\mathcal{M}}^{T}(k))^{-1}\boldsymbol{X}_{\mathcal{M}}(k)\boldsymbol{X}_{\mathcal{M}}^{T}(k) \\ &\qquad \times (\epsilon \boldsymbol{I} + \boldsymbol{X}_{\mathcal{M}}(k)\boldsymbol{X}_{\mathcal{M}}^{T}(k))^{-1}\boldsymbol{e}(k) \end{aligned} \tag{4.69}$$

Assuming sufficiently small ϵ the above minimization problem can be approximated by:

$$\min_{\boldsymbol{I}_M(k)} \boldsymbol{e}^T(k)(\boldsymbol{X}_{\mathcal{M}}(k)\boldsymbol{X}_{\mathcal{M}}^{T}(k))^{-1}\boldsymbol{e}(k) \tag{4.70}$$

Unfortunately, the full implementation of the selection criterion (4.70) is computationally very expensive. In an attempt to ease the excessive computational demand imposed by (4.70), a simplified alternative criterion with reduced complexity may be considered (Doğançay and Tanrıkulu, 2000). Using $\boldsymbol{X}_{\mathcal{M}}(k)\boldsymbol{X}_{\mathcal{M}}^{T}(k) = \boldsymbol{X}(k)\boldsymbol{I}_M(k)\boldsymbol{X}^T(k)$, rewrite (4.70) as:

$$\min_{\boldsymbol{I}_M(k)} \boldsymbol{e}^T(k) \left(\epsilon \boldsymbol{I} + \sum_{j=1}^{N} i_j(k)\boldsymbol{x}_A(k-j+1)\boldsymbol{x}_A^T(k-j+1) \right)^{-1} \boldsymbol{e}(k) \tag{4.71}$$

The simplified selection criterion is:

$$\max_{\boldsymbol{I}_M(k)} \operatorname{tr} \left(\sum_{j=1}^{N} i_j(k)\boldsymbol{x}_A(k-j+1)\boldsymbol{x}_A^T(k-j+1) \right) \tag{4.72}$$

which ranks the traces of outer products $\|\boldsymbol{x}_A(k)\|_2^2, \|\boldsymbol{x}_A(k-1)\|_2^2, \ldots, \|\boldsymbol{x}_A(k-N+1)\|_2^2$ and sets the $i_j(k)$ corresponding to the M maxima to one. This leads to the same coefficient selection matrix that was derived for the M-max APA in (4.64).

Combining (4.68) with appropriately defined $\boldsymbol{I}_M(k)$ gives the selective-partial-update APA recursion:

$$\boldsymbol{w}(k+1) = \boldsymbol{w}(k) + \mu \boldsymbol{I}_M(k)\boldsymbol{X}^T(k)(\epsilon \boldsymbol{I} + \boldsymbol{X}(k)\boldsymbol{I}_M(k)\boldsymbol{X}^T(k))^{-1}\boldsymbol{e}(k) \quad (4.73)$$

where $\boldsymbol{I}_M(k)$ is either given by (4.69) or (4.64) where the former is the computationally expensive full implementation and the latter is the simplified approximate criterion that permits complexity savings.

4.7.6 Set-membership partial-update APA

Set-membership filtering aims to maintain a bound on the output error magnitude of an adaptive filter. Adaptive filter coefficients are only updated when the *a priori* error magnitude exceeds a predetermined threshold γ. The resulting update mechanism is sparse in time and provides an important reduction in power consumption. The set-membership partial-update NLMS updates M out of N coefficients with largest regressor norm whenever $|e(k)| > \gamma$. The coefficient updates are constructed to achieve $|e_p(k)| = \gamma$.

Extending the error magnitude bound over $P \leq M$ outputs $y(k), y(k-1), \ldots, y(k-P+1)$ and using selective partial updates to limit update complexity leads to the set-membership partial-update APA. At each coefficient update the set-membership partial-update APA solves the following constrained optimization problem:

$$\min_{\boldsymbol{I}_M(k)} \min_{\boldsymbol{w}_{\mathcal{M}}(k+1)} \| \underbrace{\boldsymbol{w}_{\mathcal{M}}(k+1) - \boldsymbol{w}_{\mathcal{M}}(k)}_{\delta \boldsymbol{w}_{\mathcal{M}}(k+1)} \|_2^2 \quad (4.74a)$$

subject to:

$$\boldsymbol{e}_p(k) = \boldsymbol{\gamma}(k) \quad (4.74b)$$

where $\boldsymbol{\gamma}(k) = [\gamma_1(k), \ldots, \gamma_P(k)]^T$ is an error constraint vector with $|\gamma_i(k)| \leq \gamma$. Equation (4.74b) can be rewritten as:

$$\begin{aligned} \boldsymbol{e}(k) - \boldsymbol{e}_p(k) &= \boldsymbol{e}(k) - \boldsymbol{\gamma}(k) \\ \boldsymbol{X}_{\mathcal{M}}(k)\delta \boldsymbol{w}_{\mathcal{M}}(k+1) &= \boldsymbol{e}(k) - \boldsymbol{\gamma}(k) \end{aligned} \quad (4.75)$$

For fixed $\boldsymbol{I}_M(k)$, (4.74) is equivalent to solving the following underdetermined minimum-norm least-squares problem for $\delta\boldsymbol{w}_{\mathcal{M}}(k+1)$:

$$\boldsymbol{X}_{\mathcal{M}}(k)\delta\boldsymbol{w}_{\mathcal{M}}(k+1) = \boldsymbol{e}(k) - \boldsymbol{\gamma}(k), \quad \|\delta\boldsymbol{w}_{\mathcal{M}}(k+1)\|_2^2 \text{ is minimum} \tag{4.76}$$

with the unique solution given by:

$$\begin{aligned}\delta\boldsymbol{w}_{\mathcal{M}}(k+1) &= \boldsymbol{X}_{\mathcal{M}}^{\dagger}(k)(\boldsymbol{e}(k) - \boldsymbol{\gamma}(k)) \\ &= \boldsymbol{X}_{\mathcal{M}}^{T}(k)(\boldsymbol{X}_{\mathcal{M}}(k)\boldsymbol{X}_{\mathcal{M}}^{T}(k))^{-1}(\boldsymbol{e}(k) - \boldsymbol{\gamma}(k))\end{aligned} \tag{4.77}$$

where $\boldsymbol{X}_{\mathcal{M}}(k)\boldsymbol{X}_{\mathcal{M}}^{T}(k)$ is assumed to be invertible. Introducing a regularization parameter to avoid rank-deficient matrix inversion we finally obtain the set-membership partial-update APA recursion for fixed $\boldsymbol{I}_M(k)$:

$$\begin{aligned}\boldsymbol{w}_{\mathcal{M}}(k+1) = \boldsymbol{w}_{\mathcal{M}}(k) + \mu(k)\boldsymbol{X}_{\mathcal{M}}^{T}(k)(\epsilon\boldsymbol{I} + \boldsymbol{X}_{\mathcal{M}}(k)\boldsymbol{X}_{\mathcal{M}}^{T}(k))^{-1} \\ \times (\boldsymbol{e}(k) - \boldsymbol{\gamma}(k))\end{aligned} \tag{4.78}$$

where:

$$\mu(k) = \begin{cases} 1 & \text{if } |e(k)| > \gamma \\ 0 & \text{otherwise} \end{cases} \tag{4.79}$$

There are infinitely many possibilities for the error constraint vector $\boldsymbol{\gamma}(k)$. Two particular choices for $\boldsymbol{\gamma}(k)$ were proposed in (Werner and Diniz, 2001). The first one sets $\boldsymbol{\gamma}(k) = \boldsymbol{0}$ which leads to:

$$\boldsymbol{w}(k+1) = \boldsymbol{w}(k) + \mu(k)\boldsymbol{I}_M(k)\boldsymbol{X}^T(k)(\epsilon\boldsymbol{I} + \boldsymbol{X}(k)\boldsymbol{I}_M(k)\boldsymbol{X}^T(k))^{-1}\boldsymbol{e}(k) \tag{4.80}$$

This recursion, which we call the *set-membership partial-update APA-1*, uses the selective-partial-update APA whenever $|e(k)| > \gamma$ and applies no update otherwise. The other possibility for $\boldsymbol{\gamma}(k)$ is defined by:

$$\boldsymbol{\gamma}(k) = \begin{bmatrix} \gamma \dfrac{e(k)}{|e(k)|} \\ e_2(k) \\ \vdots \\ e_P(k) \end{bmatrix} \tag{4.81}$$

where $\boldsymbol{e}(k) = [e(k), e_2(k), \ldots, e_P(k)]^T$ with $|e_i(k)| \leq \gamma$ for $i = 2, \ldots, P$. This results in:

$$\boldsymbol{w}(k+1) = \boldsymbol{w}(k) + \mu(k)\boldsymbol{I}_M(k)\boldsymbol{X}^T(k)(\epsilon\boldsymbol{I} + \boldsymbol{X}(k)\boldsymbol{I}_M(k)\boldsymbol{X}^T(k))^{-1}\boldsymbol{u}(k),$$

$$\boldsymbol{u}(k) = \begin{bmatrix} e(k)\left(1 - \dfrac{\gamma}{|e(k)|}\right) \\ 0 \\ \vdots \\ 0 \end{bmatrix} \tag{4.82}$$

which we shall refer to as the *set-membership partial-update APA-2*.

The optimum coefficient selection matrix $\boldsymbol{I}_M(k)$ solving (4.74) is obtained from:

$$\begin{aligned} &\min_{\boldsymbol{I}_M(k)} \left\| \boldsymbol{X}_{\mathcal{M}}^T(k)(\boldsymbol{X}_{\mathcal{M}}(k)\boldsymbol{X}_{\mathcal{M}}^T(k))^{-1}(\boldsymbol{e}(k) - \boldsymbol{\gamma}(k)) \right\|_2^2 \\ &\min_{\boldsymbol{I}_M(k)} (\boldsymbol{e}(k) - \boldsymbol{\gamma}(k))^T(\boldsymbol{X}_{\mathcal{M}}(k)\boldsymbol{X}_{\mathcal{M}}^T(k))^{-1}(\boldsymbol{e}(k) - \boldsymbol{\gamma}(k)) \end{aligned} \tag{4.83}$$

which is similar to the minimization problem for the selective-partial-update APA in (4.70). Thus, a simplified approximate coefficient selection matrix for the set-membership partial-update APA is given by (4.64). Note that the full implementation of (4.83) would be computationally expensive and, therefore, we do not advocate its use.

4.7.7 Selective-regressor APA

The selective-regressor APA (Hwang and Song, 2007) is a variant of the selective-partial-update APA. Unlike partial-update affine projection algorithms it updates all adaptive filter coefficients. However, it saves some update complexity by selecting M out of N possible regressor vectors. The selective-regressor APA assumes that the hardware resources allow a maximum of M regressors to be computed in the APA adaptation process. The constrained optimization problem solved by the selective-regressor APA is:

$$\max_{\boldsymbol{I}_M(k)} \min_{\boldsymbol{w}(k+1)} \|\boldsymbol{w}(k+1) - \boldsymbol{w}(k)\|_2^2 \tag{4.84a}$$

subject to:

$$\boldsymbol{e}_{p,\mathcal{I}}(k) - \left(\boldsymbol{I} - \mu\boldsymbol{X}_{\mathcal{I}}(k)\boldsymbol{X}_{\mathcal{I}}^T(k)(\epsilon\boldsymbol{I} + \boldsymbol{X}_{\mathcal{I}}(k)\boldsymbol{X}_{\mathcal{I}}^T(k))^{-1}\right)\boldsymbol{e}_{\mathcal{I}}(k) = \boldsymbol{0} \tag{4.84b}$$

where $\boldsymbol{X}_{\mathcal{I}}(k)$ is the $M \times N$ partition of regressor vectors corresponding to the M non-zero diagonal elements of $\boldsymbol{I}_M(k)$, and the *a priori* and *a posteriori* error vectors are defined by:

$$\begin{aligned} e_{\mathcal{I}}(k) &= d_{\mathcal{I}}(k) - X_{\mathcal{I}}(k)w(k) \\ e_{p,\mathcal{I}}(k) &= d_{\mathcal{I}}(k) - X_{\mathcal{I}}(k)w(k+1) \end{aligned} \tag{4.85}$$

Here $d_{\mathcal{I}}(k)$ is an $M \times 1$ subvector of $[d(k), d(k-1), \ldots, d(k-N+1)]^T$ corresponding to the non-zero diagonal elements of $I_M(k)$. As an example, if:

$$I_M(k) = \text{diag}(1, 0, 0, 1) \tag{4.86}$$

then:

$$X_{\mathcal{I}}(k) = \begin{bmatrix} x^T(k) \\ x^T(k-3) \end{bmatrix} \tag{4.87}$$

The equivalent underdetermined set of equations is:

$$X_{\mathcal{I}}(k)\delta w(k+1) = \mu X_{\mathcal{I}}(k)X_{\mathcal{I}}^T(k)(\epsilon I + X_{\mathcal{I}}(k)X_{\mathcal{I}}^T(k))^{-1}e_{\mathcal{I}}(k) \tag{4.88}$$

and its minimum-norm solution is given by:

$$\delta w(k+1) = \mu X_{\mathcal{I}}^T(k)(\epsilon I + X_{\mathcal{I}}(k)X_{\mathcal{I}}^T(k))^{-1}e_{\mathcal{I}}(k) \tag{4.89}$$

Thus, for fixed regressors we have:

$$w(k+1) = w(k) + \mu X_{\mathcal{I}}^T(k)(\epsilon I + X_{\mathcal{I}}(k)X_{\mathcal{I}}^T(k))^{-1}e_{\mathcal{I}}(k) \tag{4.90}$$

The optimum regressor selection is determined from:

$$\max_{I_M(k)} \|\delta w(k+1)\|_2^2 \tag{4.91}$$

Here we wish to maximize $\|\delta w(k+1)\|_2^2$, because our objective is to identify the past M regressors that would make the most contribution to the APA update term in an M-max-updates sense. For sufficiently small ϵ (4.91) may be written as:

$$\max_{I_M(k)} e_{\mathcal{I}}^T(k)(X_{\mathcal{I}}(k)X_{\mathcal{I}}^T(k))^{-1}e_{\mathcal{I}}(k) \tag{4.92}$$

This minimization problem is computationally demanding. Therefore, as we have done before, we seek to simplify it by assuming $X_{\mathcal{I}}(k)X_{\mathcal{I}}^T(k)$ is diagonal, leading to an approximate implementation of (4.92):

$$\max_{\boldsymbol{I}_M(k)} \|\boldsymbol{e}_{\mathcal{I}}^T(k)\|^2_{\boldsymbol{T}_{\mathcal{I}}(k)} \tag{4.93}$$

where:

$$\boldsymbol{T}_{\mathcal{I}}(k) = \begin{bmatrix} 1/\|\boldsymbol{x}(k-\mathcal{I}_1(k))\|_2^2 & & & \mathbf{0} \\ & 1/\|\boldsymbol{x}(k-\mathcal{I}_2(k))\|_2^2 & & \\ & & \ddots & \\ \mathbf{0} & & & 1/\|\boldsymbol{x}(k-\mathcal{I}_M(k))\|_2^2 \end{bmatrix} \tag{4.94}$$

Here the $(\mathcal{I}_j(k)+1, \mathcal{I}_j(k)+1)$th entry of $\boldsymbol{I}_M(k)$ is equal to one for $j = 1, \ldots, M$. Based on (4.93) and (4.94) the simplified coefficient selection matrix is given by:

$$\boldsymbol{I}_M(k) = \begin{bmatrix} i_1(k) & 0 & \cdots & 0 \\ 0 & i_2(k) & \ddots & \vdots \\ \vdots & \ddots & \ddots & 0 \\ 0 & \cdots & 0 & i_N(k) \end{bmatrix},$$

$$i_j(k) = \begin{cases} 1 & \text{if } \dfrac{e_j^2(k)}{\|\boldsymbol{x}(k-j+1)\|_2^2} \in \max_{1 \le l \le N}\left(\dfrac{e_l^2(k)}{\|\boldsymbol{x}(k-l+1)\|_2^2}, M\right) \\ 0 & \text{otherwise} \end{cases} \tag{4.95}$$

where $\boldsymbol{e}(k) = [e_1(k), e_2(k), \ldots, e_N(k)]^T$.

4.7.8 Computational complexity

The computation of the full-update APA in (4.51) comprises the following steps at each iteration:

- PN multiplications and PN additions to compute $\boldsymbol{e}(k)$.
- P multiplications to compute $\mu \boldsymbol{e}(k)$.
- To compute $\epsilon \boldsymbol{I} + \boldsymbol{X}(k)\boldsymbol{X}^T(k)$, we can write:

$$\begin{aligned} \epsilon \boldsymbol{I} + \boldsymbol{X}(k)\boldsymbol{X}^T(k) &= \epsilon \boldsymbol{I} + \sum_{i=k-N+1}^{k} \boldsymbol{x}_A(i)\boldsymbol{x}_A^T(i) \\ &= \epsilon \boldsymbol{I} + \sum_{i=k-N}^{k-1} \boldsymbol{x}_A(i)\boldsymbol{x}_A^T(i) + \boldsymbol{x}_A(k)\boldsymbol{x}_A^T(k) \\ &\quad - \boldsymbol{x}_A(k-N)\boldsymbol{x}_A^T(k-N) \\ &= \epsilon \boldsymbol{I} + \boldsymbol{X}(k-1)\boldsymbol{X}^T(k-1) + \boldsymbol{x}_A(k)\boldsymbol{x}_A^T(k) \\ &\quad - \boldsymbol{x}_A(k-N)\boldsymbol{x}_A^T(k-N) \end{aligned} \tag{4.96}$$

Thus we only need to compute $\boldsymbol{x}_A(k)\boldsymbol{x}_A^T(k)$ at each k. Using the above relationship between $\epsilon\boldsymbol{I} + \boldsymbol{X}(k)\boldsymbol{X}^T(k)$ and $\epsilon\boldsymbol{I} + \boldsymbol{X}(k-1)\boldsymbol{X}^T(k-1)$, the matrix $\epsilon\boldsymbol{I} + \boldsymbol{X}(k)\boldsymbol{X}^T(k)$ can be computed using P^2 multiplications to obtain the outerproduct $\boldsymbol{x}_A(k)\boldsymbol{x}_A^T(k)$ and $2P^2$ additions to add $\boldsymbol{x}_A(k)\boldsymbol{x}_A^T(k)$ and $-\boldsymbol{x}_A(k-N)\boldsymbol{x}_A^T(k-N)$ to $\epsilon\boldsymbol{I} + \boldsymbol{X}(k-1)\boldsymbol{X}^T(k-1)$.

- P^3 multiplications and P^3 additions to compute the matrix inverse $(\epsilon\boldsymbol{I} + \boldsymbol{X}(k)\boldsymbol{X}^T(k))^{-1}$.
- P^2 multiplications and $P(P-1)$ additions to compute $\mu(\epsilon\boldsymbol{I} + \boldsymbol{X}(k)\boldsymbol{X}^T(k))^{-1}\boldsymbol{e}(k)$.
- PN multiplications and $N(P-1)$ additions to compute the update vector $\mu\boldsymbol{X}^T(k)(\epsilon\boldsymbol{I} + \boldsymbol{X}(k)\boldsymbol{X}^T(k))^{-1}\boldsymbol{e}(k)$.
- N additions to update $\boldsymbol{w}(k)$.

The total computational complexity of the full-update APA is $P(2N+1)+2P^2+P^3$ multiplications and $P(2N-1)+3P^2+P^3$ additions at each iteration.

The computational steps of the periodic-partial-update APA are:

- N multiplications and N additions to compute the current error signal $d(k) - \boldsymbol{x}^T(k)\boldsymbol{w}(k)$, which is done by the filtering process at every iteration.
- $(P-1)N$ multiplications and $(P-1)N$ additions over S iterations to compute the remaining entries of $\boldsymbol{e}(kS)$.
- P multiplications to compute $\mu\boldsymbol{e}(kS)$ over S iterations.
- If $S < N$, $\epsilon\boldsymbol{I} + \boldsymbol{X}(k)\boldsymbol{X}^T(k)$ is computed at each iteration at a cost of P^2 multiplications and $2P^2$ additions per iteration. If $S \geq N$, $\epsilon\boldsymbol{I} + \boldsymbol{X}(kS)\boldsymbol{X}^T(kS)$ is computed over S iterations at a cost of P^2N multiplications and $P^2(N-1)+N$ additions.
- P^3 multiplications and P^3 additions over S iterations to compute the matrix inverse $(\epsilon\boldsymbol{I} + \boldsymbol{X}(kS)\boldsymbol{X}^T(kS))^{-1}$.
- P^2 multiplications and $P(P-1)$ additions over S iterations to compute $\mu(\epsilon\boldsymbol{I} + \boldsymbol{X}(kS)\boldsymbol{X}^T(kS))^{-1}\boldsymbol{e}(kS)$.
- PN multiplications and $N(P-1)$ additions over S iterations to compute the update vector $\mu\boldsymbol{X}^T(kS)(\epsilon\boldsymbol{I} + \boldsymbol{X}(kS)\boldsymbol{X}^T(kS))^{-1}\boldsymbol{e}(kS)$.
- N additions over S iterations to update $\boldsymbol{w}(kS)$.

If $S < N$ the total computational complexity of the periodic-partial-update APA per iteration is:

$$\frac{(2P-1)N}{S} + \frac{P}{S} + P^2\left(1 + \frac{1}{S}\right) + \frac{P^3}{S} + N \quad \text{multiplications}$$

and:

$$\frac{(2P-1)N}{S} + \frac{P(P-1)}{S} + 2P^2 + \frac{P^3}{S} + N \quad \text{additions.}$$

Otherwise it is:

$$\frac{P}{S} + \frac{(2P-1)N}{S} + \frac{P^2(N+1)}{S} + \frac{P^3}{S} + N \quad \text{multiplications}$$

and:

$$\frac{2PN}{S} + \frac{P(P-1)}{S} + \frac{P^2(N-1)}{S} + \frac{P^3}{S} + N \quad \text{additions.}$$

The sequential-partial-update APA performs the following computations at each iteration:

- PN multiplications and PN additions to compute $\boldsymbol{e}(k)$.
- If $P \leq M$, P multiplications to compute $\mu\boldsymbol{e}(k)$. Otherwise delay multiplying μ with the update vector until the partial-update vector is computed. The complexity of this step is therefore $\min(P, M)$ multiplications.
- P^2 multiplications and $2P^2$ additions to compute $\epsilon\boldsymbol{I} + \boldsymbol{X}(k)\boldsymbol{X}^T(k)$.
- P^3 multiplications and P^3 additions to compute the matrix inverse $(\epsilon\boldsymbol{I} + \boldsymbol{X}(k)\boldsymbol{X}^T(k))^{-1}$.
- P^2 multiplications and $P(P-1)$ additions to compute $\mu(\epsilon\boldsymbol{I} + \boldsymbol{X}(k)\boldsymbol{X}^T(k))^{-1}\boldsymbol{e}(k)$.
- PM multiplications and $M(P-1)$ additions to compute the partial update vector $\mu\boldsymbol{I}_M(k)\boldsymbol{X}^T(k)(\epsilon\boldsymbol{I} + \boldsymbol{X}(k)\boldsymbol{X}^T(k))^{-1}\boldsymbol{e}(k)$.
- M additions to update $\boldsymbol{w}(k)$.

The total computational complexity of the sequential-partial-update APA is $\min(P, M) + P(M+N) + 2P^2 + P^3$ multiplications and $P(M+N-1) + 3P^2 + P^3$ additions at each iteration.

The stochastic-partial-update APA requires a further 2 multiplications and 2 additions per iteration to implement the linear congruential generator in (4.22) in addition to the computational requirements of the sequential-partial-update APA.

The computational complexity of the M-max APA is slightly larger than that of the sequential-partial-update APA due to M-max coefficient selection. Considering the simplified M-max APA with coefficient selection matrix given in (4.64), the additional complexity arising from M-max coefficient selection is $2\lceil \log_2 N \rceil + 2$ comparisons and $P-1$ additions. We use the sortline algorithm to rank $\|\boldsymbol{x}_A(k)\|_2^2, \ldots, \|\boldsymbol{x}_A(k-N+1)\|_2^2$. Note that the squared Euclidean norms of the $\boldsymbol{x}_A(k)$ only require $P-1$

additions per iteration since the squared entries of $\boldsymbol{x}_A(k)$ are already available from the outerproduct $\boldsymbol{x}_A(k)\boldsymbol{x}_A^T(k)$.

The selective-partial-update APA defined in (4.73) has the same computational steps as the M-max APA with the exception of $\epsilon \boldsymbol{I} + \boldsymbol{X}(k)\boldsymbol{I}_M(k)\boldsymbol{X}^T(k)$, which has the same computational complexity as $\epsilon \boldsymbol{I} + \boldsymbol{X}(k)\boldsymbol{X}^T(k)$ in the worst case. Therefore, the computational complexities of the M-max APA and selective-partial-update APA are identical.

The set-membership partial-update APA-1 has the computational complexity of $P(M+N) + 2P^2 + P^3$ multiplications, $P(M+N) - 1 + 3P^2 + P^3$ additions and $2\lceil \log_2 N \rceil + 3$ comparisons at iterations with a coefficient update. On the other hand, the set-membership partial-update APA-2 requires:

- N multiplications and N additions to compute $e(k)$.
- One comparison to check $|e(k)| > \gamma$.
- One multiplication, one division and one addition to compute $\boldsymbol{u}(k)$.
- $2\lceil \log_2 N \rceil + 2$ comparisons to obtain $\boldsymbol{I}_M(k)$.
- P^2 multiplications and $2P^2$ additions to compute $\epsilon \boldsymbol{I} + \boldsymbol{X}(k)\boldsymbol{I}_M(k)\boldsymbol{X}^T(k)$.
- P^3 multiplications and P^3 additions to compute $(\epsilon \boldsymbol{I} + \boldsymbol{X}(k)\boldsymbol{I}_M(k)\boldsymbol{X}^T(k))^{-1}$.
- P multiplications to compute $(\epsilon \boldsymbol{I} + \boldsymbol{X}(k)\boldsymbol{X}^T(k))^{-1}\boldsymbol{u}(k)$.
- PM multiplications and $M(P-1)$ additions to compute $\boldsymbol{I}_M(k)\boldsymbol{X}^T(k)(\epsilon \boldsymbol{I} + \boldsymbol{X}(k)\boldsymbol{I}_M(k)\boldsymbol{X}^T(k))^{-1}\boldsymbol{u}(k)$.
- M additions to update $\boldsymbol{w}(k)$.

at every update iteration. The total computational complexity of the set-membership partial-update APA-2 is $P(M+1) + N + 1 + P^2 + P^3$ multiplications, $MP + N + 1 + 2P^2 + P^3$ additions, one division and $2\lceil \log_2 N \rceil + 3$ comparisons at each update.

The selective-regressor APA defined in (4.90) and (4.95) requires:

- N^2 multiplications and N^2 additions to compute the $N \times 1$ error vector $\boldsymbol{e}(k)$.
- N multiplications to compute $e_i^2(k)$, $i = 1, \ldots, N$.
- One multiplication and $2N$ additions to compute $\|\boldsymbol{x}(k)\|_2^2, \ldots, \|\boldsymbol{x}(k-N+1)\|_2^2$ (here we exploit the shift structure of $\boldsymbol{x}(k)$ to reduce computation).
- N divisions to compute $e_i^2(k)/\|\boldsymbol{x}(k-j+1)\|_2^2$, $i = 1, \ldots, N$.
- $2N + 2\min(M, N-M)\log_2 N$ or $2M + 2(N-M)\log_2 M$ comparisons using heapsort to obtain $\boldsymbol{I}_M(k)$ in (4.95).
- M^2N multiplications and $M^2(N-1) + M$ additions to compute $\epsilon \boldsymbol{I} + \boldsymbol{X}_{\mathcal{I}}(k)\boldsymbol{X}_{\mathcal{I}}^T(k)$.
- M^3 multiplications and M^3 additions to compute the matrix inverse $(\epsilon \boldsymbol{I} + \boldsymbol{X}_{\mathcal{I}}(k)\boldsymbol{X}_{\mathcal{I}}^T(k))^{-1}$.
- $M^2 + M$ multiplications and $M(M-1)$ additions to compute $\mu(\epsilon \boldsymbol{I} + \boldsymbol{X}_{\mathcal{I}}(k)\boldsymbol{X}_{\mathcal{I}}^T(k))^{-1}\boldsymbol{e}_{\mathcal{I}}(k)$.

- MN multiplications and $N(M-1)$ additions to compute the update vector $\mu \boldsymbol{X}_{\mathcal{I}}^T(k)(\epsilon \boldsymbol{I} + \boldsymbol{X}_{\mathcal{I}}(k)\boldsymbol{X}_{\mathcal{I}}^T(k))^{-1}\boldsymbol{e}_{\mathcal{I}}(k)$.
- N additions to update $\boldsymbol{w}(k)$.

The total computational complexity of the selective-regressor APA is $M^3 + M(N+1)(M+1) + N^2 + N + 1$ multiplications, $M^3 + MN(M+1) + N(N+2)$ additions, N divisions, and $2N + 2\min(M, N-M)\log_2 N$ or $2M + 2(N-M)\log_2 M$ comparisons, whichever is smaller, at each iteration. Note that the data to be ranked does not have a shift structure, which rules out the use of sortline. We, therefore, employ the heapsort algorithm (see Appendix A) to compute the coefficient selection matrix $\boldsymbol{I}_M(k)$.

Table 4.3 provides a summary of the computational complexity of APA and its partial-update variants. In cases where the computational complexity varies from iteration to iteration as in set-membership partial updates, we only consider the peak computational complexity per iteration. As is clear from Table 4.3, the computational complexity of the full-update APA is $O(P^3)$, which assumes that the matrix decomposition in (4.96) is utilized to reduce the computational complexity of $\epsilon \boldsymbol{I} + \boldsymbol{X}(k)\boldsymbol{X}^T(k)$. If this matrix is computed from scratch at each iteration without exploiting the recursive structure in (4.96), the complexity of the full-update APA increases to $O(P^2N)$ (Sayed, 2003). In general, the periodic-partial-update APA has the smallest complexity out of all partial-update implementations of APA with complexity $O(P^2)$ or $O(P^3/S)$, depending on the specific values of the APA order P and update period S.

4.8 RECURSIVE LEAST SQUARE ALGORITHM

The RLS algorithm has the fastest convergence rate among all adaptive filter algorithms studied in this chapter. It is also more complicated and computationally more demanding than the previous stochastic gradient algorithms. RLS can be conceived as an instantaneous approximation to Newton's method using exponentially weighted sample averages for the input correlation matrix. In Newton's method:

$$\boldsymbol{w}(k+1) = \boldsymbol{w}(k) + \mu(k)\boldsymbol{R}^{-1}(\boldsymbol{p} - \boldsymbol{R}\boldsymbol{w}(k)), \quad k = 0, 1, \ldots \tag{4.97}$$

replacing $\boldsymbol{R}$ with:

$$\boldsymbol{R} \approx \frac{1}{k+1}\sum_{i=0}^{k} \lambda^{k-i}\boldsymbol{x}(i)\boldsymbol{x}^T(i) \tag{4.98}$$

where $0 < \lambda \leq 1$ is the exponential forgetting factor and $\boldsymbol{p} - \boldsymbol{R}\boldsymbol{w}(k)$ with its instantaneous approximation:

$$\boldsymbol{p} - \boldsymbol{R}\boldsymbol{w}(k) \approx \boldsymbol{x}(k)\underbrace{(d(k) - \boldsymbol{x}^T(k)\boldsymbol{w}(k))}_{e(k)} \tag{4.99}$$

Table 4.3 Computational complexity of APA and partial-update APA at each k

Algorithm		×	+	÷	≶
APA		$P(2N+1)+2P^2+P^3$	$P(2N-1)+3P^2+P^3$		
Per-par-upd APA	$S<N$	$\frac{(2P-1)N}{S}+\frac{P}{S}+P^2\left(1+\frac{1}{S}\right)+\frac{P^3}{S}+N$	$\frac{(2P-1)N}{S}+\frac{P(P-1)}{S}+2P^2+\frac{P^3}{S}+N$		
	$S \geq N$	$\frac{P}{S}+\frac{(2P-1)N}{S}+\frac{P^2(N+1)}{S}+\frac{P^3}{S}+N$	$\frac{2PN}{S}+\frac{P(P-1)}{S}+\frac{P^2(N-1)}{S}+\frac{P^3}{S}+N$		
Seq-par-upd APA		$\min(P,M)+P(M+N)+2P^2+P^3$	$P(M+N-1)+3P^2+P^3$		
Stoch-par-upd APA		$2+\min(P,M)+P(M+N)+2P^2+P^3$	$2+P(M+N-1)+3P^2+P^3$		
M-max APA		$\min(P,M)+P(M+N)+2P^2+P^3$	$P(M+N)-1+3P^2+P^3$		$2\lceil\log_2 N\rceil+2$
Sel-par-upd APA		$\min(P,M)+P(M+N)+2P^2+P^3$	$P(M+N)-1+3P^2+P^3$		$2\lceil\log_2 N\rceil+2$
Set-mem par-upd APA-1		$P(M+N)+2P^2+P^3$	$P(M+N)-1+3P^2+P^3$		$2\lceil\log_2 N\rceil+3$
Set-mem par-upd APA-2		$P(M+1)+N+1+P^2+P^3$	$MP+N+1+2P^2+P^3$	1	$2\lceil\log_2 N\rceil+3$
Sel-reg APA		$M^3+M(N+1)(M+1)+N^2+N+1$	$M^3+MN(M+1)+N(N+2)$	N	$\min(2N+2\min(M,N-M)\log_2 N,$ $2M+2(N-M)\log_2 M)$

we obtain:

$$\boldsymbol{w}(k+1) = \boldsymbol{w}(k) + \left(\sum_{i=0}^{k} \lambda^{k-i} \boldsymbol{x}(i)\boldsymbol{x}^T(i)\right)^{-1} \boldsymbol{x}(k)e(k), \quad k = 0, 1, \ldots \tag{4.100}$$

where $\mu(k) = 1/(k+1)$.

Writing the exponentially weighted input correlation estimate as:

$$\begin{aligned}\boldsymbol{\Theta}(k) &= \sum_{i=0}^{k} \lambda^{k-i} \boldsymbol{x}(i)\boldsymbol{x}^T(i) \\ &= \boldsymbol{x}(k)\boldsymbol{x}^T(k) + \lambda \sum_{i=0}^{k-1} \lambda^{k-1-i} \boldsymbol{x}(i)\boldsymbol{x}^T(i) \\ &= \boldsymbol{x}(k)\boldsymbol{x}^T(k) + \lambda\boldsymbol{\Theta}(k-1)\end{aligned} \tag{4.101}$$

and applying the matrix inversion lemma to $\boldsymbol{P}(k) = \boldsymbol{\Theta}^{-1}(k)$ results in:

$$\boldsymbol{P}(k) = \lambda^{-1}\boldsymbol{P}(k-1) - \frac{\lambda^{-2}\boldsymbol{P}(k-1)\boldsymbol{x}(k)\boldsymbol{x}^T(k)\boldsymbol{P}(k-1)}{1 + \lambda^{-1}\boldsymbol{x}^T(k)\boldsymbol{P}(k-1)\boldsymbol{x}(k)} \tag{4.102}$$

where $\boldsymbol{P}(k)$ is initialized to $\boldsymbol{P}(-1) = \epsilon^{-1}\boldsymbol{I}$ to generate regularized RLS estimates. The above recursion computes $\boldsymbol{\Theta}^{-1}(k)$ without any matrix inversion, saving a significant amount of computation. Using $\boldsymbol{P}(k)$ in (4.100), the RLS recursion can be written as:

$$\boldsymbol{w}(k+1) = \boldsymbol{w}(k) + \boldsymbol{P}(k)\boldsymbol{x}(k)e(k), \quad k = 0, 1, \ldots \tag{4.103}$$

Similar to the LMS and NLMS algorithms, the usual choice for initialization for the RLS algorithm is the all zeros vector (i.e. $\boldsymbol{w}(0) = \boldsymbol{0}$).

At each RLS iteration the *a priori* estimation error is related to the *a posteriori* estimation error through:

$$e_p(k) = \frac{1}{1 + \lambda^{-1}\boldsymbol{x}^T(k)\boldsymbol{P}(k-1)\boldsymbol{x}(k)} e(k) \tag{4.104}$$

where $1/(1 + \lambda^{-1}\boldsymbol{x}^T(k)\boldsymbol{P}(k-1)\boldsymbol{x}(k))$ is referred to as the *conversion factor* (Sayed, 2003). To see this, rewrite (4.104) as:

$$e(k) - e_p(k) = \left(1 - \frac{1}{1 + \lambda^{-1}\boldsymbol{x}^T(k)\boldsymbol{P}(k-1)\boldsymbol{x}(k)}\right) e(k) \tag{4.105}$$

or:

$$\boldsymbol{x}^T(k)\delta\boldsymbol{w}(k+1) = \frac{\lambda^{-1}\boldsymbol{x}^T(k)\boldsymbol{P}(k-1)\boldsymbol{x}(k)}{1+\lambda^{-1}\boldsymbol{x}^T(k)\boldsymbol{P}(k-1)\boldsymbol{x}(k)}e(k) \tag{4.106}$$

The last equation has infinitely many solutions for $\delta\boldsymbol{w}(k+1)$ if $N > 1$. A solution is immediately obtained by deleting $\boldsymbol{x}^T(k)$ from both sides of the equation, yielding:

$$\delta\boldsymbol{w}(k+1) = \frac{\lambda^{-1}\boldsymbol{P}(k-1)\boldsymbol{x}(k)}{1+\lambda^{-1}\boldsymbol{x}^T(k)\boldsymbol{P}(k-1)\boldsymbol{x}(k)}e(k) \tag{4.107}$$

Note that this solution is different to the minimum-norm solution which would have been given by:

$$\delta\boldsymbol{w}(k+1) = \frac{\boldsymbol{x}(k)}{\|\boldsymbol{x}(k)\|_2^2}\frac{\lambda^{-1}\boldsymbol{x}^T(k)\boldsymbol{P}(k-1)\boldsymbol{x}(k)}{1+\lambda^{-1}\boldsymbol{x}^T(k)\boldsymbol{P}(k-1)\boldsymbol{x}(k)}e(k) \tag{4.108}$$

Substituting $\delta\boldsymbol{w}(k+1) = \boldsymbol{w}(k+1) - \boldsymbol{w}(k)$ into (4.107) gives:

$$\boldsymbol{w}(k+1) = \boldsymbol{w}(k) + \frac{\lambda^{-1}\boldsymbol{P}(k-1)\boldsymbol{x}(k)}{1+\lambda^{-1}\boldsymbol{x}^T(k)\boldsymbol{P}(k-1)\boldsymbol{x}(k)}e(k), \quad k = 0, 1, \ldots \tag{4.109}$$

Post-multiplying both sides of (4.102) with $\boldsymbol{x}(k)$ leads to:

$$\boldsymbol{P}(k)\boldsymbol{x}(k) = \frac{\lambda^{-1}\boldsymbol{P}(k-1)\boldsymbol{x}(k)}{1+\lambda^{-1}\boldsymbol{x}^T(k)\boldsymbol{P}(k-1)\boldsymbol{x}(k)} \tag{4.110}$$

which confirms that (4.109) is equal to (4.103). We thus conclude that the RLS algorithm satisfies the error constraint in (4.104). However, it does not obey the principle of minimum disturbance since the coefficient updates do not have the minimum norm (i.e. $\delta\boldsymbol{w}(k+1)$ is not a minimum-norm solution of (4.104)).

The RLS algorithm can also be constructed as a linear least-squares estimator that solves:

$$\boldsymbol{w}(k+1) = \arg\min_{\boldsymbol{w}} \sum_{i=0}^{k} \lambda^{k-i}(d(i) - \boldsymbol{x}^T(i)\boldsymbol{w})^2 \tag{4.111}$$

The weighted least-squares cost function minimized by the RLS algorithm is:

$$J_{LS}(k) = (\boldsymbol{d}(k) - \boldsymbol{X}(k)\boldsymbol{w})^T\boldsymbol{\Lambda}(k)(\boldsymbol{d}(k) - \boldsymbol{X}(k)\boldsymbol{w}) \tag{4.112}$$

where:

$$\boldsymbol{d}(k) = \begin{bmatrix} d(k) \\ d(k-1) \\ \vdots \\ d(0) \end{bmatrix}, \quad \boldsymbol{X}(k) = \begin{bmatrix} \boldsymbol{x}^T(k) \\ \boldsymbol{x}^T(k-1) \\ \vdots \\ \boldsymbol{x}^T(0) \end{bmatrix},$$

$$\boldsymbol{\Lambda}(k) = \begin{bmatrix} 1 & & & \boldsymbol{0} \\ & \lambda & & \\ & & \ddots & \\ \boldsymbol{0} & & & \lambda^k \end{bmatrix} \tag{4.113}$$

The vector $\boldsymbol{w}$ minimizing $J_{LS}(k)$ is obtained from:

$$\left.\frac{\partial J_{LS}(k)}{\partial \boldsymbol{w}}\right|_{\boldsymbol{w}=\boldsymbol{w}(k+1)} = \boldsymbol{0} \tag{4.114}$$

which leads to:

$$\boldsymbol{w}(k+1) = (\boldsymbol{X}^T(k)\boldsymbol{\Lambda}(k)\boldsymbol{X}(k))^{-1}\boldsymbol{X}^T(k)\boldsymbol{\Lambda}(k)\boldsymbol{d}(k) \tag{4.115}$$

Expanding this expression we have:

$$\begin{aligned} \boldsymbol{w}(k+1) &= \left(\sum_{i=0}^{k} \lambda^{k-i}\boldsymbol{x}(i)\boldsymbol{x}^T(i)\right)^{-1} \sum_{i=0}^{k} \lambda^{k-i}\boldsymbol{x}(i)d(i) \\ &= \left(\boldsymbol{x}(k)\boldsymbol{x}^T(k) + \lambda \sum_{i=0}^{k-1} \lambda^{k-1-i}\boldsymbol{x}(i)\boldsymbol{x}^T(i)\right)^{-1} \\ &\quad \times \left(\boldsymbol{x}(k)d(k) + \lambda \sum_{i=0}^{k-1} \lambda^{k-1-i}\boldsymbol{x}(i)d(i)\right) \\ &= \boldsymbol{P}(k)\left(\boldsymbol{x}(k)d(k) + \lambda \sum_{i=0}^{k-1} \lambda^{k-1-i}\boldsymbol{x}(i)d(i)\right) \end{aligned} \tag{4.116}$$

Substituting (4.102) into the last equation above, we obtain:

$$
\begin{aligned}
\boldsymbol{w}(k+1) &= \left(\lambda^{-1}\boldsymbol{P}(k-1) - \frac{\lambda^{-2}\boldsymbol{P}(k-1)\boldsymbol{x}(k)\boldsymbol{x}^T(k)\boldsymbol{P}(k-1)}{1+\lambda^{-1}\boldsymbol{x}^T(k)\boldsymbol{P}(k-1)\boldsymbol{x}(k)}\right) \\
&\quad \times \left(\boldsymbol{x}(k)d(k) + \lambda\sum_{i=0}^{k-1}\lambda^{k-1-i}\boldsymbol{x}(i)d(i)\right), \quad k=0,1,\ldots \\
&= \boldsymbol{w}(k) + \lambda^{-1}\boldsymbol{P}(k-1)\boldsymbol{x}(k)d(k) \\
&\quad - \frac{\lambda^{-2}\boldsymbol{P}(k-1)\boldsymbol{x}(k)\boldsymbol{x}^T(k)\boldsymbol{P}(k-1)\boldsymbol{x}(k)d(k)}{1+\lambda^{-1}\boldsymbol{x}^T(k)\boldsymbol{P}(k-1)\boldsymbol{x}(k)} \\
&\quad - \frac{\lambda^{-1}\boldsymbol{P}(k-1)\boldsymbol{x}(k)\boldsymbol{x}^T(k)\boldsymbol{w}(k)}{1+\lambda^{-1}\boldsymbol{x}^T(k)\boldsymbol{P}(k-1)\boldsymbol{x}(k)}, \quad k=0,1,\ldots \\
&= \boldsymbol{w}(k) + \frac{\lambda^{-1}\boldsymbol{P}(k-1)\boldsymbol{x}(k)}{1+\lambda^{-1}\boldsymbol{x}^T(k)\boldsymbol{P}(k-1)\boldsymbol{x}(k)}(d(k)-\boldsymbol{x}^T(k)\boldsymbol{w}(k)), \\
&\quad k=0,1,\ldots \\
&= \boldsymbol{w}(k) + \boldsymbol{P}(k)\boldsymbol{x}(k)e(k), \quad k=0,1,\ldots
\end{aligned} \tag{4.117}
$$

which is identical to the RLS recursion given in (4.103).

For an adaptive filter with N coefficients the computational complexity of the RLS algorithm is $O(N^2)$ (see Section 4.9.7). Computationally efficient fast RLS algorithms with complexity $O(N)$ are also available (Sayed, 2003). These algorithms exploit the shift structure of the regressor vector in order to simplify the RLS update vector computation.

4.9 PARTIAL-UPDATE RLS ALGORITHMS

The RLS algorithm directly solves the weighted least-squares problem defined in (4.111). In this sense it is different from the other stochastic gradient algorithms that rely on the averaging effects of the update term. As a result, caution needs to be exercised when applying partial coefficient update methods to the RLS algorithm. From (4.115), it is seen that the normal equations for RLS are:

$$\boldsymbol{X}^T(k)\boldsymbol{\Lambda}(k)\boldsymbol{X}(k)\boldsymbol{w}(k+1) = \boldsymbol{X}^T(k)\boldsymbol{\Lambda}(k)\boldsymbol{d}(k). \tag{4.118}$$

Partial updating of coefficients results in modification of the RLS normal equations. We need to ensure that any such modification does not introduce undesirable estimation bias or increased steady-state MSE. One way to meet these requirements is to resort to the method of instrumental variables (Ljung, 1999), which allows $\boldsymbol{X}^T(k)$ to be replaced with another full-rank matrix (instrumental variable matrix) in both sides of the normal equations. The motivation behind instrumental variables is to make the resulting estimate asymptotically unbiased. In our case, however, we wish to develop partial-update RLS algorithms by means of substituting an appropriate matrix for $\boldsymbol{X}^T(k)$ in (4.118).

We propose to replace $\boldsymbol{X}^T(k)$ with the partial-update regressor matrix $\boldsymbol{X}_M^T(k)$ defined by:

$$\boldsymbol{X}_M(k) = \begin{bmatrix} \boldsymbol{x}^T(k)\boldsymbol{I}_M(k) \\ \boldsymbol{x}^T(k-1)\boldsymbol{I}_M(k-1) \\ \vdots \\ \boldsymbol{x}^T(0)\boldsymbol{I}_M(0) \end{bmatrix} \tag{4.119}$$

which results in 'partial-update' normal equations:

$$\boldsymbol{X}_M^T(k)\boldsymbol{\Lambda}(k)\boldsymbol{X}(k)\boldsymbol{w}(k+1) = \boldsymbol{X}_M^T(k)\boldsymbol{\Lambda}(k)\boldsymbol{d}(k) \tag{4.120}$$

As long as $\boldsymbol{X}_M(k)$ is full-rank, the 'partial-update' RLS solution is given by:

$$\boldsymbol{w}(k+1) = (\boldsymbol{X}_M^T(k)\boldsymbol{\Lambda}(k)\boldsymbol{X}(k))^{-1}\boldsymbol{X}_M^T(k)\boldsymbol{\Lambda}(k)\boldsymbol{d}(k) \tag{4.121}$$

The partial-update RLS solution can be written as:

$$\begin{aligned} \boldsymbol{w}(k+1) = \boldsymbol{w}(k) + &\left(\sum_{i=0}^{k} \lambda^{k-i}\boldsymbol{I}_M(i)\boldsymbol{x}(i)\boldsymbol{x}^T(i)\right)^{-1} \\ &\boldsymbol{I}_M(k)\boldsymbol{x}(k)e(k), \quad k = 0, 1, \ldots \end{aligned} \tag{4.122}$$

Defining:

$$\begin{aligned} \boldsymbol{\Theta}_M(k) &= \boldsymbol{I}_M(k)\boldsymbol{x}(k)\boldsymbol{x}^T(k) + \lambda \sum_{i=0}^{k-1} \lambda^{k-1-i}\boldsymbol{I}_M(i)\boldsymbol{x}(i)\boldsymbol{x}^T(i) \\ &= \boldsymbol{I}_M(k)\boldsymbol{x}(k)\boldsymbol{x}^T(k) + \lambda\boldsymbol{\Theta}_M(k-1) \end{aligned} \tag{4.123}$$

and applying the matrix inversion lemma to $\boldsymbol{P}_M(k) = \boldsymbol{\Theta}_M^{-1}(k)$, we obtain:

$$\begin{aligned} \boldsymbol{P}_M(k) &= \lambda^{-1}\boldsymbol{P}_M(k-1) \\ &\quad - \frac{\lambda^{-2}\boldsymbol{P}_M(k-1)\boldsymbol{I}_M(k)\boldsymbol{x}(k)\boldsymbol{x}^T(k)\boldsymbol{P}_M(k-1)}{1 + \lambda^{-1}\boldsymbol{x}^T(k)\boldsymbol{P}_M(k-1)\boldsymbol{I}_M(k)\boldsymbol{x}(k)} \end{aligned} \tag{4.124}$$

where $\boldsymbol{P}_M(-1) = \epsilon^{-1}\boldsymbol{I}$. Substituting (4.124) into (4.122) results in the *partial-update RLS algorithm*

$$\boldsymbol{w}(k+1) = \boldsymbol{w}(k) + \boldsymbol{P}_M(k)e(k)\boldsymbol{I}_M(k)\boldsymbol{x}(k), \quad k = 0, 1, \ldots \tag{4.125}$$

Strictly speaking, the partial-update RLS algorithm is not a partial-coefficient-update algorithm because it updates the entire coefficient vector $\boldsymbol{w}(k)$ at each iteration. However, it uses a subset of the regressor vector $\boldsymbol{x}(k)$ to compute the update vector, saving some complexity in the adaptation process.

We now show that the partial-update RLS algorithm in (4.125) obeys the following fundamental relationship between *a posteriori* and *a priori* errors (cf. (4.104)):

$$e_p(k) = \frac{1}{1 + \lambda^{-1}\boldsymbol{x}^T(k)\boldsymbol{P}_M(k-1)\boldsymbol{I}_M(k)\boldsymbol{x}(k)} e(k) \tag{4.126}$$

After subtracting both sides of (4.126) from $e(k)$ we have:

$$\boldsymbol{x}^T(k)\delta\boldsymbol{w}(k+1) = \frac{\lambda^{-1}\boldsymbol{x}^T(k)\boldsymbol{P}_M(k-1)\boldsymbol{I}_M(k)\boldsymbol{x}(k)}{1 + \lambda^{-1}\boldsymbol{x}^T(k)\boldsymbol{P}_M(k-1)\boldsymbol{I}_M(k)\boldsymbol{x}(k)} e(k) \tag{4.127}$$

A solution to the above under-determined equation is simply obtained by deleting $\boldsymbol{x}^T(k)$ from both sides of the equation (note that this does not give the minimum-norm solution):

$$\delta\boldsymbol{w}(k+1) = \frac{\lambda^{-1}\boldsymbol{P}_M(k-1)\boldsymbol{I}_M(k)\boldsymbol{x}(k)}{1 + \lambda^{-1}\boldsymbol{x}^T(k)\boldsymbol{P}_M(k-1)\boldsymbol{I}_M(k)\boldsymbol{x}(k)} e(k) \tag{4.128}$$

whence we have:

$$\boldsymbol{w}(k+1) = \boldsymbol{w}(k) + \frac{\lambda^{-1}\boldsymbol{P}_M(k-1)\boldsymbol{I}_M(k)\boldsymbol{x}(k)}{1 + \lambda^{-1}\boldsymbol{x}^T(k)\boldsymbol{P}_M(k-1)\boldsymbol{I}_M(k)\boldsymbol{x}(k)} e(k), \quad k = 0, 1, \ldots \tag{4.129}$$

From the definition of $\boldsymbol{P}_M(k)$ in (4.124) we have:

$$\boldsymbol{P}_M(k)\boldsymbol{I}_M(k)\boldsymbol{x}(k) = \frac{\lambda^{-1}\boldsymbol{P}_M(k-1)\boldsymbol{I}_M(k)\boldsymbol{x}(k)}{1 + \lambda^{-1}\boldsymbol{x}^T(k)\boldsymbol{P}_M(k-1)\boldsymbol{I}_M(k)\boldsymbol{x}(k)} \tag{4.130}$$

Using this equation in (4.129) we obtain:

$$\boldsymbol{w}(k+1) = \boldsymbol{w}(k) + \boldsymbol{P}_M(k)e(k)\boldsymbol{I}_M(k)\boldsymbol{x}(k), \quad k = 0, 1, \ldots$$

which is identical to the partial-update RLS algorithm in (4.125). Thus, we conclude that the partial-update RLS algorithm relates $e(k)$ to $e_p(k)$ according to (4.126).

4.9.1 Periodic-partial-update RLS algorithm

The periodic-partial-update RLS algorithm is given by the recursion:

$$\boldsymbol{w}((k+1)S) = \boldsymbol{w}(kS) + \boldsymbol{P}(kS)\boldsymbol{x}(kS)e(kS), \quad k = 0, 1, \ldots \tag{4.131}$$

where:

$$\begin{aligned}\boldsymbol{P}(kS) &= \lambda^{-1}\boldsymbol{P}((k-1)S) \\ &\quad - \frac{\lambda^{-2}\boldsymbol{P}((k-1)S)\boldsymbol{x}(kS)\boldsymbol{x}^T(kS)\boldsymbol{P}((k-1)S)}{1+\lambda^{-1}\boldsymbol{x}^T(kS)\boldsymbol{P}((k-1)S)\boldsymbol{x}(kS)}\end{aligned} \tag{4.132}$$

and $\boldsymbol{P}(-S) = \epsilon^{-1}\boldsymbol{I}$. All adaptive filter coefficients are updated every S iterations, thereby giving the adaptation process S iterations to complete its update vector computation. No partial-update regressors are used in the update vector computation. This gives us some reassurance that the correlation matrix will not become ill-conditioned, causing spikes in the MSE curve. The periodic-partial-update RLS algorithm is arguably the most stable partial-update RLS algorithm.

4.9.2 Sequential-partial-update RLS algorithm

Using (4.125), the sequential-partial-update RLS algorithm is defined by:

$$\begin{aligned}\boldsymbol{w}(k+1) &= \boldsymbol{w}(k) + \boldsymbol{P}_M(k)e(k)\boldsymbol{I}_M(k)\boldsymbol{x}(k), \quad k=0,1,\dots \\ \boldsymbol{P}_M(k) &= \lambda^{-1}\boldsymbol{P}_M(k-1) - \frac{\lambda^{-2}\boldsymbol{P}_M(k-1)\boldsymbol{I}_M(k)\boldsymbol{x}(k)\boldsymbol{x}^T(k)\boldsymbol{P}_M(k-1)}{1+\lambda^{-1}\boldsymbol{x}^T(k)\boldsymbol{P}_M(k-1)\boldsymbol{I}_M(k)\boldsymbol{x}(k)}, \\ \boldsymbol{P}_M(-1) &= \epsilon^{-1}\boldsymbol{I}\end{aligned} \tag{4.133}$$

where the coefficient selection matrix $\boldsymbol{I}_M(k)$ is given by (4.14).

4.9.3 Stochastic-partial-update RLS algorithm

The stochastic-partial-update RLS algorithm uses randomized coefficient selection and is defined by the recursion:

$$\begin{aligned}\boldsymbol{w}(k+1) &= \boldsymbol{w}(k) + \boldsymbol{P}_M(k)e(k)\boldsymbol{I}_M(k)\boldsymbol{x}(k), \quad k=0,1,\dots \\ \boldsymbol{P}_M(k) &= \lambda^{-1}\boldsymbol{P}_M(k-1) - \frac{\lambda^{-2}\boldsymbol{P}_M(k-1)\boldsymbol{I}_M(k)\boldsymbol{x}(k)\boldsymbol{x}^T(k)\boldsymbol{P}_M(k-1)}{1+\lambda^{-1}\boldsymbol{x}^T(k)\boldsymbol{P}_M(k-1)\boldsymbol{I}_M(k)\boldsymbol{x}(k)}, \\ \boldsymbol{P}_M(-1) &= \epsilon^{-1}\boldsymbol{I}\end{aligned} \tag{4.134}$$

where the coefficient selection matrix $\boldsymbol{I}_M(k)$ is given by (4.17).

4.9.4 Selective-partial-update RLS algorithm

Consider the selective-partial-update version of Newton's method:

$$\boldsymbol{w}(k+1) = \boldsymbol{w}(k) + \mu(k)\boldsymbol{R}_M^{-1}(\boldsymbol{p}_M - \boldsymbol{R}_M\boldsymbol{w}(k)), \quad k=0,1,\dots \tag{4.135}$$

We saw in Section 2.6.2 that the selective-partial-update NLMS algorithm can be derived as an instantaneous approximation to Newton's method. We adopt the same approach here. However, this time we use a better approximation for the inverse of the partial-update correlation matrix $\boldsymbol{R}_M^{-1}$.

Following the same steps as in the RLS development, we replace $\boldsymbol{R}_M$ with the exponentially weighted 'partial-update' sample average:

$$\boldsymbol{R}_M \approx \frac{1}{k+1}\sum_{i=0}^{k} \lambda^{k-i} \boldsymbol{I}_M(i)\boldsymbol{x}(i)\boldsymbol{x}^T(i) \tag{4.136}$$

where $0 < \lambda \leq 1$ is the exponential forgetting factor, and $\boldsymbol{p}_M - \boldsymbol{R}_M\boldsymbol{w}(k)$ with the instantaneous approximation:

$$\boldsymbol{p}_M - \boldsymbol{R}_M\boldsymbol{w}(k) \approx \boldsymbol{I}_M(k)\boldsymbol{x}(k)\underbrace{(d(k) - \boldsymbol{x}^T(k)\boldsymbol{w}(k))}_{e(k)} \tag{4.137}$$

An approximate selective-partial-update version of Newton's method is then given by the recursion:

$$\boldsymbol{w}(k+1) = \boldsymbol{w}(k) + \left(\sum_{i=0}^{k} \lambda^{k-i} \boldsymbol{I}_M(i)\boldsymbol{x}(i)\boldsymbol{x}^T(i)\right)^{-1} \boldsymbol{I}_M(k)\boldsymbol{x}(k)e(k), \quad k = 0, 1, \ldots$$

where

$$\mu(k) = 1/(k+1).$$

Note that this recursion is identical to the generic partial-update RLS recursion in (4.122). Therefore, the selective-partial-update RLS algorithm has the same update equation as the partial-update RLS recursion in (4.125) and it only differs from the other partial-update RLS algorithms in terms of its coefficient selection criterion.

In addition to describing the relationship between $e(k)$ and $e_p(k)$ for the partial-update RLS algorithm, (4.126) provides a criterion for the optimum coefficient selection matrix. The coefficient selection matrix for the selective-partial-update RLS algorithm minimizes the magnitude of *a posteriori* error $|e_p(k)|$ at each k. The formal definition of $\boldsymbol{I}_M(k)$ is:

$$\min_{\boldsymbol{I}_M(k)} \left| \frac{1}{1 + \lambda^{-1}\boldsymbol{x}^T(k)\boldsymbol{P}_M(k-1)\boldsymbol{I}_M(k)\boldsymbol{x}(k)} \right| \tag{4.138}$$

or:

$$\max_{\boldsymbol{I}_M(k)} \boldsymbol{x}^T(k)\boldsymbol{P}_M(k-1)\boldsymbol{I}_M(k)\boldsymbol{x}(k) \tag{4.139}$$

The full implementation of this selection criterion is computationally expensive. Assuming a stationary and approximately white input signal the maximization problem in (4.139) can be replaced by:

$$\max_{\boldsymbol{I}_M(k)} \boldsymbol{x}^T(k)\boldsymbol{I}_M(k)\boldsymbol{x}(k) \tag{4.140}$$

This simplified selection criterion is identical to the M-max LMS criterion given in (4.20).

4.9.5 Set-membership partial-update RLS algorithm

The final partial-update algorithm that we consider for RLS is the set-membership partial-update RLS algorithm which extends set membership filtering and selective partial updates into the RLS algorithm.

In the set-membership adaptive filtering context we impose the following equality constrained on the *a posteriori* error of the selective-partial-update RLS algorithm:

$$e_p(k) = \frac{\alpha(k)}{1 + \lambda^{-1}\boldsymbol{x}^T(k)\boldsymbol{P}_M(k-1)\boldsymbol{I}_M(k)\boldsymbol{x}(k)} e(k) \tag{4.141}$$

Our objective is to bound the error magnitude $|e_p(k)|$ with an appropriate selection of $\alpha(k)$. In particular, for a given bound γ we wish to have $|e_p(k)| = \gamma$ if $|e(k)| > \gamma$ and $e_p(k) = e(k)$ otherwise. Referring to (4.141) the time-varying step-size parameter $\alpha(k)$ satisfying these requirements is given by:

$$\alpha(k) = \begin{cases} \dfrac{\gamma}{|e(k)|}(1 + \lambda^{-1}\boldsymbol{x}^T(k)\boldsymbol{P}_M(k-1)\boldsymbol{I}_M(k)\boldsymbol{x}(k)) & \text{if } |e(k)| > \gamma \\ 1 + \lambda^{-1}\boldsymbol{x}^T(k)\boldsymbol{P}_M(k-1)\boldsymbol{I}_M(k)\boldsymbol{x}(k) & \text{otherwise} \end{cases} \tag{4.142}$$

Subtracting both sides of (4.141) from $e(k)$ and using the definition of $e(k)$ and $e_p(k)$, we obtain:

$$\boldsymbol{x}^T(k)\delta\boldsymbol{w}(k+1) = \frac{1 - \alpha(k) + \lambda^{-1}\boldsymbol{x}^T(k)\boldsymbol{P}_M(k-1)\boldsymbol{I}_M(k)\boldsymbol{x}(k)}{1 + \lambda^{-1}\boldsymbol{x}^T(k)\boldsymbol{P}_M(k-1)\boldsymbol{I}_M(k)\boldsymbol{x}(k)} e(k) \tag{4.143}$$

Rewriting the above equation as:

$$\begin{aligned} &\boldsymbol{x}^T(k)\delta\boldsymbol{w}(k+1) \\ &= \boldsymbol{x}^T(k)\frac{\frac{\boldsymbol{x}(k)}{\|\boldsymbol{x}(k)\|_2^2}(1 - \alpha(k)) + \lambda^{-1}\boldsymbol{P}_M(k-1)\boldsymbol{I}_M(k)\boldsymbol{x}(k)}{1 + \lambda^{-1}\boldsymbol{x}^T(k)\boldsymbol{P}_M(k-1)\boldsymbol{I}_M(k)\boldsymbol{x}(k)} e(k) \end{aligned} \tag{4.144}$$

we see that a non-unique solution for $\delta \boldsymbol{w}(k+1)$ is given by:

$$\boldsymbol{w}(k+1) = \boldsymbol{w}(k) + \frac{\frac{\boldsymbol{x}(k)}{\|\boldsymbol{x}(k)\|_2^2}(1-\alpha(k)) + \lambda^{-1}\boldsymbol{P}_M(k-1)\boldsymbol{I}_M(k)\boldsymbol{x}(k)}{1+\lambda^{-1}\boldsymbol{x}^T(k)\boldsymbol{P}_M(k-1)\boldsymbol{I}_M(k)\boldsymbol{x}(k)} e(k) \tag{4.145}$$

Using (4.130) in the above equation we obtain:

$$\boldsymbol{w}(k+1) = \boldsymbol{w}(k) + \left(\frac{\boldsymbol{x}(k)}{\|\boldsymbol{x}(k)\|_2^2} \frac{1-\alpha(k)}{1+\lambda^{-1}\boldsymbol{x}^T(k)\boldsymbol{P}_M(k-1)\boldsymbol{I}_M(k)\boldsymbol{x}(k))} + \boldsymbol{P}_M(k)\boldsymbol{I}_M(k)\boldsymbol{x}(k) \right) e(k), \quad k = 0, 1, \ldots \tag{4.146}$$

where under simplifying assumptions $\boldsymbol{I}_M(k)$ is given by (4.20). We will refer to (4.146) as the *set-membership partial-update RLS algorithm.*

The set-membership partial-update RLS algorithm updates the adaptive filter coefficients only when $|e(k)| > \gamma$. The resulting update recursion is:

$$\boldsymbol{w}(k+1) = \boldsymbol{w}(k) + \left(\frac{\boldsymbol{x}(k)}{\|\boldsymbol{x}(k)\|_2^2} \left(\frac{1}{1+\lambda^{-1}\boldsymbol{x}^T(k)\boldsymbol{P}_M(k-1)\boldsymbol{I}_M(k)\boldsymbol{x}(k)} - \frac{\gamma}{|e(k)|} \right) + \boldsymbol{P}_M(k)\boldsymbol{I}_M(k)\boldsymbol{x}(k) \right) e(k), \quad \text{if } |e(k)| > \gamma \tag{4.147}$$

4.9.6 Partial-update RLS simulations

Consider a system identification problem where the unknown system has transfer function:

$$H(z) = 1 - 0.8z^{-1} + 0.3z^{-2} + 0.2z^{-3} + 0.1z^{-4} - 0.2z^{-5} + 0.1z^{-6} + 0.4z^{-7} - 0.2z^{-8} + 0.1z^{-9} \tag{4.148}$$

The input signal $x(k)$ is an AR(1) Gaussian process with $a = -0.8$. The system output is measured in additive zero-mean white Gaussian noise with variance $\sigma_n^2 = 10^{-4}$. The RLS parameters are set to $N = 10$, $\lambda = 0.999$ and $\epsilon = 10^{-4}$. The convergence performance of the RLS algorithm and its partial-update versions has been measured by *misalignment*, which is defined by:

$$\frac{\|\boldsymbol{w}(k) - \boldsymbol{h}\|_2^2}{\|\boldsymbol{h}\|_2^2} \tag{4.149}$$

where $\boldsymbol{h}$ is the true system impulse response.

Figures 4.1 and 4.2 show the time evolution of misalignment for full-update and partial-update RLS algorithms for one realization of input and noise signals. The partial-update parameters are $S = 10$ for periodic partial updates and $M = 1$ for the other partial-update algorithms. The error magnitude bound for the set-membership partial-update RLS algorithm is set to $\gamma = 0.02$. The selective-partial-update and set-membership partial-update RLS algorithms use the simplified coefficient selection criterion. From Figure 4.1 we observe that the initial convergence rate of the periodic-partial-update RLS algorithm is roughly $S = 10$ times slower than the full-update RLS algorithm as expected. The sequential and stochastic-partial-update RLS algorithms both exhibit divergent behaviour as evidenced by spikes in their misalignment curves. These spikes occur whenever the time-averaged partial-update correlation matrix $\boldsymbol{R}_M$ becomes ill-conditioned due to the use of sparse partial-update regressor vectors $\boldsymbol{I}_M(k)\boldsymbol{x}(k)$ in the update of $\boldsymbol{R}_M$. The likelihood of ill-conditioning increases with smaller M and λ (in this case $M = 1$).

Even though the selective-partial-update and set-membership partial-update RLS algorithms are not entirely immune from ill-conditioning of $\boldsymbol{R}_M$, the data-dependent nature of coefficient selection is likely to reduce the frequency of ill-conditioning. In fact, in Figure 4.2 there are no visible misalignment spikes that can be attributed to ill-conditioning of $\boldsymbol{R}_M$. The selective-partial-update and set-membership partial-update RLS algorithms appear to perform well compared with the sequential and stochastic-partial-update RLS algorithms. Figure 4.2(c) shows the sparse time updates (iterations at which partial-update coefficients are updated) of the set-membership partial-update RLS algorithm, which is a feature of set-membership adaptive filters leading to reduced power consumption.

4.9.7 Computational complexity

At each iteration k, the RLS algorithm performs two main tasks; viz. update the adaptive filter coefficients $\boldsymbol{w}(k)$ and compute $\boldsymbol{P}(k)$ for the next coefficient update. The computational complexity analysis of the full-update RLS algorithm in (4.103) is presented below:

- N multiplications and N additions to compute $e(k)$.
- $N(N+1)/2$ multiplications* to compute $\lambda^{-1}\boldsymbol{P}(k-1)$.
- N^2 multiplications and $N(N-1)$ additions to compute $\lambda^{-1}\boldsymbol{P}(k-1)\boldsymbol{x}(k)$.
- N multiplications and $N-1$ additions to compute $\lambda^{-1}\boldsymbol{x}^T(k)\boldsymbol{P}(k-1)\boldsymbol{x}(k)$.
- One addition to compute $1 + \lambda^{-1}\boldsymbol{x}^T(k)\boldsymbol{P}(k-1)\boldsymbol{x}(k)$.

---

*Here we make use of the fact that $\boldsymbol{P}(k)$ is a symmetric matrix and therefore it is sufficient multiply λ^{-1} with the upper or lower diagonal entries of $\boldsymbol{P}(k-1)$. The remaining entries of $\lambda^{-1}\boldsymbol{P}(k-1)$ are simply copies of the computed entries.

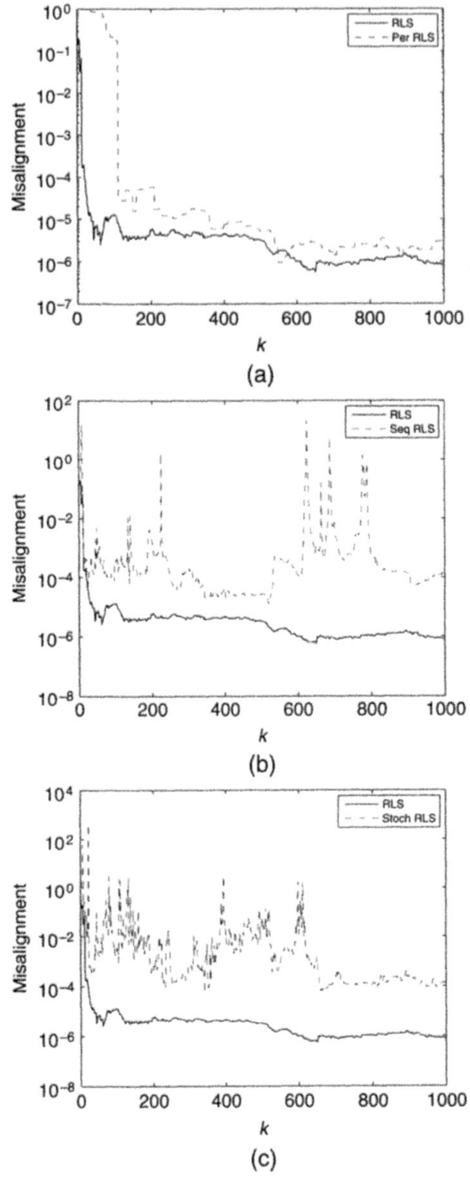

Figure 4.1 Time evolution of misalignment for data-independent partial-update RLS algorithms: (a) Periodic-partial-update RLS; (b) sequential-partial-update RLS; (c) stochastic-partial-update RLS.

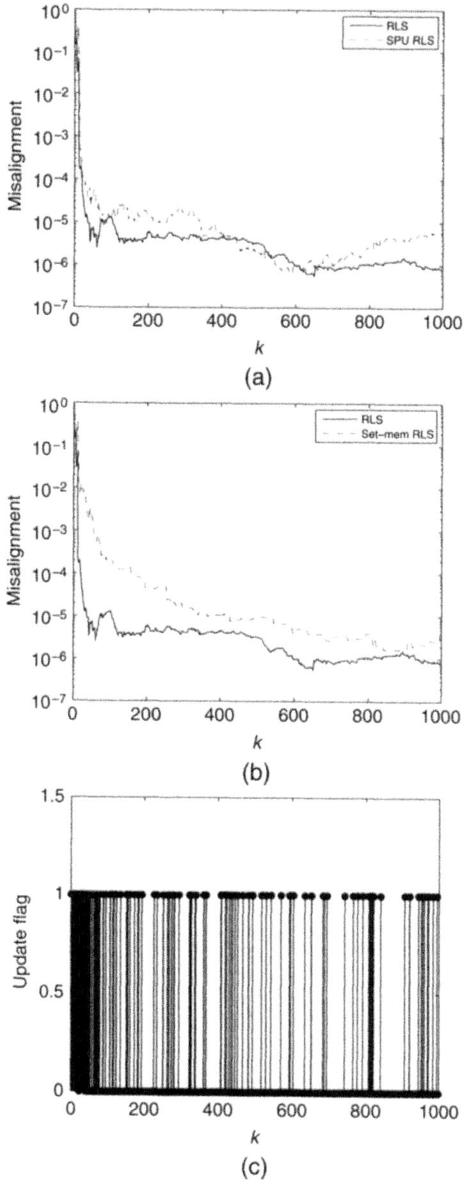

Figure 4.2 Time evolution of misalignment for data-dependent partial-update RLS algorithms: (a) Selective-partial-update RLS; (b) set-membership partial-update RLS; (c) coefficient update indicator for set-membership partial-update RLS.

- One division to compute $1/(1+\lambda^{-1}\boldsymbol{x}^T(k)\boldsymbol{P}(k-1)\boldsymbol{x}(k))$.
- N multiplications to compute $\lambda^{-1}\boldsymbol{P}(k-1)\boldsymbol{x}(k)/(1+\lambda^{-1}\boldsymbol{x}^T(k)\boldsymbol{P}(k-1)\boldsymbol{x}(k))$.
- $N(N+1)/2$ multiplications to compute $\lambda^{-2}\boldsymbol{P}(k-1)\boldsymbol{x}(k)\boldsymbol{x}^T(k)\boldsymbol{P}(k-1)/(1+\lambda^{-1}\boldsymbol{x}^T(k)\boldsymbol{P}(k-1)\boldsymbol{x}(k))$.
- $N(N+1)/2$ additions to compute $\boldsymbol{P}(k)$.
- N multiplications to compute $\boldsymbol{x}(k)e(k)$.
- N^2 multiplications and $N(N-1)$ additions to compute $\boldsymbol{P}(k)\boldsymbol{x}(k)e(k)$.
- N additions to update $\boldsymbol{w}(k)$.

The total computational complexity of the full-update RLS algorithm is $3N^2+5N$ multiplications, $\frac{5}{2}N^2+\frac{3}{2}N$ additions and one division at each iteration. The full-update RLS algorithm, therefore, has computational complexity of $O(N^2)$.

The periodic-partial-update RLS algorithm effectively operates at a rate S times slower than the input signal rate. The only exception to this is the filtering process which operates at the same rate as the input signal. Therefore, the computational complexity of the periodic-partial-update RLS algorithm per iteration is $(3N^2+4N)/S+N$ multiplications, $(5N^2+N)/(2S)+N$ additions and $1/S$ divisions.

The sequential-partial-update RLS algorithm uses a subset of the regressor vector in update vector computation. A detailed analysis of its computational complexity is provided below:

- N multiplications and N additions to compute $e(k)$.
- N^2 multiplications* to compute $\lambda^{-1}\boldsymbol{P}_M(k-1)$.
- N^2 multiplications and $N(N-1)$ additions to compute $\lambda^{-1}\boldsymbol{P}_M(k-1)\boldsymbol{x}(k)$ and $\lambda^{-1}\boldsymbol{P}_M(k-1)\boldsymbol{I}_M(k)\boldsymbol{x}(k)$.
- M multiplications and $M-1$ additions to compute $\lambda^{-1}\boldsymbol{x}^T(k)\boldsymbol{P}_M(k-1)\boldsymbol{I}_M(k)\boldsymbol{x}(k)$.
- One addition to compute $1+\lambda^{-1}\boldsymbol{x}^T(k)\boldsymbol{P}_M(k-1)\boldsymbol{I}_M(k)\boldsymbol{x}(k)$.
- One division to compute $1/(1+\lambda^{-1}\boldsymbol{x}^T(k)\boldsymbol{P}_M(k-1)\boldsymbol{I}_M(k)\boldsymbol{x}(k))$.
- N multiplications to compute $\lambda^{-1}\boldsymbol{P}_M(k-1)\boldsymbol{I}_M(k)\boldsymbol{x}(k)/(1+\lambda^{-1}\boldsymbol{x}^T(k)\boldsymbol{P}_M(k-1)\boldsymbol{I}_M(k)\boldsymbol{x}(k))$.
- N^2 multiplications to compute $\lambda^{-2}\boldsymbol{P}_M(k-1)\boldsymbol{I}_M(k)\boldsymbol{x}(k)\boldsymbol{x}^T(k)\boldsymbol{P}_M(k-1)/(1+\lambda^{-1}\boldsymbol{x}^T(k)\boldsymbol{P}_M(k-1)\boldsymbol{I}_M(k)\boldsymbol{x}(k))$.
- N^2 additions to compute $\boldsymbol{P}_M(k)$.
- M multiplications to compute $e(k)\boldsymbol{I}_M(k)\boldsymbol{x}(k)$.
- NM multiplications and $N(M-1)$ additions to compute $\boldsymbol{P}_M(k)e(k)\boldsymbol{I}_M(k)\boldsymbol{x}(k)$.
- N additions to update $\boldsymbol{w}(k)$.

*The matrix $\boldsymbol{P}_M(k-1)$ is not symmetric even though $\boldsymbol{R}_M$ is for stationary $x(k)$. As a result we must multiply λ^{-1} with all entries of $\boldsymbol{P}_M(k-1)$.

The total computational complexity of the sequential-partial-update RLS algorithm is $3N^2 + 2N + 2M + NM$ multiplications, $2N^2 + NM + M$ additions and one division per iteration. The non-symmetry of $\boldsymbol{P}_M(k)$ which is a consequence of computing a partial-update correlation matrix appears to increase the complexity of the sequential-partial-update RLS algorithm.

The stochastic-partial-update RLS algorithm requires a further 2 multiplications and 2 additions in addition to the computational requirements of the sequential-partial-update RLS algorithm. The additional complexity arises from the linear congruential generator in (4.22) which is used to generate random numbers for the coefficient selection matrix.

The computational complexity of the selective-partial-update RLS algorithm is slightly larger than that of the sequential-partial-update RLS algorithm due to data sorting required by data-dependent coefficient selection. For the simplified selection criterion in (4.140) the computational complexity of constructing $\boldsymbol{I}_M(k)$ is $2\lceil \log_2 N \rceil + 2$ comparisons using the sortline algorithm (see Appendix A).

The set-membership partial-update RLS algorithm combines selective partial coefficient updates with sparse time updates. Whether or not an update decision is made, the matrix $\boldsymbol{P}_M(k)$ needs to be computed at each iteration in preparation for future updates and to keep track of variations in signal statistics. Referring to the update equation in (4.147), we see that the set-membership partial-update RLS algorithm requires the following vector to be computed in its update term in addition to what is required by the selective-partial-update RLS algorithm:

$$\frac{\boldsymbol{x}(k)}{\|\boldsymbol{x}(k)\|_2^2}\left(\frac{1}{1+\lambda^{-1}\boldsymbol{x}^T(k)\boldsymbol{P}_M(k-1)\boldsymbol{I}_M(k)\boldsymbol{x}(k)} - \frac{\gamma}{|e(k)|}\right)e(k) \qquad (4.150)$$

The additional complexity imposed by this vector is $2N + 1$ multiplications, N additions and 2 divisions.

Table 4.4 compares the computational complexity of RLS and its partial-update versions. The method of periodic partial updates achieves the most dramatic complexity reduction. The computational complexity of the set-membership partial-update RLS algorithm is the peak complexity per iteration, which corresponds to the complexity of an update whenever it is decided. The average complexity is usually much smaller thanks to sparse time updates.

4.10 TRANSFORM-DOMAIN LEAST-MEAN-SQUARE ALGORITHM

The convergence rate of the LMS algorithm depends on the eigenvalue spread of the input signal correlation matrix $\boldsymbol{R} = E\{\boldsymbol{x}(k)\boldsymbol{x}^T(k)\}$. The fastest convergence rate is achieved when the eigenvalue spread is unity (i.e. the input signal is white

Table 4.4 Computational complexity of RLS algorithm and partial-update RLS algorithms at each k

Algorithm	$\times$	$+$	$\div$	$\lessgtr$
RLS	$3N^2+5N$	$\frac{5}{2}N^2+\frac{3}{2}N$	1	
Per-par-upd RLS	$(3N^2+4N)/S+N$	$(5N^2+N)/(2S)+N$	$1/S$	
Seq-par-upd RLS	$3N^2+2N+2M$ $+NM$	$2N^2+NM+M$	1	
Stoch-par-upd RLS	$3N^2+2N+2M$ $+NM+2$	$2N^2+NM+M+2$	1	
Sel-par-upd RLS	$3N^2+2N+2M$ $+NM$	$2N^2+NM+M$	1	$2\lceil\log_2 N\rceil$ $+2$
Set-mem par-upd RLS	$3N^2+4N+2M$ $+NM+1$	$2N^2+NM+M+N$	3	$2\lceil\log_2 N\rceil$ $+2$

and stationary). For strongly correlated input signals with large eigenvalue spread, the LMS algorithm exhibits significantly slower convergence. The convergence rate of LMS can be greatly improved by decorrelating and power normalizing the input regressor as illustrated in Figure 4.3. The input regressor vector $\boldsymbol{x}(k)$ is firstly applied to an $N \times N$ unitary transform $\boldsymbol{T}$ that aims to decorrelate the regressor entries:

$$\boldsymbol{v}(k) = \begin{bmatrix} v_1(k) \\ v_2(k) \\ \vdots \\ v_N(k) \end{bmatrix} = \boldsymbol{T}\boldsymbol{x}(k) \tag{4.151}$$

where $\boldsymbol{T}\boldsymbol{T}^H = \boldsymbol{I}$ and $\boldsymbol{T}^H\boldsymbol{T} = \boldsymbol{I}$:

$$E\{\boldsymbol{v}^*(k)\boldsymbol{v}^T(k)\} = \boldsymbol{T}^*\boldsymbol{R}\boldsymbol{T}^T = \boldsymbol{\Sigma} \tag{4.152}$$

and $\boldsymbol{\Sigma}$ is a real-valued diagonal matrix:

$$\boldsymbol{\Sigma} = \begin{bmatrix} \sigma_{11} & & & \mathbf{0} \\ & \sigma_{22} & & \\ & & \ddots & \\ \mathbf{0} & & & \sigma_{NN} \end{bmatrix} \tag{4.153}$$

Since $\boldsymbol{T}$ is in general complex-valued we use the Hermitian (denoted ^{H}) and complex conjugate (denoted *) in the above definitions. It is clear from (4.152) that $\boldsymbol{T}$ diagonalizes the input correlation matrix $\boldsymbol{R}$. Even though the entries of $\boldsymbol{v}(k)$ may be uncorrelated as a result of diagonalization, the eigenvalue spread of $E\{\boldsymbol{v}^*(k)\boldsymbol{v}^T(k)\}$

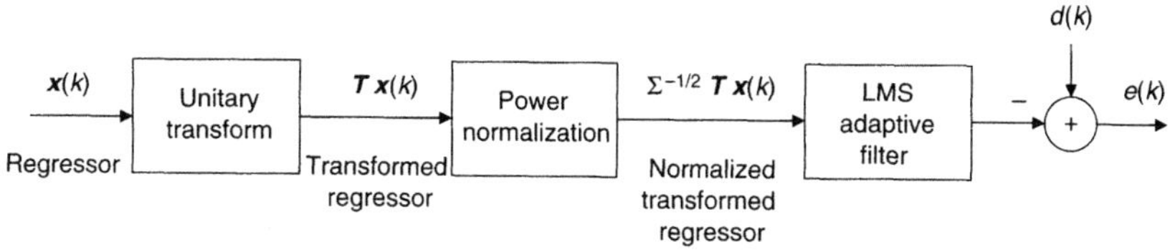

Figure 4.3 Transform-domain whitening of regressor vector for LMS adaptive filtering.

is still the same as that of $\boldsymbol{R}$. Power normalizing the entries of $\boldsymbol{v}(k)$ by:

$$\boldsymbol{v}_n(k) = \boldsymbol{\Sigma}^{-1/2}\boldsymbol{v}(k) \tag{4.154}$$

results in $E\{\boldsymbol{v}_n^*(k)\boldsymbol{v}_n^T(k)\} = \boldsymbol{\Sigma}^{-1/2}\boldsymbol{\Sigma}\boldsymbol{\Sigma}^{-1/2} = \boldsymbol{I}$ which has unity eigenvalue spread as desired. Thus the vector $\boldsymbol{v}_n(k)$ obtained from the original regressor $\boldsymbol{x}(k)$ by unitary transform $\boldsymbol{T}$ and power normalization has uncorrelated entries with unity variance. In other words, $x(k)$ has been pre-whitened. Thus the setup in Figure 4.3 is expected to yield improved convergence rate for correlated input signals. For stationary white input signals the whitening approach described above does not offer any convergence advantages.

The LMS algorithm for the whitened input signal can be implemented as an approximation to Newton's method:

$$\boldsymbol{w}(k+1) = \boldsymbol{w}(k) + \mu\boldsymbol{R}^{-1}(\boldsymbol{p} - \boldsymbol{R}\boldsymbol{w}(k)), \quad k = 0, 1, \ldots \tag{4.155}$$

Substituting the decorrelated transform output $\boldsymbol{v}(k)$ for $\boldsymbol{x}(k)$ and using the instantaneous approximation for $\boldsymbol{p} - \boldsymbol{R}\boldsymbol{w}(k)$ we obtain:

$$\begin{aligned}
\boldsymbol{w}(k+1) &= \boldsymbol{w}(k) + \mu\boldsymbol{\Sigma}^{-1}(\boldsymbol{v}^*(k)d(k) - \boldsymbol{v}^*(k)\boldsymbol{v}^T(k)\boldsymbol{w}(k)), \quad k = 0, 1, \ldots \\
\boldsymbol{w}(k+1) &= \boldsymbol{w}(k) + \mu\boldsymbol{\Sigma}^{-1}\boldsymbol{v}^*(k)(d(k) - \boldsymbol{v}^T(k)\boldsymbol{w}(k)), \quad k = 0, 1, \ldots \qquad (4.156) \\
\boldsymbol{w}(k+1) &= \boldsymbol{w}(k) + \mu\boldsymbol{\Sigma}^{-1}\boldsymbol{v}^*(k)e(k), \quad k = 0, 1, \ldots
\end{aligned}$$

The last recursion above is known as the *transform-domain LMS algorithm*. The structure of the transform-domain LMS adaptive filter is shown in Figure 4.4. Recall that the NLMS algorithm uses a more crude instantaneous estimate of $\boldsymbol{R}$ given by $\epsilon\boldsymbol{I} + \boldsymbol{x}(k)\boldsymbol{x}^T(k)$ [see (4.31)].

An alternative derivation for the transform-domain LMS algorithm follows from direct implementation of the LMS algorithm for the whitened regressor vector $\boldsymbol{v}_n(k)$:

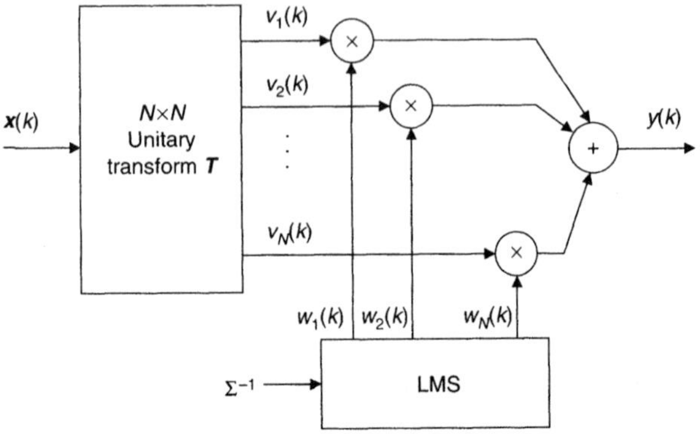

Figure 4.4 Implementation of transform-domain LMS adaptive filter.

$$
\begin{aligned}
\boldsymbol{w}_n(k+1) &= \boldsymbol{w}_n(k) + \mu \boldsymbol{v}_n^*(k)(d(k) - \boldsymbol{v}_n^T(k)\boldsymbol{w}_n(k)), \quad k = 0, 1, \ldots \\
\boldsymbol{\Sigma}^{-1/2}\boldsymbol{w}_n(k+1) &= \boldsymbol{\Sigma}^{-1/2}\boldsymbol{w}_n(k) + \mu \boldsymbol{\Sigma}^{-1/2}\boldsymbol{v}_n^*(k)(d(k) \\
&\quad - \boldsymbol{v}^T(k)\boldsymbol{\Sigma}^{-1/2}\boldsymbol{w}_n(k)), \quad k = 0, 1, \ldots \qquad (4.157) \\
\boldsymbol{w}(k+1) &= \boldsymbol{w}(k) + \mu \boldsymbol{\Sigma}^{-1}\boldsymbol{v}^*(k)(d(k) - \boldsymbol{v}^T(k)\boldsymbol{w}(k)), \quad k = 0, 1, \ldots \\
\boldsymbol{w}(k+1) &= \boldsymbol{w}(k) + \mu \boldsymbol{\Sigma}^{-1}\boldsymbol{v}^*(k)e(k), \quad k = 0, 1, \ldots
\end{aligned}
$$

where $\boldsymbol{w}(k) = \boldsymbol{\Sigma}^{-1/2}\boldsymbol{w}_n(k)$. Rewriting the error signal as:

$$
\begin{aligned}
e(k) &= d(k) - \boldsymbol{v}^T(k)\boldsymbol{w}(k) \\
&= d(k) - \boldsymbol{x}^T(k) \underbrace{\boldsymbol{T}^T \boldsymbol{w}(k)}_{\boldsymbol{w}_{\text{eq}}(k)}
\end{aligned} \tag{4.158}
$$

we note that the equivalent *time-domain* coefficient vector for the transform-domain adaptive filter coefficients $\boldsymbol{w}(k)$ is:

$$\boldsymbol{w}_{\text{eq}}(k) = \boldsymbol{T}^T \boldsymbol{w}(k) \tag{4.159}$$

Premultiplying both sides of the above equation with $\boldsymbol{T}^*$ we obtain $\boldsymbol{w}(k) = \boldsymbol{T}^* \boldsymbol{w}_{\text{eq}}(k)$. From this the optimum solution for a transform-domain adaptive filter is:

$$\boldsymbol{w}_o = \boldsymbol{T}^* \boldsymbol{R}^{-1} \boldsymbol{p} \tag{4.160}$$

According to (4.152) the unitary transform matrix $\boldsymbol{T}$ is defined by:

$$\boldsymbol{R} = \boldsymbol{T}^T \boldsymbol{\Sigma} \boldsymbol{T}^* \tag{4.161}$$

Since $\boldsymbol{R}$ is a symmetric matrix, this can be interpreted as the eigendecomposition of $\boldsymbol{R}$ with eigenvalues $\lambda_1, \ldots, \lambda_N$ and eigenvectors $\boldsymbol{u}_1, \ldots, \boldsymbol{u}_N$. Then $\boldsymbol{T}$ is given by:

$$\boldsymbol{T} = [\boldsymbol{u}_1, \boldsymbol{u}_2, \ldots, \boldsymbol{u}_N]^T \tag{4.162}$$

and $\boldsymbol{\Sigma}$ is:

$$\boldsymbol{\Sigma} = \begin{bmatrix} \lambda_1 & & & \mathbf{0} \\ & \lambda_2 & & \\ & & \ddots & \\ \mathbf{0} & & & \lambda_N \end{bmatrix} \tag{4.163}$$

The transform matrix $\boldsymbol{T}$ defined above is known as the *Karhunen-Loève transform* (KLT). While optimum in terms of achieving perfect decorrelation of the input signal, the KLT requires prior knowledge of $\boldsymbol{R}$ and is not suitable for non-stationary input signals with time-varying correlation properties. In the face of these difficulties, the preferred choice for $\boldsymbol{T}$ is usually a fixed data-independent transform realized using one of the discrete transforms such as the discrete Fourier transform (DFT), discrete cosine transform (DCT), Hartley transform, etc. The DFT and DCT asymptotically converge to the KLT. Therefore for long filters they provide a particularly attractive solution. However, unlike the KLT, the fixed transforms do not result in perfect diagonalization of $\boldsymbol{R}$.

In the case of fixed transforms the matrix $\boldsymbol{\Sigma}$ needs to be estimated from the observed transform outputs $v_i(k)$. A commonly used method for online estimation of the diagonal entries of $\boldsymbol{\Sigma}$ is based on exponentially weighted averaging:

$$\sigma_i^2(k) = (1 - \lambda)|v_i(k)|^2 + \lambda\sigma_i^2(k-1), \quad k = 0, 1, \ldots \tag{4.164}$$

where $0 < \lambda < 1$ is the exponential forgetting factor. The quantity $1/(1 - \lambda)$ is a measure of the memory of the sliding exponential window (Haykin, 1996). For non-stationary input signals, (4.164) enables tracking of changes in the transform output power levels. Suppose that $v_i(k)$ is stationary. Then taking the expectation of both sides of (4.164) results in:

$$E\{\sigma_i^2(k)\} = (1 - \lambda^{k+1})E\{v_i^2(k)\} \tag{4.165}$$

For sufficiently large k this can be approximated by:

$$E\{\sigma_i^2(k)\} \approx E\{v_i^2(k)\} \tag{4.166}$$

To save some computation the factor $1 - \lambda$ multiplying $|v_i(k)|^2$ in (4.164) may be dropped, leading to another sliding exponential window power estimate:

$$\sigma_i^2(k) = |v_i(k)|^2 + \lambda\sigma_i^2(k-1), \quad k = 0, 1, \dots \tag{4.167}$$

For large k the mean power estimate of (4.167) is:

$$E\{\sigma_i^2(k)\} \approx \frac{1}{1-\lambda} E\{|v_i(k)|^2\} \tag{4.168}$$

The factor $1 - \lambda$ may be absorbed into the step-size parameter μ, making the sliding exponential window power estimates in (4.164) and (4.167) effectively identical when appropriately scaled step-sizes are used. The transform-domain LMS algorithm with sliding exponential window power estimate is defined by:

$$\begin{aligned}
&\boldsymbol{w}(k+1) = \boldsymbol{w}(k) + \mu\boldsymbol{\Sigma}^{-1}(k)\boldsymbol{v}^*(k)e(k), \quad k = 0, 1, \dots \\
&\boldsymbol{\Sigma}(k) = \begin{bmatrix} \sigma_1^2(k) & & & \mathbf{0} \\ & \sigma_2^2(k) & & \\ & & \ddots & \\ \mathbf{0} & & & \sigma_N^2(k) \end{bmatrix}, \quad \sigma_i^2(k) = |v_i(k)|^2 + \lambda\sigma_i^2(k-1), \\
&\sigma_i^2(-1) = \epsilon
\end{aligned} \tag{4.169}$$

where ϵ is a small positive constant.

The DFT-based transform matrix $\boldsymbol{T}$ is defined by:

$$v_i(k) = \frac{1}{\sqrt{N}} \sum_{l=0}^{N-1} x(k-l) \exp\left(-\frac{j2\pi l(i-1)}{N}\right) \tag{4.170}$$

where the factor $1/\sqrt{N}$ is included so that $\boldsymbol{T}\boldsymbol{T}^H = \boldsymbol{T}^H\boldsymbol{T} = \boldsymbol{I}$. In practice, it may be dropped by absorbing it into μ. The transform output $v_i(k)$ corresponds to the response of an FIR filter to $x(k)$ with transfer function:

$$\begin{aligned}
H_i(z) = \frac{1}{\sqrt{N}} \Bigg(1 &+ \exp\left(-\frac{j2\pi(i-1)}{N}\right) z^{-1} + \exp\left(-\frac{j4\pi(i-1)}{N}\right) z^{-2} + \cdots \\
&+ \exp\left(-\frac{j2(N-1)\pi(i-1)}{N}\right) z^{-(N-1)}\Bigg)
\end{aligned} \tag{4.171}$$

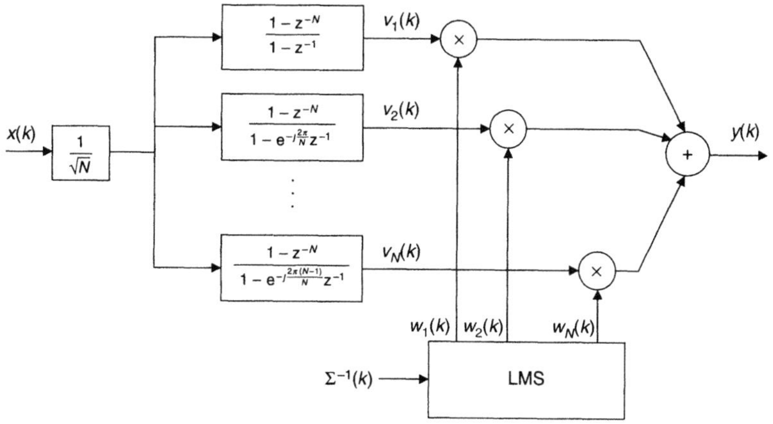

Figure 4.5 DFT-LMS structure implemented using a bank of bandpass comb filters.

Rewriting $H_i(z)$ as a difference of two infinite impulse response (IIR) filters we obtain a recursive filter implementation:

$$\begin{aligned} H_i(z) &= \frac{1}{\sqrt{N}}\left(\sum_{l=0}^{\infty}\exp\left(-\frac{\jmath 2\pi l(i-1)}{N}\right)z^{-l}\right. \\ &\qquad \left. - z^{-N}\sum_{l=0}^{\infty}\exp\left(-\frac{\jmath 2\pi l(i-1)}{N}\right)z^{-l}\right) \\ &= \frac{1-z^{-N}}{1-\exp\left(-\frac{\jmath 2\pi(i-1)}{N}\right)z^{-1}} \end{aligned} \tag{4.172}$$

Thus, $H_i(z)$ can be interpreted as an IIR bandpass comb filter with passband centred at $\omega = \frac{2\pi(i-1)}{N}$ rad. Figure 4.5 shows the comb filter implementation of the DFT-LMS algorithm. Referring to (4.172), we see that $H_i(z)$ has N zeros equally spaced on the unit circle (Nth roots of unity), $N-1$ poles at the origin and a single pole at $e^{-\jmath 2\pi(i-1)/N}$. This pole is cancelled by one of the N zeros.

For the DCT-based $\boldsymbol{T}$ we have:

$$\begin{aligned} v_i(k) &= a_i \sum_{l=0}^{N-1} x(k-l)\cos\left(\frac{\pi(2l+1)(i-1)}{2N}\right), \\ a_i &= \begin{cases} \sqrt{\dfrac{1}{N}} & \text{if } i = 1 \\ \sqrt{\dfrac{2}{N}} & \text{otherwise} \end{cases} \end{aligned} \tag{4.173}$$

This is referred to as the DCT-2 (Oppenheim *et al.*, 1999). The DCT-based transform also leads to a frequency sampling implementation involving a bank of bandpass filters with each bandpass filter given by:

$$H_i(z) = a_i \sum_{l=0}^{N-1} \cos\left(\frac{\pi(2l+1)(i-1)}{2N}\right) z^{-l} \tag{4.174a}$$

$$= \frac{a_i}{2} \sum_{l=0}^{N-1} \left(\exp\left(\jmath \frac{\pi(2l+1)(i-1)}{2N}\right) + \exp\left(-\jmath \frac{\pi(2l+1)(i-1)}{2N}\right)\right) z^{-l} \tag{4.174b}$$

$$= \frac{a_i}{2}(1-(-1)^{i-1}z^{-N}) \left(\frac{\exp\left(\jmath \frac{\pi(i-1)}{2N}\right)}{1-\exp\left(\jmath \frac{\pi(i-1)}{N}\right) z^{-1}} + \frac{\exp\left(-\jmath \frac{\pi(i-1)}{2N}\right)}{1-\exp\left(-\jmath \frac{\pi(i-1)}{N}\right) z^{-1}}\right) \tag{4.174c}$$

$$= a_i \cos\left(\frac{\pi(i-1)}{2N}\right) \frac{(1-z^{-1})(1-(-1)^{i-1}z^{-N})}{1-2\cos\frac{\pi(i-1)}{N}z^{-1}+z^{-2}} \tag{4.174d}$$

Being real-valued, the DCT offers the advantage of reduced computational complexity. An inspection of the poles and zeros of $H_i(z)$ in the last equation above reveals that $H_i(z)$ has $N+1$ zeros on the unit circle $|z| = 1$, $N-1$ poles at the origin, and a complex conjugate pair of poles at $e^{\pm \jmath\pi(i-1)/N}$. These poles are cancelled by the corresponding complex conjugate zero pair, creating the passband for the IIR comb filter.

While the frequency sampling interpretation of the DFT and DCT leads to a low-complexity bandpass comb filter implementation, the poles on the unit circle and pole-zero cancellation at the centre of passband can cause numerical problems in finite-precision implementations. If the poles are slightly outside the unit circle due to finite-precision quantization, the filters become unstable. The stability problem arising from the poles being pushed outside the unit circle can be fixed by multiplying z^{-1} with β where $0 \ll \beta < 1$ (Shynk, 1992), leading to:

$$H_i(z) = \frac{1-\beta^N z^{-N}}{1-\beta\exp\left(-\frac{\jmath 2\pi(i-1)}{N}\right) z^{-1}}, \quad \text{DFT} \tag{4.175}$$

$$H_i(z) = a_i \cos\left(\frac{\pi(i-1)}{2N}\right) \frac{(1-\beta z^{-1})(1-(-1)^{i-1}\beta^N z^{-N})}{1-2\beta\cos\frac{\pi(i-1)}{N}z^{-1}+\beta^2 z^{-2}}, \quad \text{DCT} \tag{4.176}$$

These transfer functions now have poles and zeros on a circle with radius $|z| = \beta$ which is inside the unit circle. If a pole-zero cancellation does not occur as a result of finite-precision arithmetic, this may impact the peak magnitude response of the comb filters in the passband. The impact is minimal if the pole and zero to be cancelled have approximately the same distance from the unit circle, i.e.:

$$\frac{1 - |z_l|}{1 - |p_l|} \approx 1 \tag{4.177}$$

where z_l and p_l denote the zero and pole, respectively, that survived cancellation due to finite-precision implementation.

4.10.1 Power normalization

In the context of transform-domain LMS adaptive filtering defined in (4.169) power normalization refers to the operation $\mathbf{\Sigma}^{-1}(k)\boldsymbol{v}^*(k)$. Strictly speaking, this operation does not normalize the variances of $v_i(k)$. Power normalization occurs when $\mathbf{\Sigma}^{-1/2}\boldsymbol{v}(k)$ is computed [see (4.154)]. However, the change of variables from $\boldsymbol{w}_n(k)$ to $\boldsymbol{w}(k)$ in (4.157) brings about the additional factor $\mathbf{\Sigma}^{-1/2}$; hence the reference to $\mathbf{\Sigma}^{-1}\boldsymbol{v}^*(k)$ as power normalization.

Power normalization is an important component of transform-domain adaptive filters. It produces whitened regressor entries by normalizing the variances of decorrelated regressor entries $v_i(k)$. This normalization would ideally require prior knowledge of $E\{|v_i(k)|^2\}$. However, in practical applications this information is usually not available. Moreover, for non-stationary input signals the variances of $v_i(k)$ may change with time. Thus $E\{|v_i(k)|^2\}$ needs to be estimated and tracked. The computational complexity of power normalization can be quite significant especially for large N because of the division operations involved in computing $\mathbf{\Sigma}^{-1}\boldsymbol{v}^*(k)$ for which the DSP processors are not optimized (Eyre and Bier, 2000).

Consider the sliding exponential window power estimate for $v_i(k)$ given in (4.167). Unlike (4.164) this estimate does not employ multiplication of $|\sigma_i(k)|^2$ with $1 - \lambda$ and, therefore, saves one multiplication. The computational effort required by this power estimate is 2 multiplications (one for squaring modulus of $v_i(k)$ and another to compute $\lambda\sigma_i^2(k-1)$) for each transform output $v_i(k)$. Power normalization requires one division for each $v_i^*(k)$ to compute:

$$\frac{v_i^*(k)}{\sigma_i^2(k)}, \quad i = 1, \ldots, N \tag{4.178}$$

Thus, the total computational complexity for power normalization using (4.167) is $2N$ multiplications and N divisions.

The large number of divisions in power normalization is a major obstacle to the adoption of the transform-domain LMS algorithm in practical applications (Storn, 1996). An alternative method for power normalization that aims to reduce the number of divisions is to combine power estimate update and normalization into a single step by means of the matrix inversion lemma:

$$(\boldsymbol{A} + \boldsymbol{B}\boldsymbol{C}\boldsymbol{D})^{-1} = \boldsymbol{A}^{-1} - \boldsymbol{A}^{-1}\boldsymbol{B}(\boldsymbol{C}^{-1} + \boldsymbol{D}\boldsymbol{A}^{-1}\boldsymbol{B})^{-1}\boldsymbol{D}\boldsymbol{A}^{-1} \tag{4.179}$$

In power normalization we aim to determine $\boldsymbol{\Sigma}^{-1}(k)$, i.e., $1/\sigma_i^2(k)$, $i = 1, \ldots, N$, from $\boldsymbol{v}(k)$ and $\boldsymbol{\Sigma}(k-1)$:

$$\begin{aligned}\boldsymbol{\Sigma}^{-1}(k) &= \begin{bmatrix} 1/\sigma_1^2(k) & & & \boldsymbol{0} \\ & 1/\sigma_1^2(k) & & \\ & & \ddots & \\ \boldsymbol{0} & & & 1/\sigma_N^2(k) \end{bmatrix} \\ &= \left(\lambda \boldsymbol{\Sigma}(k-1) + \begin{bmatrix} |v_1(k)|^2 & & \boldsymbol{0} \\ & \ddots & \\ \boldsymbol{0} & & |v_N(k)|^2 \end{bmatrix} \right)^{-1}\end{aligned} \tag{4.180}$$

An inspection of (4.179) reveals that complexity reduction is only possible if the term $\boldsymbol{C}^{-1} + \boldsymbol{D}\boldsymbol{A}^{-1}\boldsymbol{B}$ in the right-hand side of (4.179) is a scalar so that matrix inversion is avoided. For (4.180) this can only be achieved if we set:

$$\boldsymbol{A} = \lambda\boldsymbol{\Sigma}(k-1), \quad \boldsymbol{B} = \boldsymbol{D}^H = \boldsymbol{v}(k) \quad \text{and} \quad \boldsymbol{C} = 1$$

leading to the following approximation of (4.180):

$$\begin{aligned}(\boldsymbol{A} + \boldsymbol{B}\boldsymbol{C}\boldsymbol{D})^{-1} &= (\lambda\boldsymbol{\Sigma}(k-1) + \boldsymbol{v}(k)\boldsymbol{v}^H(k))^{-1} \\ &\approx \boldsymbol{\Sigma}^{-1}(k).\end{aligned} \tag{4.181}$$

This is considered to be a good approximation because the unitary transform $\boldsymbol{T}$ approximately decorrelates the regressor vector. Thus, $E\{\boldsymbol{v}(k)\boldsymbol{v}^H(k)\}$ is approximately a diagonal matrix, and sample averaging of $\boldsymbol{v}(k)\boldsymbol{v}^H(k)$ and $\text{diag}(|v_1(k)|^2, \ldots, |v_N(k)|^2)$ should asymptotically converge to almost identical auto-correlation matrices. In what follows we therefore replace the above approximation with equality. The application of the matrix inversion lemma to (4.181) now yields:

$$\Sigma^{-1}(k) = \frac{1}{\lambda}\Sigma^{-1}(k-1) - \frac{1}{\lambda^2}\Sigma^{-1}(k-1)\boldsymbol{v}(k)\boldsymbol{v}^H(k)\Sigma^{-1}(k-1) \times \left(1 + \frac{1}{\lambda}\boldsymbol{v}^H(k)\Sigma^{-1}(k-1)\boldsymbol{v}(k)\right)^{-1}$$
$$= \frac{1}{\lambda}\Sigma^{-1}(k-1) - \frac{\Sigma^{-1}(k-1)\boldsymbol{v}(k)\boldsymbol{v}^H(k)\Sigma^{-1}(k-1)}{\lambda^2 + \lambda\boldsymbol{v}^H(k)\Sigma^{-1}(k-1)\boldsymbol{v}(k)} \tag{4.182}$$

where explicit inverse power calculation for $v_i(k)$ is no longer required. The second term in the right-hand side of (4.182) is not diagonal. In order to calculate only the diagonal entries, we modify (4.182) to:

$$\Sigma^{-1}(k) = \frac{1}{\lambda}\Sigma^{-1}(k-1) - \frac{1}{\lambda^2 + \lambda\boldsymbol{v}^H(k)\Sigma^{-1}(k-1)\boldsymbol{v}(k)} \times \Sigma^{-1}(k-1)\begin{bmatrix} |v_0(k)|^2 & & \mathbf{0} \\ & \ddots & \\ \mathbf{0} & & |v_{N-1}(k)|^2 \end{bmatrix}\Sigma^{-1}(k-1) \tag{4.183}$$

Let $\phi_i(k) = 1/\sigma_i^2(k)$. Then (4.183) may be conveniently rewritten as:

$$\begin{bmatrix} \phi_1(k) \\ \vdots \\ \phi_N(k) \end{bmatrix} = \frac{1}{\lambda}\begin{bmatrix} \phi_1(k-1) \\ \vdots \\ \phi_N(k-1) \end{bmatrix} - \frac{1}{1 + \frac{1}{\lambda}\sum_{i=1}^{N}|v_i(k)|^2\phi_i(k-1)} \times \begin{bmatrix} |v_1(k)|^2\left(\dfrac{\phi_0(k-1)}{\lambda}\right)^2 \\ \vdots \\ |v_{N-1}(k)|^2\left(\dfrac{\phi_N(k-1)}{\lambda}\right)^2 \end{bmatrix} \tag{4.184}$$

We refer to the above recursion as the single-division power normalization algorithm. The computational complexity associated with the single-division power normalization algorithm is detailed in Table 4.5.

The single-division power normalization algorithm (4.184) generates the reciprocal of the transform output power values recursively. However, the reciprocal power values must be multiplied with the transform outputs to achieve the required power normalization. If this was done explicitly, it would increase the number of multiplications by N. In order to save N multiplications, we use the second row in Table 4.5, from which delayed power normalized transform outputs may be obtained as:

Table 4.5 Computational complexity of single-division power normalization in (4.184)

Computation	×	+	÷
$\lambda^{-1}\phi_i(k-1)$	N		
$(\lambda^{-1}\phi_i(k-1))v_i(k)v_i^*(k)$	$2N$		
$(\lambda^{-1}\phi_i(k-1)\mid v_i(k)\mid^2)(\lambda^{-1}\phi_i(k-1))$	N		
$\frac{1}{1+\lambda^{-1}\sum\cdots}\left[\;\right]$	N	N	1
$[\phi_1(k),\ldots,\phi_N(k)]^T$		N	
Total	$5N$	$2N$	1

$$\lambda^{-1}\phi_i(k-1)v_i^*(k) = \lambda^{-1}\frac{v_i^*(k)}{\sigma_i^2(k-1)}, \quad i = 1, \ldots, N \tag{4.185}$$

This expression differs from (4.178) in two aspects; viz. (4.185) has a factor λ^{-1} and uses delayed power estimates for normalization. The constant factor can be absorbed into the step-size parameter μ without any effect on power normalization. The one-sample delay for the power estimate in fact corresponds to the following transform-domain LMS recursion:

$$\boldsymbol{w}(k+1) = \boldsymbol{w}(k) + \mu\boldsymbol{\Sigma}^{-1}(k-1)\boldsymbol{v}^*(k)e(k) \tag{4.186}$$

The delayed normalization of $v_i^*(k)$ will not have noticeable impact on the performance of the transform-domain LMS algorithm since its tracking performance will hardly be affected by this delay. Thus, (4.185) may be used instead of (4.178) as the power normalized transform output.

As shown in Table 4.5, the total computational complexity of (4.184) is $5N$ multiplications and one division. Compared with the power normalization method based on (4.167) and (4.178), the single-division power normalization algorithm in (4.184) reduces the number of divisions from N to one at the expense of a 2.5-fold increase in the number of multiplications. DSP processors are highly optimized for the multiply-and-accumulate (MAC) operation used in linear filtering and convolution. For example, on Texas Instruments DSP TMS320C62x, a MAC operation takes half a cycle while a division operation can take anywhere between 16–41 cycles. Therefore, reducing the number of divisions from N to one while increasing the number of multiplications from $2N$ to $5N$ in fact does reduce the cycle count of power normalization. The reduction in complexity is particularly significant for large N (Doğançay, 2003b). The single-division power normalization algorithm in (4.184) is, therefore, likely to save a significant amount of complexity for long adaptive filters such as those typically encountered in acoustic echo cancellation applications.

4.10.2 Comparison of power normalization algorithms

In this section we show by simulation the approximate equivalence of the reciprocal power estimates produced by the traditional sliding exponential window power estimation algorithm in (4.167), and the single-division power normalization algorithm in (4.184) which is derived from the matrix inversion lemma using some simplifying approximations.

In the simulations presented, the input signal $x(k)$ is an AR(1) Gaussian signal with zero mean. The AR parameter is set to $a = -0.9$ for $k = 0, \ldots, 5000$ and $a = -0.5$ for $k = 5001, \ldots, 10000$. Thus, the second-order statistics of the input signal undergoes a sudden change at time instant $k = 5001$. The unitary transform $\boldsymbol{T}$ is taken to be the 8-point DCT (i.e. $N = 8$). A small N has been chosen in order to allow graphical illustration of the power estimates produced by the two algorithms. Both algorithms are initialized to $\sigma_i^2(-1) = \epsilon$ with $\epsilon = 0.02$. For $\lambda = 0.995$ the reciprocal power estimates $1/\sigma_i^2(k)$ and $\phi_i(k)$, $i = 1, \ldots, 8$, generated by the two algorithms are shown in Figure 4.6(a). Following an initial transient period, the two estimates settle on approximately the same power levels and behave similarly when the signal statistics change at $k = 5001$. A small offset between the two estimates is observed. This is attributed to the simplifying assumptions that we had to make in the derivation of the single-division power normalization algorithm from the matrix inversion lemma. To ascertain whether this offset is uniform across the transform outputs, we have also calculated the product $\sigma_i^2(k)\phi_i(k)$, $i = 1, \ldots, 8$, versus k, which is shown in Figure 4.6(b).

If the outputs of the two power estimation algorithms were identical, we would have:

$$\sigma_i^2(k)\phi_i(k) = 1, \quad 1 \leq i \leq N \tag{4.187}$$

If the power estimates are simply related by the equation $\phi_i(k) = \alpha/\sigma_i^2(k)$, $i = 1, \ldots, N$, where α is a constant, then we would have:

$$\sigma_i^2(k)\phi_i(k) = \alpha, \quad 0 \leq i \leq N - 1 \tag{4.188}$$

which implies that the outputs of the two power estimation algorithms only differ by a scaling factor. This constant scaling factor can be absorbed into the step-size parameter, thereby allowing us to treat the two power estimation algorithms as equivalent. Whether or not the scaling factor α is uniform for all transform outputs as prescribed by (4.188) can be checked simply by inspecting the plot of the products $\sigma_i^2(k)\phi_i(k)$, $i = 1, \ldots, N$. If these plots form identical horizontal lines when plotted against k, then we conclude that (4.188) is satisfied and therefore the two estimation algorithms are equivalent. In Figure 4.6(b) we observe that the products $\sigma_i^2(k)\phi_i(k)$ form approximately identical horizontal lines with $\alpha \approx 1.04$ as

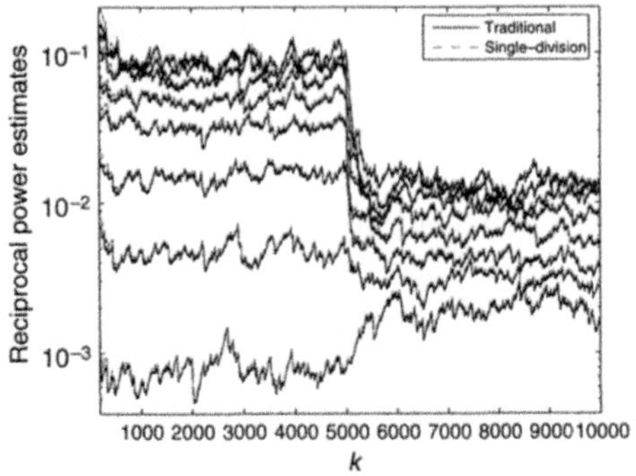

(a)

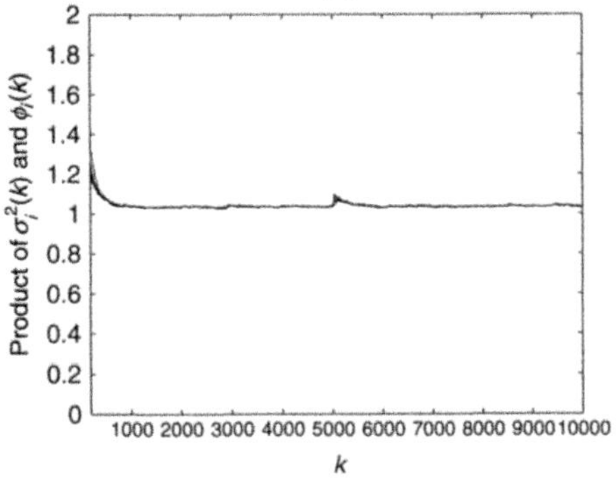

(b)

Figure 4.6 Plot of (a) $1/\sigma_i^2(k)$ and $\phi_i(k)$ versus k, and (b) $\sigma_i^2(k)\phi_i(k)$ versus k for $\lambda = 0.995$ and $N = 8$. Note that $\sigma_i^2(k)\phi_i(k)$ should ideally yield horizontal lines at the same level for all i if the two estimates are equivalent.

k increases, confirming the approximate equivalence of the two power estimates. The previous simulations were repeated for $\lambda = 0.999$. Figure 4.7 shows the reciprocal power estimates produced by the two power estimates and the product $\sigma_i^2(k)\phi_i(k)$, $i = 1, \ldots, 8$ for $\lambda = 0.999$. Note that α is now closer to one.

The two power estimation algorithms have also been compared for a longer adaptive filter with $N = 32$ and all other parameters remaining unchanged. Figure 4.8 shows a plot of $\sigma_i^2(k)\phi_i(k)$ for $\lambda = 0.995$ and $\lambda = 0.999$. In Figure 4.8(a) (4.188) roughly holds with $\alpha \approx 1.20$ for large k, again confirming the approximate equivalence of the two power estimates. As λ approaches one, the two algorithms produce closer results as can be seen from Figure 4.8(b).

While the sliding exponential window power estimation algorithm and the single-division power normalization algorithm appear to generate similar power estimates for the $v_i(k)$ upon convergence, their convergence behaviour during initial transient period and tracking of changes in signal statistics are not necessarily similar. In practical applications, the single-division power normalization algorithm may need to be run for several iterations to allow it to converge before the coefficient updates start. During statistical changes, the single-division power normalization algorithm also seems to underestimate the power levels, as evidenced by spikes in $\sigma_i^2(k)\phi_i(k)$, particularly when λ is smaller. In view of this for non-stationary input signals the equivalence of the two algorithms may break down.

4.11 PARTIAL-UPDATE TRANSFORM-DOMAIN LMS ALGORITHMS

4.11.1 Periodic-partial-update transform-domain LMS algorithm

The periodic-partial-update transform-domain LMS algorithm is defined by the recursion:

$$
\begin{aligned}
\boldsymbol{w}((k+1)S) &= \boldsymbol{w}(kS) + \mu \boldsymbol{\Sigma}^{-1}(kS)\boldsymbol{v}^*(kS)e(kS), \quad k = 0, 1, \ldots \\
\boldsymbol{\Sigma}(kS) &= \begin{bmatrix} \sigma_1^2(kS) & & & \mathbf{0} \\ & \sigma_2^2(kS) & & \\ & & \ddots & \\ \mathbf{0} & & & \sigma_N^2(kS) \end{bmatrix}, \quad \sigma_i^2(kS) = |v_i(kS)|^2 + \lambda\sigma_i^2((k-1)S)
\end{aligned}
\tag{4.189}
$$

where $\sigma_i^2(-1) = \epsilon$. The entire coefficient vector is updated every S iterations, resulting in sparse time updates. The adaptation algorithm has S iterations to compute the update vector.

4.11.2 Sequential-partial-update transform-domain LMS algorithm

The sequential-partial-update transform-domain LMS algorithm takes the following form:

$$
\begin{aligned}
\boldsymbol{w}(k+1) &= \boldsymbol{w}(k) + \mu e(k)\boldsymbol{I}_M(k)\boldsymbol{\Sigma}^{-1}(k)\boldsymbol{v}^*(k), \quad k = 0, 1, \ldots \\
\boldsymbol{\Sigma}(k) &= \begin{bmatrix} \sigma_1^2(k) & & & \mathbf{0} \\ & \sigma_2^2(k) & & \\ & & \ddots & \\ \mathbf{0} & & & \sigma_N^2(k) \end{bmatrix}, \quad \sigma_i^2(k) = |v_i(k)|^2 + \lambda\sigma_i^2(k-1)
\end{aligned}
\tag{4.190}
$$

An M-subset of the coefficient vector $\boldsymbol{w}(k)$ is selected at each iteration by $\boldsymbol{I}_M(k)$ in a deterministic round-robin manner. The coefficient selection matrix $\boldsymbol{I}_M(k)$ is given in (4.14). The application of sequential partial updates to the sliding exponential window power estimate may also be considered.

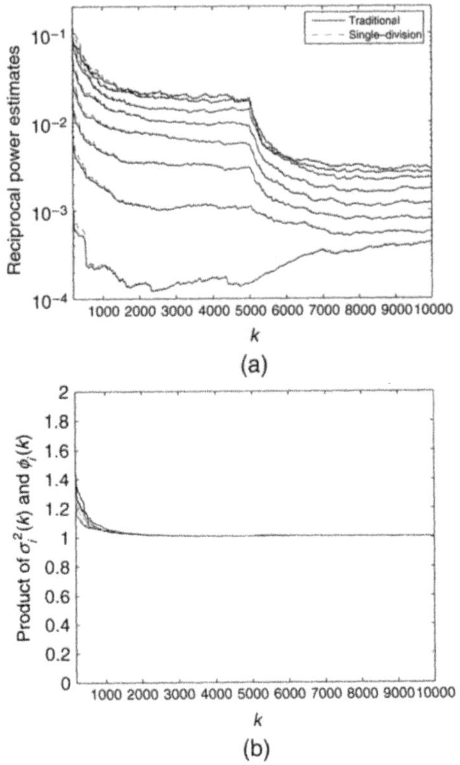

Figure 4.7 Plot of (a) $1/\sigma_i^2(k)$ and $\phi_i(k)$ versus k, and (b) $\sigma_i^2(k)\phi_i(k)$ versus k for $\lambda = 0.999$ and $N = 8$.

4.11.3 Stochastic-partial-update transform-domain LMS algorithm

The stochastic-partial-update transform-domain LMS algorithm employs a randomized coefficient selection process rather than a deterministic periodic selection. It is defined by the same recursion as the sequential-partial-update transform-domain LMS algorithm:

$$\boldsymbol{w}(k+1) = \boldsymbol{w}(k) + \mu e(k)\boldsymbol{I}_M(k)\boldsymbol{\Sigma}^{-1}(k)\boldsymbol{v}^*(k), \quad k = 0, 1, \ldots$$

$$\boldsymbol{\Sigma}(k) = \begin{bmatrix} \sigma_1^2(k) & & & \mathbf{0} \\ & \sigma_2^2(k) & & \\ & & \ddots & \\ \mathbf{0} & & & \sigma_N^2(k) \end{bmatrix}, \quad \sigma_i^2(k) = |v_i(k)|^2 + \lambda\sigma_i^2(k-1) \quad (4.191)$$

with the exception of a different $\boldsymbol{I}_M(k)$ given in (4.17).

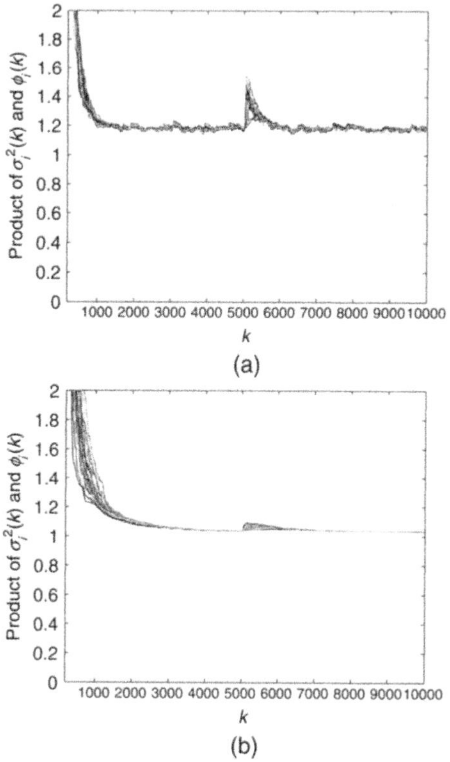

Figure 4.8 Plot of $\sigma_i^2(k)\phi_i(k)$ versus k for (a) $\lambda = 0.995$ and (b) $\lambda = 0.999$ ($N = 32$).

4.11.4 M-max transform-domain LMS algorithm

The M-max transform-domain LMS algorithm is a data-dependent partial-update transform-domain LMS algorithm defined by the recursion:

$$\boldsymbol{w}(k+1) = \boldsymbol{w}(k) + \mu e(k)\boldsymbol{I}_M(k)\boldsymbol{\Sigma}^{-1}(k)\boldsymbol{v}^*(k), \quad k = 0, 1, \ldots$$
$$\boldsymbol{\Sigma}(k) = \begin{bmatrix} \sigma_1^2(k) & & & \boldsymbol{0} \\ & \sigma_2^2(k) & & \\ & & \ddots & \\ \boldsymbol{0} & & & \sigma_N^2(k) \end{bmatrix}, \quad \sigma_i^2(k) = |v_i(k)|^2 + \lambda\sigma_i^2(k-1) \tag{4.192}$$

where the coefficient selection matrix $\boldsymbol{I}_M(k)$ solves

$$\max_{\boldsymbol{I}_M(k)} \boldsymbol{v}_n^H(k)\boldsymbol{I}_M(k)\boldsymbol{v}_n(k) \tag{4.193}$$

Here $\boldsymbol{v}_n(k) = \boldsymbol{\Sigma}^{-1/2}(k)\boldsymbol{v}(k)$ is the whitened regressor vector (cf. (4.154)) which is subsequently applied to the LMS algorithm as illustrated in Figure 4.3. In the selection of M-max coefficients we, therefore, rank the magnitude-squared entries of $\boldsymbol{\Sigma}^{-1/2}(k)\boldsymbol{v}(k)$ rather than those of $\boldsymbol{\Sigma}^{-1}(k)\boldsymbol{v}(k)$. The use of the latter will result in the selection of wrong coefficients.

Taking the square magnitude of the entries of $\boldsymbol{\Sigma}^{-1/2}(k)\boldsymbol{v}(k)$ we have:

$$\left[\frac{|v_1(k)|^2}{\sigma_1^2(k)}, \frac{|v_2(k)|^2}{\sigma_2^2(k)}, \ldots, \frac{|v_N(k)|^2}{\sigma_N^2(k)}\right]^T \tag{4.194}$$

The computation of the above vector requires an additional N multiplications and is therefore not attractive. However, referring to the second row of Table 4.5 we realize that the single-division power normalization algorithm computes a version of (4.194) with delayed power normalization:

$$\left[\frac{|v_1(k)|^2}{\sigma_1^2(k-1)}, \frac{|v_2(k)|^2}{\sigma_2^2(k-1)}, \ldots, \frac{|v_N(k)|^2}{\sigma_N^2(k-1)}\right]^T \tag{4.195}$$

In practical applications power normalization with one-sample delay will not have any adverse impact on the convergence performance. Moreover, (4.195) does not require additional complexity overheads since it is already computed by the single-division power normalization algorithm. Thus, based on (4.195) the coefficient selection matrix $\boldsymbol{I}_M(k)$ for the M-max transform-domain LMS algorithm can be defined as:

$$\boldsymbol{I}_M(k) = \begin{bmatrix} i_1(k) & 0 & \cdots & 0 \\ 0 & i_2(k) & \ddots & \vdots \\ \vdots & \ddots & \ddots & 0 \\ 0 & \cdots & 0 & i_N(k) \end{bmatrix},$$

$$i_j(k) = \begin{cases} 1 & \text{if } |v_j(k)|^2/\sigma_j^2(k-1) \in \max_{1 \le l \le N}(|v_l(k)|^2/\sigma_l^2(k-1), M) \\ 0 & \text{otherwise} \end{cases} \tag{4.196}$$

4.11.5 Computational complexity

The computational complexity of a transform-domain adaptive filter comprises complexities associated with the unitary transform, power normalization, adaptation of coefficients and data-dependent partial updates, if any. The unitary transforms can be implemented as IIR filter banks (Bondyopadhyay, 1988; Narayan *et al.*, 1983) or as sliding window transforms (Farhang-Boroujeny, 1995) with computational complexity of $O(N)$. The complexity analysis presented here assumes the use of the N-point

DCT to construct a unitary transform $\boldsymbol{T}$. The transform is implemented as a bank of IIR filters with transfer functions given in (4.174d). The computational complexity of transform outputs $v_i(k)$ is seen to be $5N$ multiplications and $5N$ additions per iteration. For large N the IIR filter bank implementation is computationally more efficient than FFT-based implementations.

For power normalization we have two choices, viz. exponential sliding window power normalization in (4.167) and (4.178) with complexity $2N$ multiplications, N additions, and N divisions, and single-division power normalization in (4.184) with complexity $5N$ multiplications, $2N$ additions and one division. Recall that single-division power normalization implies one-sample delay power normalization.

The full-update transform-domain LMS algorithm performs the following operations at each iteration:

- N multiplications and N additions to compute $e(k)$.
- One multiplication to compute $\mu e(k)$.
- $5N$ multiplications and $5N$ additions to compute $\boldsymbol{v}(k)$.
- Power normalization:
  - $2N$ multiplications, N additions, and N divisions for exponential sliding window power normalization.
  - $5N$ multiplications, $2N$ additions and one division for single-division power normalization.
- N multiplications to compute $\mu\boldsymbol{\Sigma}^{-1}(k)\boldsymbol{v}^*(k)e(k)$.
- N additions to update $\boldsymbol{w}(k)$.

The total computational complexity of the transform-domain LMS algorithm per iteration is $9N+1$ multiplications, $8N$ additions and N divisions if exponential sliding window power normalization is used, and $12N+1$ multiplications, $9N$ additions and one division if single-division power normalization is used.

The periodic-partial-update transform-domain LMS algorithm decimates update computations by S while the filtering process operates at the input signal rate. Thus, the average computational complexity of the periodic-partial-update transform-domain LMS algorithm per iteration is $(8N+1)/S+N$ multiplications, $7N/S+N$ additions and N/S divisions if exponential sliding window power normalization is used, and $(11N+1)/S+N$ multiplications, $8N/S+N$ additions and $1/S$ divisions if single-division power normalization is used.

The sequential-partial-update transform-domain LMS algorithm replaces the last two steps of the full-update transform-domain LMS algorithm with:

- M multiplications to compute $\mu e(k)\boldsymbol{I}_M(k)\boldsymbol{\Sigma}^{-1}(k)\boldsymbol{v}^*(k)$.
- M additions to update $\boldsymbol{w}(k)$.

This provides a saving of $N - M$ multiplications and $N - M$ additions. As a result the total computational complexity of the sequential-partial-update transform-domain LMS algorithm per iteration is $8N + M + 1$ multiplications, $7N + M$ additions and N divisions if exponential sliding window power normalization is used, and $11N + M + 1$ multiplications, $8N + M$ additions and one division if single-division power normalization is used.

Assuming that the linear congruential generator in (4.22) is used to randomize coefficient selection, the computational complexity of the stochastic-partial-update transform-domain LMS algorithm is 2 multiplications and 2 additions more than that of the sequential-partial-update transform-domain LMS algorithm.

The computational complexity of the M-max transform-domain LMS algorithm is given by the computational complexity of the sequential-partial-update transform-domain LMS algorithm plus the complexity of M-max selection. In the case of transform-domain LMS algorithm the M-max coefficient selection requires ranking of $|v_i(k)|^2/\sigma_i^2(k)$. The computation of these numbers requires N multiplications if exponential sliding window power normalization is used, and no complexity overheads if single-division power normalization is used, provided that delayed power normalization in (4.195) is acceptable. The numbers to be ranked $|v_i(k)|^2/\sigma_i^2(k)$, $i = 1, \ldots, N$, do not possess a shift structure. In other words, at every iteration we get a new set of $|v_i(k)|^2/\sigma_i^2(k)$ with no common numbers to past iterations. Therefore, we apply the heapsort algorithm (see Appendix A) to find M maxima of $|v_i(k)|^2/\sigma_i^2(k)$. The minimum complexity of finding M maxima is given by:

$$\min(2N + 2\min(M, N - M)\log_2 N, 2M + 2(N - M)\log_2 M).$$

The total computational complexity of M-max transform-domain LMS algorithm is $9N + M + 1$ multiplications, $7N + M$ additions, N divisions and $\min(2N + 2\min(M, N - M)\log_2 N, 2M + 2(N - M)\log_2 M)$ comparisons if exponential sliding window power normalization is used, and $11N + M + 1$ multiplications, $8N + M$ additions, one division and $\min(2N + 2\min(M, N - M)\log_2 N, 2M + 2(N - M)\log_2 M)$ comparisons if single-division power normalization is used.

Tables 4.6 and 4.7 summarize the computational complexity of the transform-domain LMS algorithm and its partial-update implementations using exponential sliding window power normalization and single-division power normalization, respectively. The complexity is dominated by transform computation and power normalization (the number of divisions). The division operation is computationally very expensive on DSP systems. Single-division power normalization trades divisions for multiplications, which is a good strategy for most DSP implementations and can lead to a significant computational complexity reduction. We observe from Table 4.6 that the M-max transform-domain LMS algorithm with exponential sliding window power normalization is not a computationally attractive choice because its computational

Table 4.6 Computational complexity of transform-domain LMS and its partial-update implementations at each k using exponential sliding window power normalization

Algorithm	×	+	÷	⋚
TD-LMS	$9N+1$	$8N$	N	
Per-par-upd TD-LMS	$N+(8N+1)/S$	$N+7N/S$	N/S	
Seq-par-upd TD-LMS	$8N+M+1$	$7N+M$	N	
Stoch-par-upd TD-LMS	$8N+M+3$	$7N+M+2$	N	
M-max TD-LMS	$9N+M+1$	$7N+M$	N	$\min(2N+2\min(M,N-M)\log_2 N, 2M+2(N-M)\log_2 M)$

Table 4.7 Computational complexity of transform-domain LMS and its partial-update implementations at each k using single-division power normalization

Algorithm	×	+	÷	⋚
TD-LMS	$12N+1$	$9N$	1	
Per-par-upd TD-LMS	$N+(11N+1)/S$	$N+8N/S$	$1/S$	
Seq-par-upd TD-LMS	$11N+M+1$	$8N+M$	1	
Stoch-par-upd TD-LMS	$11N+M+3$	$8N+M+2$	1	
M-max TD-LMS	$11N+M+1$	$8N+M$	1	$\min(2N+2\min(M,N-M)\log_2 N, 2M+2(N-M)\log_2 M)$

complexity exceeds that of the full-update transform-domain LMS algorithm. The periodic-partial-update transform-domain LMS algorithm has by far the smallest complexity requirements. Given that periodic and sequential partial updates have similar convergence properties, the periodic-partial-update transform-domain LMS algorithm comes out as the winner among data-independent partial-update algorithms.

4.12 GENERALIZED-SUBBAND-DECOMPOSITION LEAST-MEAN-SQUARE ALGORITHM

The implementation of the generalized-subband-decomposition least-mean-square (GSD-LMS) adaptive filter is shown in Figure 4.9 (Petraglia and Mitra, 1993).

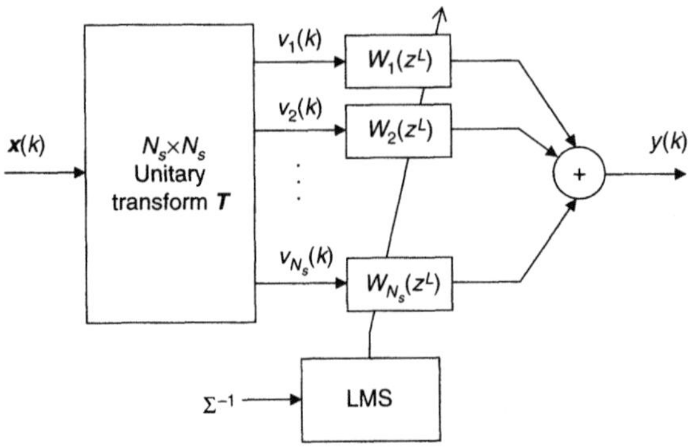

Figure 4.9 Implementation of generalized-subband-decomposition LMS algorithm.

Like the transform-domain LMS algorithm, the GSD-LMS algorithm uses a unitary transform matrix $\boldsymbol{T}$, which acts as an analysis filterbank. However, as different from the transform-domain LMS algorithm, the size of the unitary transform N_s is smaller than N. The $N_s \times 1$ transform output vector is given by:

$$\boldsymbol{v}(k) = \boldsymbol{T}\boldsymbol{x}(k) \tag{4.197}$$

where $\boldsymbol{x}(k) = [x(k), x(k-1), \ldots, x(k-N_s+1)]^T$ is the $N_s \times 1$ input regressor vector. The transform outputs $v_i(k)$ are applied to *sparse* adaptive FIR subfilters with transfer function $W_i(z^L)$, $i = 1, \ldots, N_s$, where $L \leq N_s$ (see Figure 4.9). Each subfilter has K non-zero coefficients $w_{i,1}(k), w_{i,2}(k), \ldots, w_{i,K}(k)$ spaced by the sparsity factor L. At time k, the transfer functions of the adaptive subfilters are given by:

$$W_i(z^L) = \sum_{j=1}^{K} w_{i,j}(k) z^{-(j-1)L}, \quad i = 1, 2, \ldots, N_s.$$

For $L \leq N_s$, the relationship between the equivalent time-domain adaptive filter length N and the GSD-LMS parameters is given by:

$$N = (K-1)L + N_s. \tag{4.198}$$

The total number of adaptive filter coefficients is KN_s. If $L < N_s$, the equivalent length N of the GSD-LMS adaptive filter is smaller than the number of adaptive coefficients KN_s.

The output of the GSD adaptive filter is:

$$y(k) = \sum_{j=1}^{K} \boldsymbol{v}^T(k-(j-1)L)\boldsymbol{w}_j(k) \tag{4.199}$$

where $\boldsymbol{w}_j(k)$ is the $N_s \times 1$ partition of adaptive filter coefficients at time delay $(j-1)L$, $j = 1, \ldots, K$:

$$\boldsymbol{w}_j(k) = \begin{bmatrix} w_{1,j}(k) \\ w_{2,j}(k) \\ \vdots \\ w_{N_s,j}(k) \end{bmatrix} \quad (4.200)$$

By augmenting $\boldsymbol{v}^T(k-(j-1)L)$ and $\boldsymbol{w}_j(k)$, (4.199) can be more compactly written as:

$$y(k) = \boldsymbol{v}_a^T(k)\boldsymbol{w}_a(k) \quad (4.201)$$

where:

$$\boldsymbol{v}_a(k) = \begin{bmatrix} \boldsymbol{v}(k) \\ \boldsymbol{v}(k-L) \\ \vdots \\ \boldsymbol{v}(k-(K-1)L) \end{bmatrix}_{KN_s \times 1}, \quad \boldsymbol{w}_a(k) = \begin{bmatrix} \boldsymbol{w}_1(k) \\ \boldsymbol{w}_2(k) \\ \vdots \\ \boldsymbol{w}_K(k) \end{bmatrix}_{KN_s \times 1} \quad (4.202)$$

Figure 4.10 shows the equivalent GSD adaptive filter structure that implements (4.201). In the next section we will use this equivalent GSD structure to gain more insight into the role played by the sparse adaptive subfilters.

The GSD-LMS algorithm is defined by (Petraglia and Mitra, 1993):

$$\boldsymbol{w}_a(k+1) = \boldsymbol{w}_a(k) + \mu \boldsymbol{\Sigma}_a^{-1}\boldsymbol{v}_a^*(k)e(k), \quad k = 0, 1, \ldots \quad (4.203)$$

where $e(k) = d(k) - y(k)$ is the error signal and $\boldsymbol{\Sigma}_a$ is the $KN_s \times KN_s$ augmented diagonal power matrix of transform outputs, which for stationary $\boldsymbol{v}(k)$ takes the form:

$$\boldsymbol{\Sigma}_a = \begin{bmatrix} \boldsymbol{\Sigma} & & \boldsymbol{0} \\ & \ddots & \\ \boldsymbol{0} & & \boldsymbol{\Sigma} \end{bmatrix}_{KN_s \times KN_s} \quad (4.204)$$

Here the $N_s \times N_s$ power matrix $\boldsymbol{\Sigma}$ is defined by $\boldsymbol{\Sigma} = \text{diag}(\sigma_1^2, \sigma_2^2, \ldots, \sigma_{N_s}^2)$ where $\sigma_i^2 = E\{|v_i(k)|^2\}$. When the input signal $x(k)$ is non-stationary or its second-order statistics are unknown, we may use a sliding exponential window estimate to compute the diagonal entries of $\boldsymbol{\Sigma}$:

$$\sigma_i^2(k) = \lambda\sigma_i^2(k-1) + |v_i(k)|^2, \quad i = 1, \ldots, N_s \quad (4.205)$$

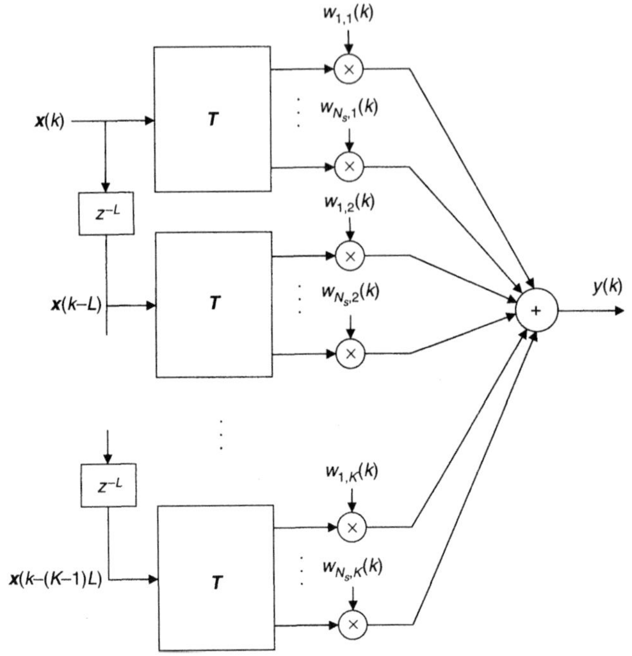

Figure 4.10 Equivalent GSD adaptive filter.

where $0 < \lambda < 1$ is the exponential forgetting factor. Replacing $\boldsymbol{\Sigma}_a$ with the online estimate in (4.205) we have:

$$\boldsymbol{\Sigma}_a(k) = \begin{bmatrix} \boldsymbol{\Sigma}(k) & & & \mathbf{0} \\ & \boldsymbol{\Sigma}(k-L) & & \\ & & \ddots & \\ \mathbf{0} & & & \boldsymbol{\Sigma}(k-(K-1)L) \end{bmatrix}_{KN_s \times KN_s} \tag{4.206}$$

where:

$$\boldsymbol{\Sigma}(k) = \begin{bmatrix} \sigma_1^2(k) & & & \mathbf{0} \\ & \sigma_2^2(k) & & \\ & & \ddots & \\ \mathbf{0} & & & \sigma_{N_s}^2(k) \end{bmatrix} \tag{4.207}$$

The diagonal matrix partitions of $\boldsymbol{\Sigma}_a(k)$ are perfectly time-aligned with the partitions of $\boldsymbol{v}_a(k)$ [see (4.202)]. An important complexity implication of this is that power normalization needs to be applied only to the current transform output $\boldsymbol{v}(k)$ at iteration k since $\boldsymbol{v}(k-L), \ldots, \boldsymbol{v}(k-(K-1)L)$ have already been power normalized at previous iterations.

The transform-domain LMS algorithm may be considered to be a special case of the GSD-LMS algorithm since setting $N_s = N$ and $K = 1$ reduces the GSD-LMS to the transform-domain LMS algorithm. As we saw in Section 4.11.5, the computational complexity of the transform-domain LMS algorithm is dominated by transform computation and power normalization. The GSD-LMS algorithm alleviates the high complexity of the transform-domain LMS algorithm by using a smaller transform of size N_s. The GSD-LMS algorithm not only has a lower computational complexity than the transform-domain LMS algorithm, but it can also achieve a convergence performance very close to that of the transform-domain LMS algorithm (Petraglia and Mitra, 1993).

4.12.1 Relationship between GSD-LMS coefficients and equivalent time-domain response

Using (4.197) in (4.201), the output of the GSD adaptive filter can be expressed as

$$y(k) = \boldsymbol{x}_a^T(k)\,\boldsymbol{T}_a^T\,\boldsymbol{w}_a(k) \tag{4.208}$$

where:

$$\boldsymbol{x}_a(k) = \begin{bmatrix} \boldsymbol{x}(k) \\ \boldsymbol{x}(k-L) \\ \vdots \\ \boldsymbol{x}(k-(K-1)L) \end{bmatrix}_{KN_s \times 1}, \quad \boldsymbol{T}_a = \begin{bmatrix} \boldsymbol{T} & & \boldsymbol{0} \\ & \boldsymbol{T} & \\ & & \ddots & \\ \boldsymbol{0} & & & \boldsymbol{T} \end{bmatrix}_{KN_s \times KN_s}.$$

The corresponding time-domain equation is

$$y(k) = \underbrace{[x(k), x(k-1), \ldots, x(k-N+1)]}_{\boldsymbol{x}_{\text{eq}}^T(k)}\,\boldsymbol{w}_{\text{eq}}(k) \tag{4.209}$$

where $\boldsymbol{w}_{\text{eq}}(k)$ is the $N \times 1$ coefficient vector of the equivalent time-domain adaptive filter.

Suppose that $\boldsymbol{w}_{\text{eq}}(k)$ is given and we wish to find the corresponding GSD coefficients $\boldsymbol{w}_a(k)$. Firstly we obtain $\boldsymbol{x}_a(k)$ from $\boldsymbol{x}_{\text{eq}}(k)$. If $L = N_s$ (i.e. the sparsity factor is identical to the transform size) then from (4.198) we have $K = N/N_s$ and:

$$\boldsymbol{x}_a(k) = \boldsymbol{x}_{\text{eq}}(k) \tag{4.210}$$

Thus, the GSD coefficient vector $\boldsymbol{w}_a(k)$ corresponding to $\boldsymbol{w}_{\text{eq}}(k)$ can be obtained from:

$$\boldsymbol{T}_a^T\,\boldsymbol{w}_a(k) = \boldsymbol{w}_{\text{eq}}(k) \tag{4.211}$$

This equation has the following unique solution:

$$\boldsymbol{w}_a(k) = \boldsymbol{T}_a^* \boldsymbol{w}_{\text{eq}}(k) \tag{4.212}$$

Thus, for $L = N_s$ the GSD coefficient vector $\boldsymbol{w}_a(k)$ that represents the equivalent time-domain filter with coefficients $\boldsymbol{w}_{\text{eq}}(k)$ is uniquely determined from (4.212). With reference to Figure 4.10, we observe that the case of $L = N_s$ may be considered a special case of the transform-domain adaptive filter with the $N \times N$ transform replaced by K smaller $N_s \times N_s$ transforms applied to adjacent blocks of the input regressor.

For $L < N_s$, $\boldsymbol{w}_a(k)$ is no longer uniquely related to $\boldsymbol{w}_{\text{eq}}(k)$. This is best demonstrated by an example. Consider a GSD adaptive filter with $N_s = 3$, $L = 2$, and $K = 3$ which results in $N = 7$. Using (4.209) we have:

$$y(k) = \sum_{i=1}^{7} x(k-i+1) w_{\text{eq},i}(k) \tag{4.213}$$

where:

$$\boldsymbol{w}_{\text{eq}}(k) = \begin{bmatrix} w_{\text{eq},1}(k) \\ w_{\text{eq},2}(k) \\ \vdots \\ w_{\text{eq},N}(k) \end{bmatrix}$$

From (4.208) we obtain:

$$\begin{aligned} y(k) &= \underbrace{x(k)h_1(k) + x(k-1)h_2(k) + x(k-2)h_3(k)}_{\boldsymbol{x}^T(k)\boldsymbol{h}_1(k)} \\ &\quad + \underbrace{x(k-2)h_4(k) + x(k-3)h_5(k) + x(k-4)h_6(k)}_{\boldsymbol{x}^T(k-L)\boldsymbol{h}_2(k)} \\ &\quad + \underbrace{x(k-4)h_7(k) + x(k-5)h_8(k) + x(k-6)h_9(k)}_{\boldsymbol{x}^T(k-2L)\boldsymbol{h}_3(k)} \\ &= x(k)h_1(k) + x(k-1)h_2(k) + (h_3(k) + h_4(k))x(k-2) \\ &\quad + x(k-3)h_5(k) + (h_6(k) + h_7(k))x(k-4) + x(k-5)h_8(k) \\ &\quad + x(k-6)h_9(k) \end{aligned} \tag{4.214}$$

where:

$$\boldsymbol{h}(k) = \begin{bmatrix} h_1(k) \\ h_2(k) \\ \vdots \\ h_{KN_s}(k) \end{bmatrix} = \begin{bmatrix} \boldsymbol{h}_1(k) \\ \boldsymbol{h}_2(k) \\ \vdots \\ \boldsymbol{h}_K(k) \end{bmatrix} = \boldsymbol{T}_a^T \boldsymbol{w}_a(k) \tag{4.215}$$

The corresponding GSD coefficient vector is:

$$\boldsymbol{w}_a(k) = \boldsymbol{T}_a^* \boldsymbol{h}(k) \tag{4.216}$$

Comparing (4.213) and (4.214) we see that:

$$\boldsymbol{w}_{\text{eq}}(k) = \begin{bmatrix} h_1(k) \\ h_2(k) \\ h_3(k) + h_4(k) \\ h_5(k) \\ h_6(k) + h_7(k) \\ h_8(k) \\ h_9(k) \end{bmatrix} \tag{4.217}$$

where we are free to choose $h_3(k)$, $h_4(k)$, $h_6(k)$ and $h_7(k)$ arbitrarily as long as:

$$\begin{aligned} h_3(k) + h_4(k) &= w_{\text{eq},3}(k) \\ h_6(k) + h_7(k) &= w_{\text{eq},5}(k) \end{aligned} \tag{4.218}$$

Thus, it follows from (4.216) that the GSD coefficients $\boldsymbol{w}_a(k)$ are not uniquely related to the equivalent time-domain coefficients $\boldsymbol{w}_{\text{eq}}(k)$. The non-uniqueness of $\boldsymbol{w}_a(k)$ arises from the overlap of the entries of $\boldsymbol{x}(k), \boldsymbol{x}(k-L), \ldots, \boldsymbol{x}(k-(K-1)L)$ whenever $L < N_s$. In Figure 4.10 we also observe that the condition $L < N_s$ causes the $N_s \times N_s$ transforms to be applied to overlapping regressor vectors.

4.12.2 Eigenvalue spread of GSD input correlation matrix

As with all adaptive filters, the convergence rate of the GSD-LMS algorithm is sensitive to the eigenvalue spread (condition number) of the autocorrelation matrix of the power normalized transform output. Suppose that the input signal $x(k)$ is stationary. Then referring to Figure 4.10 we see that the power normalized transform output vector is $\boldsymbol{\Sigma}_a^{-1/2} \boldsymbol{T}_a \boldsymbol{x}_a(k)$ which has the autocorrelation matrix:

$$\begin{aligned} &E\{\boldsymbol{\Sigma}_a^{-1/2} \boldsymbol{T}_a^* \boldsymbol{x}_a(k) \boldsymbol{x}_a^T(k) \boldsymbol{T}_a^T \boldsymbol{\Sigma}_a^{-1/2}\} \\ &= \boldsymbol{\Sigma}_a^{-1/2} \boldsymbol{T}_a^* E\{\boldsymbol{x}_a(k) \boldsymbol{x}_a^T(k)\} \boldsymbol{T}_a^T \boldsymbol{\Sigma}_a^{-1/2} \end{aligned} \tag{4.219}$$

If $L = N_s$, we have $\boldsymbol{x}_a(k) = \boldsymbol{x}_{\text{eq}}(k)$ and $E\{\boldsymbol{x}_a(k)\boldsymbol{x}_a^T(k)\} = \boldsymbol{R}$ where:

$$\boldsymbol{R} = E\left\{\begin{bmatrix} x(k) \\ x(k-1) \\ \vdots \\ x(k-N+1) \end{bmatrix} \begin{bmatrix} x(k) & x(k-1) & \cdots & x(k-N+1) \end{bmatrix}\right\} \tag{4.220}$$

Thus, the autocorrelation matrix of the power normalized transform output is:

$$E\{\boldsymbol{\Sigma}_a^{-1/2}\boldsymbol{T}_a^*\boldsymbol{x}_a(k)\boldsymbol{x}_a^T(k)\boldsymbol{T}_a^T\boldsymbol{\Sigma}_a^{-1/2}\} = \boldsymbol{\Sigma}_a^{-1/2}\boldsymbol{T}_a^*\boldsymbol{R}\boldsymbol{T}_a^T\boldsymbol{\Sigma}_a^{-1/2} \tag{4.221}$$

The $N_s \times N_s$ transform $\boldsymbol{T}$ can only decorrelate $E\{\boldsymbol{x}(k)\boldsymbol{x}^T(k)\}$ which is an $N_s \times N_s$ submatrix of $\boldsymbol{R}$. Therefore, in general $\boldsymbol{T}_a^*\boldsymbol{R}\boldsymbol{T}_a^T$ is not a diagonal matrix. However, for weakly correlated input signals it may be approximated by a diagonal matrix:

$$\begin{aligned} E\{\boldsymbol{\Sigma}_a^{-1/2}\boldsymbol{T}_a^*\boldsymbol{x}_a(k)\boldsymbol{x}_a^T(k)\boldsymbol{T}_a^T\boldsymbol{\Sigma}_a^{-1/2}\} &\approx \boldsymbol{\Sigma}_a^{-1/2}\boldsymbol{\Sigma}_a\boldsymbol{\Sigma}_a^{-1/2} \\ &\approx \boldsymbol{I} \end{aligned} \tag{4.222}$$

The power normalized transformed output has approximately unity eigenvalue spread only for weakly correlated input signals. If $E\{x(k)x(k-\tau)\} \gg 0$ for $\tau \neq 0$, it is not possible to achieve the desired diagonalization by means of $\boldsymbol{T}_a$. Therefore, the GSD-LMS algorithm does not perform as well as the transform-domain LMS algorithm for strongly correlated input signals.

If $L < N_s$, the autocorrelation of the augmented regressor vector can be written as:

$$\begin{aligned} E\{\boldsymbol{x}_a(k)\boldsymbol{x}_a^T(k)\} &= E\{\boldsymbol{A}\boldsymbol{x}_{\text{eq}}(k)\boldsymbol{x}_{\text{eq}}^T(k)\boldsymbol{A}^T\} \\ &= \boldsymbol{A}\boldsymbol{R}\boldsymbol{A}^T \end{aligned} \tag{4.223}$$

where:

$$\boldsymbol{A} = \left[\begin{array}{ccccccccccc} 1 & & & & & & & & & & \\ & 1 & & & & & & & & & \\ & & \ddots & & & & & & & & \\ & & & 1 & & & & & & & \\ \hdashline & 1 & & & & & & & & & \\ & & & 1 & & & & & & & \\ & & & & \ddots & & & & & & \\ & & & & & 1 & & & & & \\ \hdashline & & & & \vdots & & & & & & \\ \hdashline & & & & & & 1 & & & & \\ & & & & & & & 1 & & & \\ & & & & & & & & \ddots & & \\ & & & & & & & & & & 1 \end{array}\right]_{KN_s \times N} . \tag{4.224}$$

The matrix $\boldsymbol{A}$ generates $\boldsymbol{x}_a(k)$ from $\boldsymbol{x}(k)$, and for $L < N_s$ it is a rectangular matrix with identity submatrices of size $N_s \times N_s$. These submatrices are not diagonally aligned unless $L = N_s$. We observe that $E\{\boldsymbol{x}_a(k)\boldsymbol{x}_a^T(k)\}$ is rank-deficient by $KN_s - N$ since only N rows of $E\{\boldsymbol{x}_a(k)\boldsymbol{x}_a^T(k)\}$ are linearly independent, while the remaining rows are identical copies of the $KN_s - N$ independent rows. The rank deficiency of $E\{\boldsymbol{x}_a(k)\boldsymbol{x}_a^T(k)\}$ implies that the autocorrelation matrix of the power normalized transform output vector will also be rank deficient with $KN_s - N$ zero eigenvalues and infinite eigenvalue spread. This is not surprising since we have already established that for $L < N_s$ the GSD adaptive filter has infinitely many solutions for a given equivalent time-domain filter, suggesting a lack of persistent excitation or zero eigenvalues for the GSD adaptive filter input (Doğançay, 2003a).

The lack of persistent excitation is known to cause the coefficient drift phenomenon in practical implementations of adaptive filters (Lee and Messerschmitt, 1994). The reason for coefficient drifts is the existence of a continuum of stable points on the adaptive filter cost function surface. This is caused by the zero eigenvalues that accommodates all possible solutions of the adaptive filter with the same MSE performance. Therefore, the filter coefficients are free to move on this flat hyperdimensional cost function surface without affecting the MSE for the filter output. The movement of the coefficients can be triggered by small perturbations due to, for example, quantization errors in finite-precision implementations.

When investigating the convergence rate of an adaptive filter, the eigenvalue spread of the input autocorrelation matrix is determined using the minimum *non-zero* eigenvalue (i.e. zero eigenvalues are disregarded). Therefore, having a zero eigenvalue or multiple solutions does not necessarily degrade the convergence performance as long as the largest and smallest non-zero eigenvalues are comparable. Intuitively this can be explained as follows. Since a zero eigenvalue does not contribute to the convergence of the adaptive coefficients, the convergence rate is dominated by the smallest non-zero eigenvalue. If the largest and smallest eigenvalues are close to each other, then all coefficients converge at the same rate, attaining the maximum convergence rate.

We demonstrate the non-uniqueness of the GSD-LMS solution by way of a simulation example. In the simulation example we apply the GSD-LMS algorithm to adaptive identification of an unknown system with 64 coefficients. The system input is a zero-mean white Gaussian signal with unity variance. The additive noise at the system output is zero-mean Gaussian with variance $\sigma_n^2 = 0.001$. The GSD-LMS parameters are set to $N_s = 8$, $L = 7$, $K = 9$ with $N = 64$ (note that $L < N_s$ and the number or GSD-LMS adaptive coefficients is $KN_s = 72$). The 8-point DCT is used as the fixed transform. The GSD-LMS algorithm is initialized to two different settings, viz. all zeros and random values drawn from a zero-mean Gaussian random process. The time-averaged learning curves and the GSD coefficients to which the

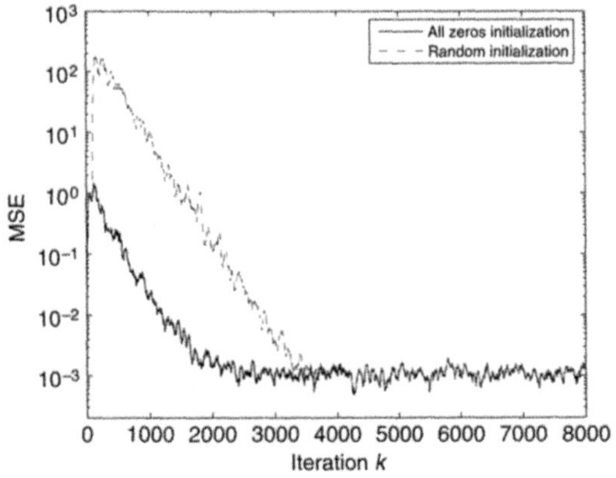

Figure 4.11 Time-averaged learning curves for GSD-LMS algorithm for all zeros and random initialization ($L < N_s$).

GSD-LMS algorithm converges are shown in Figures 4.11 and 4.12, respectively. Both converged solutions attain a steady-state MSE of −29.6 dB. Note the significant difference between the two solutions, verifying the existence of multiple GSD-LMS solutions.

Setting $L = N_s = 8$ renders the GSD-LMS solution unique. Figure 4.13 shows the solutions to which the GSD-LMS algorithm converges after being initialized to an all zeros vector and a random setting. The two solutions are identical, confirming the uniqueness of the GSD-LMS solution.

4.13 PARTIAL-UPDATE GSD-LMS ALGORITHMS

4.13.1 Periodic-partial-update GSD-LMS algorithm

The periodic-partial-update GSD-LMS algorithm is defined by:

$$\begin{aligned}
\boldsymbol{w}_a((k+1)S) &= \boldsymbol{w}_a(kS) + \mu \boldsymbol{\Sigma}_a^{-1}(kS)\boldsymbol{v}_a^*(kS)e(kS), \quad k = 0, 1, \ldots \\
\boldsymbol{\Sigma}_a(kS) &= \begin{bmatrix} \boldsymbol{\Sigma}(kS) & & & \mathbf{0} \\ & \boldsymbol{\Sigma}(kS-L) & & \\ & & \ddots & \\ \mathbf{0} & & & \boldsymbol{\Sigma}(kS-(K-1)L) \end{bmatrix} \\
\sigma_i^2(kS-jL) &= \lambda\sigma_i^2((k-1)S-jL) + |v_i(kS-jL)|^2, \\
& \qquad i = 1, \ldots, N_s, \quad j = 0, \ldots, K-1
\end{aligned} \tag{4.225}$$

The entire coefficient vector $\boldsymbol{w}_a(k)$ has sparse time updates, giving the adaptation process S iterations to compute each update.

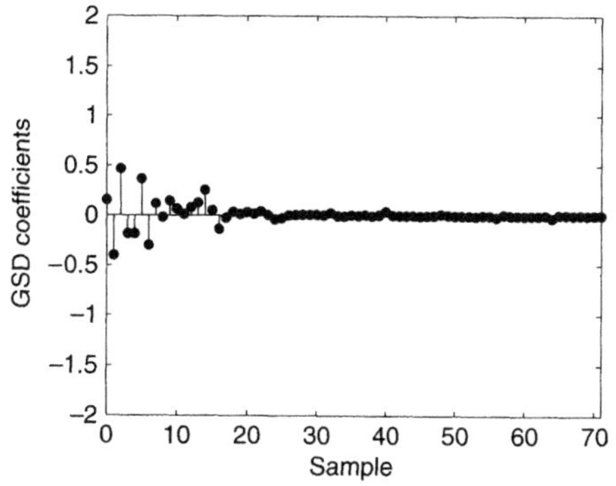

(a)

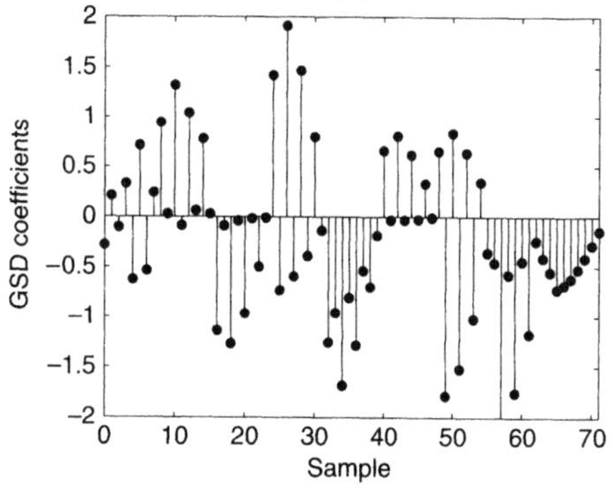

(b)

Figure 4.12 Converged GSD-LMS solutions from: (a) all zeros initialization; and (b) random initialization ($L < N_s$). The GSD-LMS algorithm does not have a unique solution.

4.13.2 Sequential-partial-update GSD-LMS algorithm

The sequential-partial-update GSD-LMS algorithm is defined by:

$$\begin{aligned}
\boldsymbol{w}_a(k+1) &= \boldsymbol{w}_a(k) + \mu e(k)\boldsymbol{I}_M(k)\boldsymbol{\Sigma}_a^{-1}(k)\boldsymbol{v}_a^*(k), \quad k = 0, 1, \ldots \\
\boldsymbol{\Sigma}_a(k) &= \begin{bmatrix} \boldsymbol{\Sigma}(k) & & & \boldsymbol{0} \\ & \boldsymbol{\Sigma}(k-L) & & \\ & & \ddots & \\ \boldsymbol{0} & & & \boldsymbol{\Sigma}(k-(K-1)L) \end{bmatrix} \\
\sigma_i^2(k) &= \lambda\sigma_i^2(k-1) + |v_i(k)|^2, \quad i = 1, \ldots, N_s
\end{aligned} \tag{4.226}$$

The coefficient selection matrix $\boldsymbol{I}_M(k)$ for the sequential-partial-update GSD-LMS algorithm is given in (4.14). A subset of M coefficients out of KN_s coefficients are selected for update at each iteration in a periodic manner.

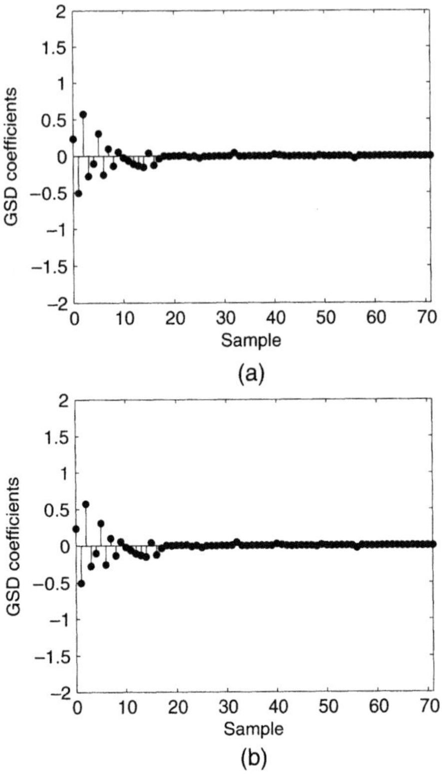

Figure 4.13 Converged GSD-LMS solutions from: (a) all zeros initialization; and (b) random initialization ($L = N_s$). The GSD-LMS now has a unique solution.

4.13.3 Stochastic-partial-update GSD-LMS algorithm

The stochastic-partial-update GSD-LMS algorithm uses a randomized coefficient selection process rather than a deterministic periodic selection:

$$\begin{aligned}
\boldsymbol{w}_a(k+1) &= \boldsymbol{w}_a(k) + \mu e(k)\boldsymbol{I}_M(k)\boldsymbol{\Sigma}_a^{-1}(k)\boldsymbol{v}_a^*(k), \quad k = 0, 1, \ldots \\
\boldsymbol{\Sigma}_a(k) &= \begin{bmatrix} \boldsymbol{\Sigma}(k) & & & \boldsymbol{0} \\ & \boldsymbol{\Sigma}(k-L) & & \\ & & \ddots & \\ \boldsymbol{0} & & & \boldsymbol{\Sigma}(k-(K-1)L) \end{bmatrix} \\
\sigma_i^2(k) &= \lambda\sigma_i^2(k-1) + |v_i(k)|^2, \quad i = 1, \ldots, N_s
\end{aligned} \tag{4.227}$$

where the random coefficient selection matrix $\boldsymbol{I}_M(k)$ is given by (4.17).

4.13.4 M-max GSD-LMS algorithm

The M-max GSD-LMS algorithm is defined by:

$$w_a(k+1) = w_a(k) + \mu e(k) I_M(k) \Sigma_a^{-1}(k) v_a^*(k), \quad k = 0, 1, \ldots$$
$$\Sigma_a(k) = \begin{bmatrix} \Sigma(k) & & & \mathbf{0} \\ & \Sigma(k-L) & & \\ & & \ddots & \\ \mathbf{0} & & & \Sigma(k-(K-1)L) \end{bmatrix} \tag{4.228}$$
$$\sigma_i^2(k) = \lambda \sigma_i^2(k-1) + |v_i(k)|^2, \quad i = 1, \ldots, N_s$$

where the data-dependent coefficient selection matrix $I_M(k)$ is determined from:

$$\max_{I_M(k)} v_n^H(k) I_M(k) v_n(k) \tag{4.229}$$

Here:

$$v_n(k) = \Sigma_a^{-1/2}(k) v_a(k) \tag{4.230}$$

is the whitened regressor vector applied to the sparse LMS adaptive filter. The selection of M-max coefficients is based on ranking of the magnitude-squared entries of $\Sigma_a^{-1/2}(k) v_a(k)$:

$$\left[\frac{|v_1(k)|^2}{\sigma_1^2(k)}, \ldots, \frac{|v_{N_s}(k)|^2}{\sigma_N^2(k)}, \frac{|v_1(k-L)|^2}{\sigma_1^2(k-L)}, \ldots, \frac{|v_{N_s}(k-L)|^2}{\sigma_N^2(k-L)}, \ldots, \right.$$
$$\left. \frac{|v_1(k-(K-1)L)|^2}{\sigma_1^2(k-(K-1)L)}, \ldots, \frac{|v_{N_s}(k-(K-1)L)|^2}{\sigma_N^2(k-(K-1)L)} \right]^T \tag{4.231}$$

The additional computational complexity imposed by the above vector can be alleviated by employing the single-division power normalization algorithm summarized in Table 4.5. The single-division power normalization algorithm generates the following version of (4.231) which employs delayed power normalization:

$$q(k) = \left[\frac{|v_1(k)|^2}{\sigma_1^2(k-1)}, \ldots, \frac{|v_{N_s}(k)|^2}{\sigma_N^2(k-1)}, \frac{|v_1(k-L)|^2}{\sigma_1^2(k-1-L)}, \ldots, \frac{|v_{N_s}(k-L)|^2}{\sigma_N^2(k-1-L)}, \ldots, \right.$$
$$\left. \frac{|v_1(k-(K-1)L)|^2}{\sigma_1^2(k-1-(K-1)L)}, \ldots, \frac{|v_{N_s}(k-(K-1)L)|^2}{\sigma_N^2(k-1-(K-1)L)} \right]^T \tag{4.232}$$

The one-sample delay in power normalization is not likely to have an adverse impact on the convergence performance. Using (4.232) the coefficient selection matrix $I_M(k)$

for the M-max GSD-LMS algorithm is defined by:

$$\boldsymbol{I}_M(k) = \begin{bmatrix} i_1(k) & 0 & \cdots & 0 \\ 0 & i_2(k) & \ddots & \vdots \\ \vdots & \ddots & \ddots & 0 \\ 0 & \cdots & 0 & i_{KN_s}(k) \end{bmatrix},$$

$$i_j(k) = \begin{cases} 1 & \text{if } q_j(k) \in \max_{1 \le l \le KN_s}(q_l(k), M) \\ 0 & \text{otherwise} \end{cases} \qquad (4.233)$$

where $\boldsymbol{q}(k) = [q_1(k), q_2(k), \ldots, q_{KN_s}(k)]^T$

4.13.5 Computational complexity

We assume that the N_s-point DCT is used to construct a unitary transform $\boldsymbol{T}$ and that the transform is implemented as a bank of IIR filters with computational complexity of $5N_s$ multiplications and $5N_s$ additions per iteration.

For power normalization we either use exponential sliding window power normalization in (4.167) and (4.178) with complexity $2N_s$ multiplications, N_s additions, and N_s divisions, or single-division power normalization in (4.184) with complexity $5N_s$ multiplications, $2N_s$ additions and one division.

The full-update GSD-LMS algorithm performs the following computations at each k:

- KN_s multiplications and KN_s additions to compute $e(k)$.
- One multiplication to compute $\mu e(k)$.
- $5N_s$ multiplications and $5N_s$ additions to compute $\boldsymbol{v}(k)$.
- Power normalization of $\boldsymbol{v}(k)$:
  - $2N_s$ multiplications, N_s additions, and N_s divisions for exponential sliding window power normalization.
  - $5N_s$ multiplications, $2N_s$ additions and one division for single-division power normalization.
- KN_s multiplications to compute $\mu \boldsymbol{\Sigma}_a^{-1}(k)\boldsymbol{v}_a^*(k)e(k)$.
- KN_s additions to update $\boldsymbol{w}(k)$.

The total computational complexity of the GSD-LMS algorithm per iteration is $2KN_s+7N_s+1$ multiplications, $2KN_s+6N_s$ additions and N_s divisions if exponential sliding window power normalization is used, and $2KN_s + 10N_s + 1$ multiplications, $2KN_s+7N_s$ additions and one division if single-division power normalization is used.

In the periodic-partial-update GSD-LMS algorithm all update operations are computed over S iterations with the exception of the filtering process which operates at the input signal rate. Thus no complexity reduction is available for the $e(kS)$ computation. We have two options for power normalization:

1. Spread power normalization computation for the augmented transform output $v_a(kS)$ over S iterations.
2. Power normalize $v(k)$ at each iteration.

Option 1 has the average computational complexity of KC_p/S per iteration, and option 2 C_p per iteration, where C_p denotes the complexity of power normalization of $v(k)$ at each iteration. Clearly option 1 has a lower complexity if $S > K$. Therefore, for $S > K$ the average computational complexity of the periodic-partial-update GSD-LMS algorithm per iteration is $(3KN_s+5N_s+1)/S+KN_s$ multiplications, $(2KN_s+5N_s)/S+KN_s$ additions and KN_s/S divisions if exponential sliding window power normalization is used; and $(6KN_s+5N_s+1)/S+KN_s$ multiplications, $(3KN_s+5N_s)/S+KN_s$ additions and K/S divisions if single-division power normalization is used. If $S \leq K$, which makes option 2 computationally cheaper, the average computational complexity of the periodic-partial-update GSD-LMS algorithm per iteration is $(KN_s+5N_s+1)/S+KN_s+2N_s$ multiplications, $(KN_s+5N_s)/S+KN_s+N_s$ additions and N_s divisions if exponential sliding window power normalization is used; and $(KN_s+5N_s+1)/S+KN_s+5N_s$ multiplications, $(KN_s+5N_s)/S+KN_s+2N_s$ additions and one division if single-division power normalization is used.

The sequential-partial-update GSD-LMS algorithm replaces the last two steps of the full-update GSD-LMS algorithm with:

- M multiplications to compute $\mu e(k)\boldsymbol{I}_M(k)\boldsymbol{\Sigma}_a^{-1}(k)\boldsymbol{v}_a^*(k)$.
- M additions to update $\boldsymbol{w}(k)$.

This provides a saving of $KN_s - M$ multiplications and $KN_s - M$ additions. The total computational complexity of the sequential-partial-update GSD-LMS algorithm per iteration is $KN_s + 7N_s + M + 1$ multiplications, $KN_s + 6N_s + M$ additions and N_s divisions if exponential sliding window power normalization is used; and $KN_s + 10N_s + M + 1$ multiplications, $KN_s + 7N_s + M$ additions and one division if single-division power normalization is used.

The use of the linear congruential generator in (4.22) to randomize coefficient selection for the stochastic-partial-update GSD-LMS algorithm adds 2 multiplications and 2 additions to the complexity of the sequential-partial-update GSD-LMS algorithm.

The computational complexity of the M-max GSD-LMS algorithm is given by the computational complexity of the sequential-partial-update GSD-LMS algorithm and the complexity of M-max selection. The number of comparisons required by heapsort (see Appendix A) to determine the M coefficients to be updated is:

$$\min(2KN_s + 2\min(M, KN_s - M)\log_2 KN_s, 2M + 2(KN_s - M)\log_2 M).$$

The total computational complexity of M-max GSD-LMS algorithm is $KN_s + 8N_s + M + 1$ multiplications, $KN_s + 6N_s + M$ additions, N_s divisions and $\min(2KN_s + 2\min(M, KN_s - M)\log_2 KN_s, 2M + 2(KN_s - M)\log_2 M)$ comparisons if exponential sliding window power normalization is used; and $KN_s + 10N_s + M + 1$ multiplications, $KN_s + 7N_s + M$ additions, one division and $\min(2KN_s + 2\min(M, KN_s - M)\log_2 KN_s, 2M + 2(KN_s - M)\log_2 M)$ comparisons if single-division power normalization is used. Note that exponential sliding window power normalization requires an additional N_s multiplications per iteration to compute the vector whose entries are to be ranked.

Tables 4.8 and 4.9 summarize the computational complexity of the GSD-LMS algorithm and its partial-update implementations using exponential sliding window power normalization and single-division power normalization, respectively. The use of a smaller transform than the transform-domain LMS algorithm evidently eases the pressure coming from transform and power normalization computations. For large K, the partial-update algorithms make more impact on complexity reduction. The periodic-partial-update GSD-LMS algorithm has, once again, the smallest computational complexity.

4.14 SIMULATION EXAMPLES: CHANNEL EQUALIZATION

In this section we present simulation examples to demonstrate the convergence performance of some of the partial-update adaptive filters studied in this chapter. Suppose that a communication channel with transfer function

$$H(z) = \frac{1}{1 - 1.1z^{-1} + 0.3z^{-2} + 0.2z^{-3} + 0.1z^{-4} - 0.2z^{-5} + 0.1z^{-6} + 0.05z^{-7} - 0.01z^{-8} + 0.02z^{-9}} \tag{4.234}$$

is equalized by an adaptive equalizer with coefficients $\boldsymbol{\theta}(k) = [\theta_1(k), \ldots, \theta_N(k)]^T$ as shown in Figure 4.14. The channel frequency response is depicted in Figure 4.15. The channel input $x(k)$ is an uncorrelated binary sequence taking on values ± 1 and the channel output is corrupted by additive zero-mean white Gaussian noise with variance $\sigma_n^2 = 0.0001$. The channel output signal-no-noise ratio (SNR) is approximately 45 dB. The adaptive equalizer has length $N = 10$, which allows perfect zero-forcing equalization, and is implemented using one of the adaptive filters investigated in this chapter. The eigenvalue spread of the regressor autocorrelation matrix is 75. The objective of adaptive equalization is to learn $\boldsymbol{\theta}(k)$ so that $E\{e^2(k)\}$ is minimized where $e(k)$ is the error signal given by:

$$e(k) = x(k - D) - \underbrace{\boldsymbol{r}^T(k)\boldsymbol{\theta}(k)}_{y(k)} \tag{4.235}$$

Table 4.8 Computational complexity of GSD-LMS and partial-update GSD-LMS algorithms at each k using exponential sliding window power normalization.

Algorithm		×	+	÷	≶
GSD-LMS		$2KN_s + 7N_s + 1$	$2KN_s + 6N_s$	N_s	
Per-par-upd GSD-LMS	$S > K$	$(3KN_s + 5N_s + 1)/S + KN_s$	$(2KN_s + 5N_s)/S + KN_s$	KN_s/S	
	$S \leq K$	$(KN_s + 5N_s + 1)/S + KN_s + 2N_s$	$(KN_s+5N_s)/S+KN_s+N_s$	N_s	
Seq-par-upd GSD-LMS		$KN_s + 7N_s + M + 1$	$KN_s + 6N_s + M$	N_s	
Stoch-par-upd GSD-LMS		$KN_s + 7N_s + M + 3$	$KN_s + 6N_s + M + 2$	N_s	
M-max GSD-LMS		$KN_s + 8N_s + M + 1$	$KN_s + 6N_s + M$	N_s	$\min(2KN_s + 2\min(M, KN_s - M)\log_2 KN_s, 2M + 2(KN_s - M)\log_2 M)$

Table 4.9 Computational complexity of GSD-LMS and partial-update GSD-LMS algorithms at each k using single-division power normalization

Algorithm		×	+	÷	≶
GSD-LMS		$2KN_s + 10N_s + 1$	$2KN_s + 7N_s$	1	
Per-par-upd GSD-LMS	$S > K$	$(6KN_s + 5N_s + 1)/S + KN_s$	$(3KN_s + 5N_s)/S + KN_s$	K/S	
	$S \leq K$	$(KN_s + 5N_s + 1)/S + KN_s + 5N_s$	$(KN_s + 5N_s)/S + KN_s + 2N_s$	1	
Seq-par-upd GSD-LMS		$KN_s + 10N_s + M + 1$	$KN_s + 7N_s + M$	1	
Stoch-par-upd GSD-LMS		$KN_s + 10N_s + M + 3$	$KN_s + 7N_s + M + 2$	1	
M-max GSD-LMS		$KN_s + 10N_s + M + 1$	$KN_s + 7N_s + M$	1	$\min(2KN_s + 2\min(M, KN_s - M)\log_2 KN_s, 2M + 2(KN_s - M)\log_2 M)$

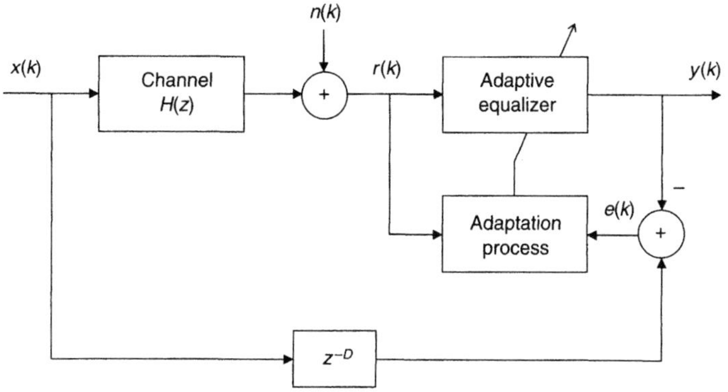

Figure 4.14 Adaptive channel equalization. The adaptive equalizer aims to remove intersymbol interference caused by linear channel $H(z)$.

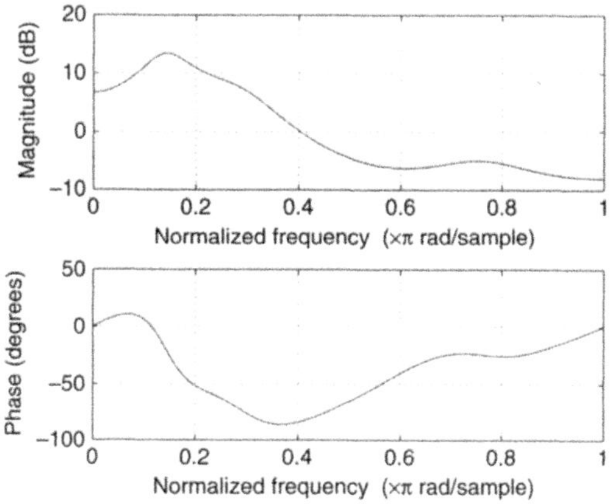

Figure 4.15 Frequency response of channel.

We assume $D = 0$ since the channel $H(z)$ is minimum-phase and its causal inverse is a stable FIR system. The minimization of $E\{e^2(k)\}$ implies alleviation of the intersymbol interference (ISI) caused by the dispersive linear channel.

We first compare the converge rates of full-update adaptive equalizers that update all available equalizer coefficients at each iteration. Figure 4.16 shows the learning curves of the full-update adaptive equalizers ensemble averaged over 400 independent realizations. The adaptive filter parameters used in this simulation are summarized in Table 4.10. Clearly, the best convergence performance is achieved by the RLS algorithm while the LMS algorithm results in the slowest convergence rate. The other adaptive filters perform between these two extremes with APA and transform-domain LMS having similar convergence rates.

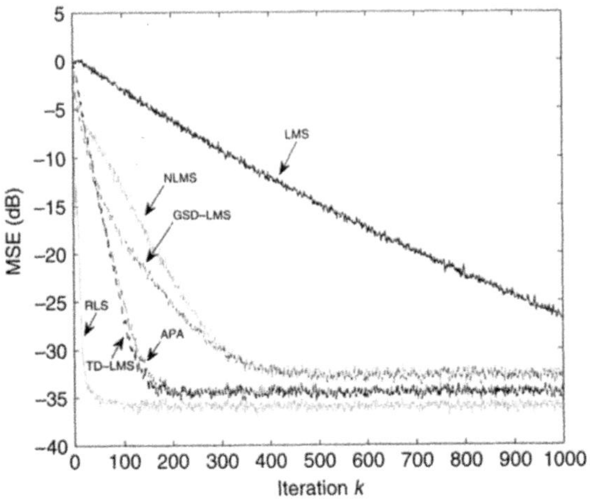

Figure 4.16 Ensemble averaged learning curves of full-update adaptive equalizers.

Table 4.10 Full-update adaptive equalizer parameters

Algorithm	μ	ϵ	λ	P	N_s	K	L
LMS	0.01						
NLMS	0.8	0.01					
APA	0.08	0.01		5			
RLS		0.01	0.99				
TD-LMS	5	0.01	0.99				
GSD-LMS	8	0.01	0.99		5	2	5

Figure 4.17 compares the ensemble averaged learning curves of partial-update LMS equalizers for $M = 1$ and $S = N/M = 10$. The step-size parameters of the algorithms are chosen to obtain the same steady-state MSE. The adaptive filter parameters used in this simulation are summarized in Table 4.11. The M-max LMS algorithm outperforms the full-update LMS algorithm despite updating only one out of 10 coefficients at each iteration. The periodic-partial-update and sequential-partial-update LMS algorithms have similar convergence rates and they both converge roughly 10 times slower than the full-update LMS algorithm.

Figure 4.18 shows the ensemble averaged learning curves of partial-update NLMS equalizers. The number of coefficients to be updated is set to $M = 1$ (which corresponds to $S = 10$) for all partial-update algorithms, with the exception of the set-membership partial-update NLMS algorithm for which $M = 2$ in order to avoid stability problems. Table 4.12 gives the adaptive filter parameters used in the simulation. The set-membership partial-update NLMS and full-update NLMS

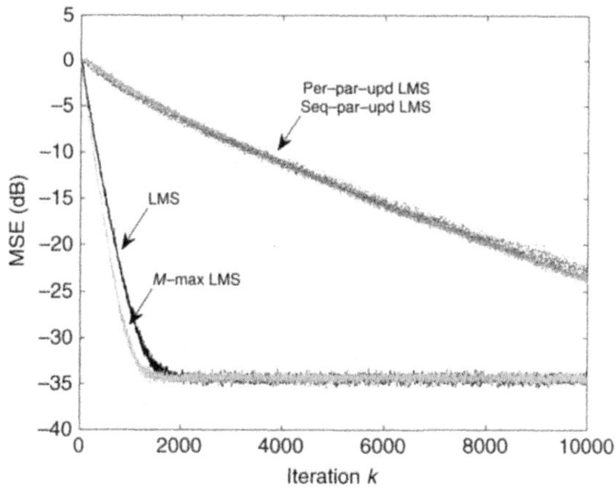

Figure 4.17 Ensemble averaged learning curves of partial-update LMS equalizers.

Table 4.11 Partial-update LMS equalizer parameters ($M = 1$)

Algorithm	μ
LMS	0.010
Per-par-upd LMS	0.010
Seq-par-upd LMS	0.011
M-max LMS	0.022

Table 4.12 Partial-update NLMS equalizer parameters

Algorithm	μ	M	S	ϵ	γ
NLMS	0.80			0.01	
Per-par-upd NLMS	0.80		10	0.01	
Seq-par-upd NLMS	0.90	1		0.01	
M-max NLMS	1.25	1		0.01	
Set-mem par-upd NLMS		2		0.01	0.05

algorithms have the same initial convergence rates and yet the set-membership partial-update NLMS algorithm achieves a better steady-state MSE. The M-max NLMS algorithm is somewhat slower than the full-update NLMS algorithm. The convergence rate of the periodic-partial-update and sequential-partial-update NLMS algorithms is significantly slower than the other data-dependent partial-update algorithms.

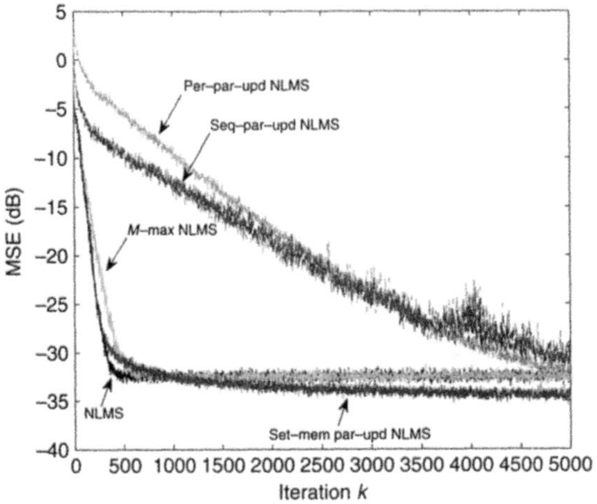

Figure 4.18 Ensemble averaged learning curves of partial-update NLMS equalizers.

Table 4.13 Partial-update APA equalizer parameters

Algorithm	μ	M	P	ϵ	γ
APA	0.08		5	0.01	
Seq-par-upd APA	0.20	1	5	0.01	
M-max APA	0.08	1	5	0.01	
Simplified M-max APA	0.22	1	5	0.01	
Sel-par-upd APA	0.24	8	2	0.01	
Set-mem par-upd APA-1		8	2	0.01	0.08
Set-mem par-upd APA-2		8	2	0.01	0.03

Figure 4.19 shows the ensemble averaged learning curves of partial-update APA equalizers. The number of coefficients to be updated and the APA order are set to $M = 1$ and $P = 5$, respectively, for all partial-update algorithms except for the selective-partial-update APA, and set-membership partial-update APA-1 and APA-2, for which $M = 8$ and $P = 2$ in order to avoid stability problems (recall that we require $P \leq M$ for selective-partial-update and set-membership partial-update APA). Table 4.13 lists the adaptive filter parameters used in the simulation. The set-membership partial-update APA-2 and full-update APA are seen to perform similarly. The full-update APA has order $P = 5$ while the set-membership partial-update APA-2 has order $P = 2$ and updates $M = 8$ coefficients. In addition, the set-membership partial-update APA-2 uses sparse time-domain updates, which helps lower the power consumption. We also observe from Figure 4.19 that the sequential-partial-update APA performs

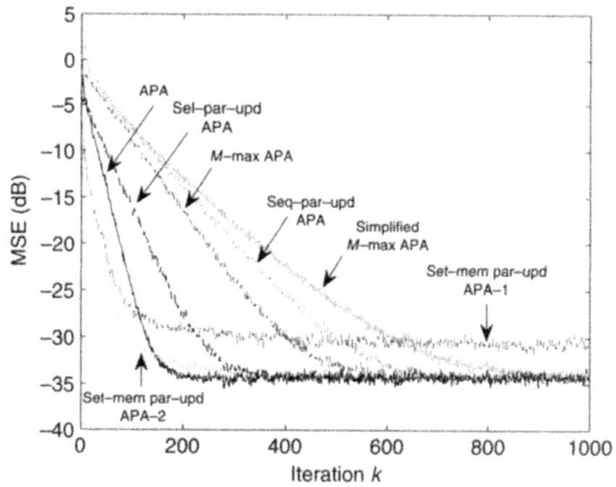

Figure 4.19 Ensemble averaged learning curves of partial-update APA equalizers.

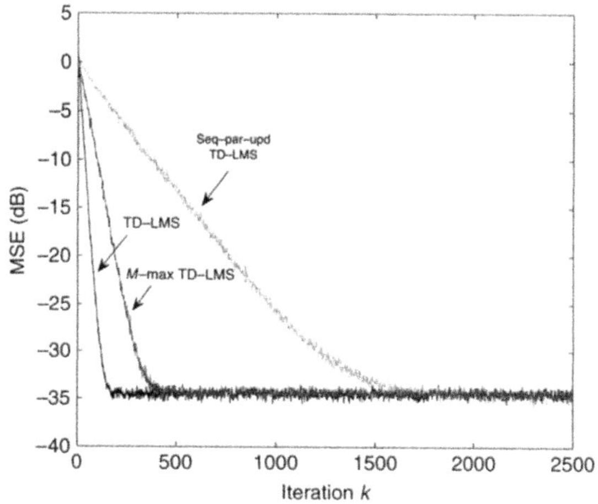

Figure 4.20 Ensemble averaged learning curves of partial-update transform-domain LMS equalizers.

well in this equalization problem, outperforming the simplified M-max APA. The computationally expensive M-max APA, that uses the full implementation of M-max coefficient selection, converges faster than the sequential-partial-update APA by a relatively small margin. The selective-partial-update APA which is implemented using the simplified approximate selection criterion converges faster than the M-max APA. This is not surprising since the selective-partial-update APA employs a correlation estimate that is consistent with the partial-update regressor vector.

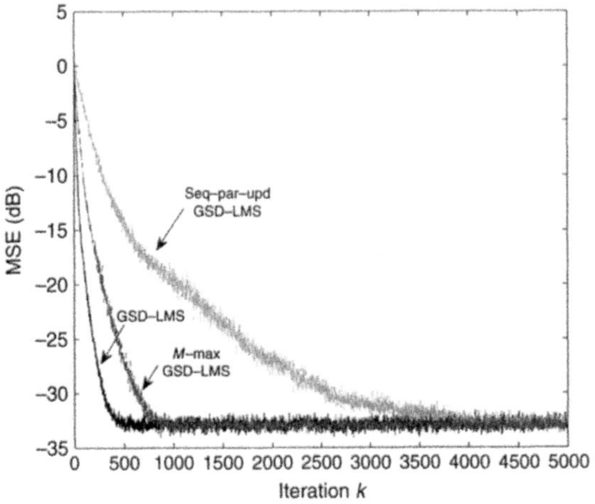

Figure 4.21 Ensemble averaged learning curves of partial-update GSD-LMS equalizers.

Table 4.14 Partial-update transform-domain LMS equalizer parameters ($M = 1$)

Algorithm	μ	ϵ	λ
TD-LMS	5	0.01	0.99
Seq-par-upd TD-LMS	5	0.01	0.99
M-max TD-LMS	6	0.01	0.99

Table 4.15 Partial-update GSD-LMS equalizer parameters ($M = 1$)

Algorithm	μ	ϵ	λ
GSD-LMS	8	0.01	0.99
Seq-par-upd GSD-LMS	10	0.01	0.99
M-max GSD-LMS	10	0.01	0.99

Figure 4.20 shows the ensemble averaged learning curves of partial-update transform-domain LMS equalizers. The number of coefficients to be updated is $M = 1$. The fixed transform $\boldsymbol{T}$ is implemented using the N-point DCT. The transform output vector is power normalized using the single-division power normalization algorithm. The M-max coefficient selection also uses the single-division power normalization algorithm. Table 4.14 lists the adaptive filter parameters used.

Figure 4.21 shows the ensemble averaged MSE curves of partial-update GSD-LMS equalizers. The number of coefficients to be updated is set to $M = 1$. The GSD parameters are $N_s = 5$, $K = 2$ and $L = 5$. The fixed transform $\boldsymbol{T}$ is implemented using the N_s-point DCT. The single-division power normalization algorithm is used to power-normalize the transform outputs. The M-max coefficient selection also uses the single-division power normalization algorithm. Table 4.15 lists the GSD-LMS adaptive filter parameters used in the simulation.

Chapter | five

Selected applications

5.1 INTRODUCTION

We present applications of partial-update adaptive signal processing to acoustic/network echo cancellation, blind fractionally spaced channel equalization, and blind adaptive linear multiuser detection. These are not the only application areas for partial-update adaptive signal processing. Any adaptive signal processing problem can potentially benefit from partial coefficient updates in terms of computational complexity reduction. It is also possible for partial-update adaptive filters to achieve performance improvement at reduced complexity. In this chapter we will demonstrate performance improvements achieved by selective partial updates in blind channel equalization and linear multiuser detection.

5.2 ACOUSTIC ECHO CANCELLATION

The acoustic echo cancellation problem arises whenever a coupling between a loudspeaker and a microphone occurs in such applications as hands-free telephone and teleconferencing. This coupling results in the far-end talker's signal being fed back to the the far-end taker creating annoying echoes and in some instances instability. The key to reducing the undesirable echoes electrically is to generate a replica of the microphone signal and subtract it from the actual microphone signal. This is illustrated in Figure 5.1. The sound waves emanating from the loudspeaker propagate through the echo path of the acoustic environment. The echo path is *a priori* unknown and also time-varying. Even a slight movement of furniture or people in the acoustic environment can lead to drastic changes in the echo path. As a result, the adaptive echo canceller has the task of not only estimating the echo path, but also keeping track of changes in it.

The adaptive identification of the echo path is a challenging problem for two main reasons: (i) the input to the adaptive filter is usually a speech signal with strong

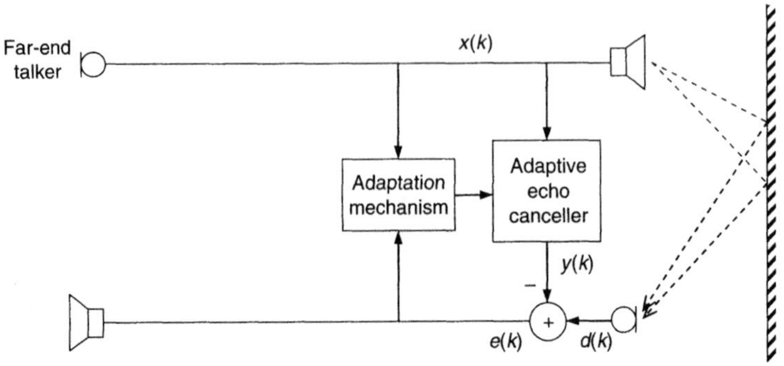

Figure 5.1 Adaptive acoustic echo cancellation.

correlation and time-varying statistics, (ii) the unknown echo path has a long impulse response even if the input signal is sampled at a rate as low as 8 kHz. The adaptive filter structure employed in acoustic echo cancellation is an FIR filter because of its guaranteed stability. This means that, in order to model a given echo path, the adaptive filter is likely to require thousands of coefficients. The large number of adaptive filter coefficients places a strong demand on the computational resources. In this context, partial-update adaptive filters offer an attractive alternative to reduce the overall adaptation complexity.

The echo path is modelled as a time-varying IIR system with impulse response $h_i(k)$. In terms of $h_i(k)$ the microphone signal is given by:

$$d(k) = \sum_{i=0}^{\infty} h_i(k)x(k-i) + n(k) \tag{5.1}$$

where $x(k)$ is the loudspeaker signal and $n(k)$ is the additive noise at the microphone. The echo signal is the noise-free version of the microphone signal:

$$r(k) = \sum_{i=0}^{\infty} h_i(k)x(k-i) \tag{5.2}$$

The adaptive echo canceller attempts to create a replica of the echo signal $y(k)$ from the loudspeaker signal:

$$y(k) = \boldsymbol{x}^T(k)\boldsymbol{w}(k) \tag{5.3}$$

The difference between the microphone signal and the adaptive filter output is the error signal which is also referred to as the residual echo signal:

$$e(k) = d(k) - y(k) \tag{5.4}$$

The adaptive echo canceller uses this error signal to update its coefficients so that the difference between the microphone signal and the echo signal synthesized by the adaptive filter is minimized.

A measure of the effectiveness of an echo canceller is given by the *echo return loss enhancement* (ERLE), which is formally defined as (Breining *et al.*, 1999)

$$\text{ERLE} = \frac{E\{r^2(k)\}}{E\{(r(k) - y(k))^2\}} \tag{5.5}$$

Another measure is *misalignment* of adaptive filter coefficients:

$$\frac{\|\boldsymbol{w}(k) - \boldsymbol{h}(k)\|_2^2}{\|\boldsymbol{h}(k)\|_2^2} \tag{5.6}$$

where $\boldsymbol{h}(k)$ is the echo path impulse response. Since $\boldsymbol{h}(k)$ has infinite length, a truncated finite-length approximation of $\boldsymbol{h}(k)$ is used in applications of the misalignment measure.

The update process of the adaptive echo canceller is inhibited whenever the near-end and far-end talkers are active simultaneously, which is referred to as *double-talk*. Double-talk can lead to divergence of the adaptive echo canceller effectively due to bursts in the additive noise variance at the microphone output.

The method of selective partial updates (Doğançay and Tanrıkulu, 2001a) has been applied to acoustic echo cancellation with considerable success (Doğançay and Naylor, 2005; Doğançay, 2003b; Doğançay and Tanrıkulu, 2000, 2001a,b, 2002). We now present computer simulations to compare the performance of partial-update NLMS algorithms in an acoustic echo cancellation setup. In the simulations, we use real speech sampled at 8 kHz as the loudspeaker signal. The acoustic echo path is a measured car echo impulse response shown in Figure 5.2. The sampled echo path impulse response has 256 taps. The echo path is assumed to be time-invariant. The return echo $r(k)$ is corrupted by zero-mean additive white Gaussian noise. The signal-to-noise ratio at the microphone output is 30 dB.

The adaptive filters have $N = 256$ coefficients. The partial-update NLMS filters update $M = 64$ coefficients at each iteration out of $N = 256$. The periodic-partial-update NLMS algorithm updates the entire coefficient vector every $S = 4$ iterations, which amounts to updating 64 coefficients per iteration on average. The step-size parameter for all adaptive filters is $\mu = 0.22$ except for the selective-partial-update NLMS algorithm which uses $\mu = 0.16$. The error magnitude bound for the set-membership partial-update NLMS algorithm is $\gamma = 0.012$. Figure 5.3 shows the time evolution of misalignment for the full-update and partial-update NLMS algorithms. As can be seen from Figure 5.3(a) the periodic-partial-update and sequential-partial-update NLMS algorithms exhibit slow convergence compared with

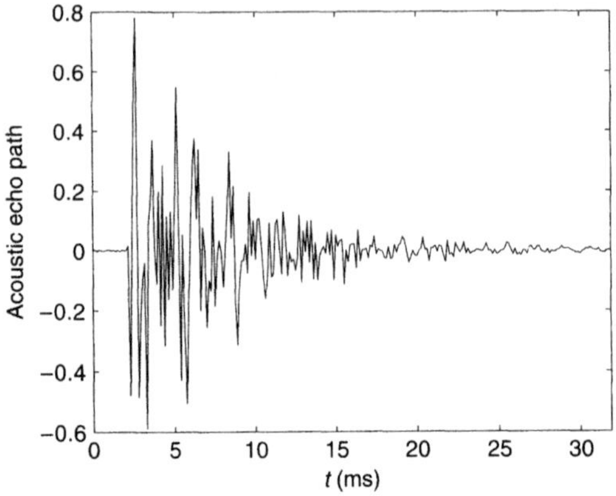

Figure 5.2 Acoustic echo path impulse response measured in a car.

the full-update NLMS algorithm. On the other hand, the data-dependent partial-update NLMS algorithms appear to perform very well (see Figure 5.3(b)), almost having the same convergence performance as the full-update NLMS algorithm. The set-membership partial-update NLMS algorithm stands out as the fasting converging data-dependent partial-update NLMS algorithm. Note that in addition to updating a quarter of the adaptive coefficients, the set-membership partial-update NLMS algorithm updates only 8% of the time on average in this case.

5.3 NETWORK ECHO CANCELLATION

A simplified long-distance telephone connection between two subscribers is illustrated in Figure 5.4. The local subscriber loops at both ends use two-wire circuits carrying both transmitted and received speech over the same pair of wires. Long-distance transmission of speech necessitates amplification using separate four-wire trunks for each direction of transmission. The cost and cabling requirements rule out the use of four-wire trunks directly between subscribers. Therefore, the two-wire local loop is converted to four-wire trunks by the so-called hybrid located at local exchanges.

Referring to Figure 5.4 the task of hybrid 1 is to convert the four-wire circuit carrying the signal of S2 to the two-wire circuit for S1 without allowing it to return to S2 through the four-wire transmission trunk. Hybrid 2 performs the same task for the signal coming from S1. However, due to impedance mismatches at the hybrid the signal received from the four-wire trunk is leaked to the four-wire transmission trunk, resulting in some echoes to be fed back to the subscribers. The delay of the echo depends on the distance between the hybrids. For short delays up to 30 ms the echo is usually hard to distinguish from the sidetone and can in fact improve the perceived quality of speech. If the delay is in excess of 30 ms the echo becomes annoying and may lead to interruption of conversation. The network echo problem has

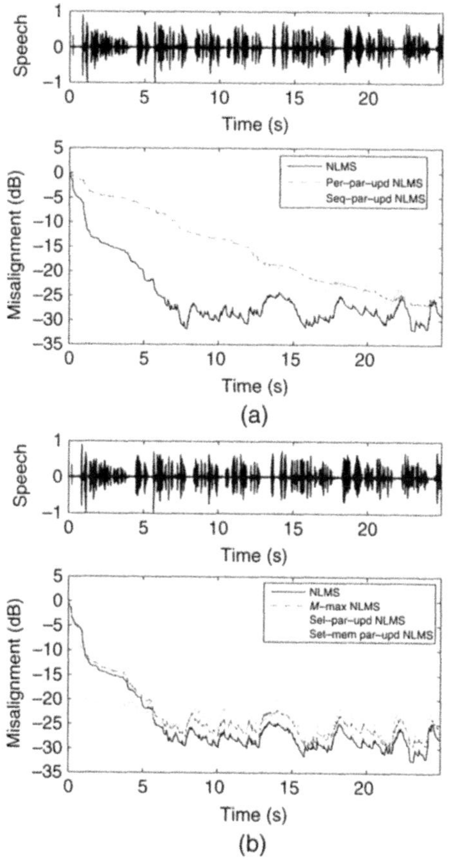

Figure 5.3 Misalignment of full-update and partial-update NLMS algorithms for speech signal: (a) data-independent partial-update NLMS algorithms; (b) data-dependent partial-update NLMS algorithms.

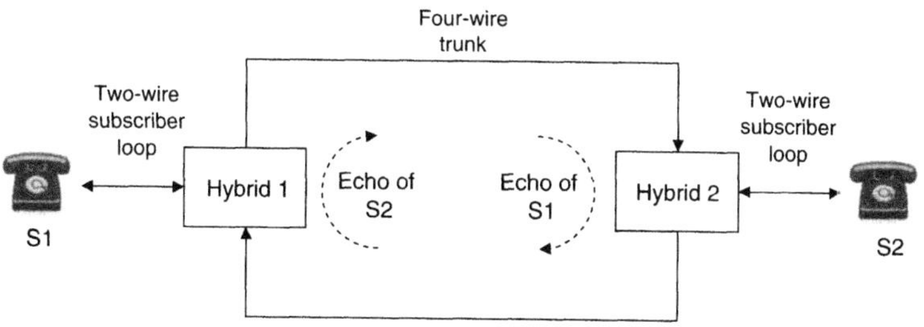

Figure 5.4 Simplified model for long-distance telephone connection.

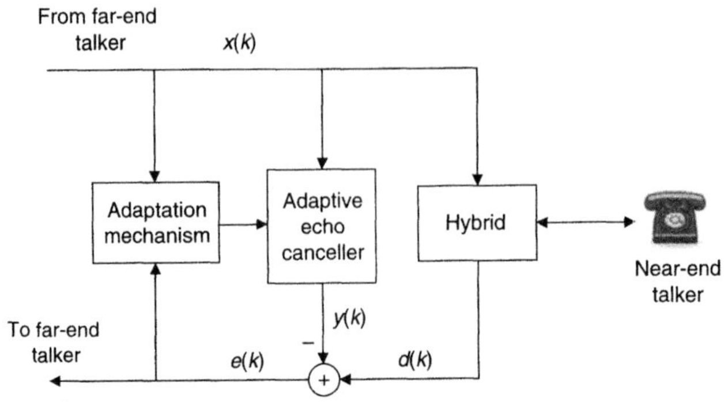

Figure 5.5 Adaptive network echo cancellation.

become particularly acute with the advent of geostationary satellites for long-distance telephone connections due to round-trip delays of around 500 ms. Even though considerably shorter round-trip delays are achieved now by optical communications, the coding delays introduced by modern wireless communication systems still make network echo cancellation an important practical problem. Figure 5.5 depicts network echo cancellation by an adaptive filter. Similar to the acoustic echo cancellation problem, the adaptive filter aims to identify the echo path impulse response of the hybrid which is modelled as an FIR filter. Network echo cancellation problems are characterized by long echo path impulse responses of the order of 128 ms, of which only a small portion (4–22 ms) is actually non-zero. Adaptive echo cancellers are designed to synthesize the full length of the echo path because: (i) the location of the non-zero portion of the echo path (the flat delay) is unknown; and (ii) the flat delay varies from call to call. The flat delay is due to the long distance between the echo canceller and the hybrid/local-loop circuit.

NLMS is the algorithm of choice in most implementations of adaptive echo canceller because of its simplicity. A recently proposed modification of NLMS—the proportionate normalized least-mean-square (PNLMS) algorithm (Duttweiler, 2000; Gay, 1998)—exploits the sparseness of the echo path impulse response to speed up the initial convergence of the conventional NLMS algorithm. PNLMS essentially adapts the non-zero portion of the echo path by weighting the update terms of the adaptive filter with the magnitude of the estimated echo path impulse response. This effectively results in a shorter adaptive filter that converges faster than the full-length adaptive filter. The key to PNLMS is the introduction of weighting for adaptive filter coefficients. A particular choice for the weighting gives rise to PNLMS.

There are some disadvantages to PNLMS, namely, an increase in computational complexity by 50% and the slow convergence of the adaptive filter coefficients after the fast initial convergence. The latter is due to the slow convergence of small coefficients after the convergence of the large coefficients over the 'active' portion of the echo path impulse response. It is possible to reduce the increased computational complexity by way of selective partial updating of the adaptive filter coefficients (Tanrıkulu and Doğançay, 2002). PNLMS is a coefficient-selective algorithm in that it weights the coefficients to be updated in direct proportion to the magnitude of the coefficient estimates. The method of selective partial updates introduces a data-selective flavour to PNLMS by selecting the regressor vector entries to be used in the update term. This in turn reduces the number of multiplications in the update term. The slow convergence of PNLMS after the initial fast convergence can be alleviated by providing a transition from PNLMS to NLMS during the adaptation process (Tanrıkulu and Doğançay, 2002). This can be achieved by changing one of the PNLMS parameters as new data samples come in. Variants of PNLMS have been proposed to improve the convergence speed, to provide robustness in the face of undetected double-talk, and to reduce the computational complexity (Benesty *et al.*, 2001; Gay, 1998). Of particular interest is the μ-law PNLMS algorithm proposed in (Deng and Doroslovački, 2006) which is capable of maintaining the fast initial convergence of PNLMS throughout the convergence process.

5.3.1 PNLMS and μ-law PNLMS with selective partial updates

In this section we derive the PNLMS and μ-law PNLMS algorithms with selective partial updates from a weighted cost function. The algorithm development is cast into the framework of a constrained optimization problem, which is different to the approach taken in (Duttweiler, 2000).

PNLMS

To start with, let us show how PNLMS can be derived from the solution of a constrained minimization problem in a similar way to the NLMS algorithm. Referring to Figure 5.5 the output $y(k)$ of the adaptive network echo canceller is given by:

$$y(k) = \boldsymbol{x}^T(k)\boldsymbol{w}(k) \tag{5.7}$$

where $\boldsymbol{w}(k) = [w_1(k), w_2(k), \ldots, w_N(k)]^T$ is the $N \times 1$ adaptive coefficient vector and $\boldsymbol{x}(k) = [x(k), x(k-1), \ldots, x(k-N+1)]^T$ is the $N \times 1$ regressor vector of the far-end signal that excites the echo path. Introducing a weighting in the constrained optimization problem for the NLMS algorithm we obtain:

$$\min_{\boldsymbol{w}(k+1)} \|\boldsymbol{w}(k+1) - \boldsymbol{w}(k)\|^2_{\boldsymbol{G}^{-1}(k)} \tag{5.8a}$$

subject to:

$$e_p(k) = \left(1 - \frac{\mu\|\boldsymbol{x}(k)\|^2_{\boldsymbol{G}(k)}}{\epsilon + \|\boldsymbol{x}(k)\|^2_{\boldsymbol{G}(k)}}\right) e(k) \tag{5.8b}$$

where $\boldsymbol{G}(k) = \text{diag}\{g_1(k), g_2(k), \ldots, g_N(k)\}$ is a diagonal coefficient weighting matrix with positive diagonal elements, and $e_p(k) = d(k) - \boldsymbol{x}^T(k)\boldsymbol{w}(k+1)$ and $e(k) = d(k) - \boldsymbol{x}^T(k)\boldsymbol{w}(k)$ are the *a posteriori* and *a priori* errors, respectively. Here $d(k)$ is the desired filter response which is the echo plus the near-end noise.

Define the cost function:

$$\begin{aligned} J(\boldsymbol{w}(k+1), \lambda) &= \|\boldsymbol{w}(k+1) - \boldsymbol{w}(k)\|^2_{\boldsymbol{G}^{-1}(k)} \\ &\quad + \lambda\left(\boldsymbol{x}^T(k)(\boldsymbol{w}(k+1) - \boldsymbol{w}(k)) - \frac{\mu\|\boldsymbol{x}(k)\|^2_{\boldsymbol{G}(k)}}{\epsilon + \|\boldsymbol{x}(k)\|^2_{\boldsymbol{G}(k)}} e(k)\right) \end{aligned} \tag{5.9}$$

where λ is a Lagrange multiplier. Setting $\partial J(\boldsymbol{w}(k+1), \lambda)/\partial \boldsymbol{w}(k+1) = \boldsymbol{0}$ and $\partial J(\boldsymbol{w}(k+1), \lambda)/\partial \lambda = 0$ we obtain:

$$2(\boldsymbol{w}(k+1) - \boldsymbol{w}(k)) + \lambda \boldsymbol{G}(k)\boldsymbol{x}(k) = \boldsymbol{0} \tag{5.10a}$$

$$\boldsymbol{x}^T(k)(\boldsymbol{w}(k+1) - \boldsymbol{w}(k)) - \frac{\mu\|\boldsymbol{x}(k)\|^2_{\boldsymbol{G}(k)}}{\epsilon + \|\boldsymbol{x}(k)\|^2_{\boldsymbol{G}(k)}} e(k) = 0 \tag{5.10b}$$

Solving (5.10a) for $\boldsymbol{w}(k+1) - \boldsymbol{w}(k)$ and substituting it into (5.10b) we have:

$$\lambda = -2\mu \frac{e(k)}{\epsilon + \|\boldsymbol{x}(k)\|^2_{\boldsymbol{G}(k)}} \tag{5.11}$$

Plugging this back into (5.10a) we finally obtain the PNLMS algorithm:

$$\boldsymbol{w}(k+1) = \boldsymbol{w}(k) + \mu \frac{\boldsymbol{G}(k)\boldsymbol{x}(k)e(k)}{\epsilon + \boldsymbol{x}^T(k)\boldsymbol{G}(k)\boldsymbol{x}(k)} \tag{5.12}$$

In the above derivation we have not defined the diagonal weighting matrix $\boldsymbol{G}(k)$ specifically. Indeed, the algorithm derivation is independent of $\boldsymbol{G}(k)$ as long as it is positive definite, and as we shall see later different definitions of $\boldsymbol{G}(k)$ can lead to improved versions of PNLMS. The PNLMS algorithm employs a particular weighting matrix which results in its diagonal entries approximating the magnitude of the estimated echo path impulse response. This intuitive selection of diagonal entries results in an uneven distribution of the available energy in the regressor vector over the adaptive filter coefficients. If we set $\boldsymbol{G}(k) = \boldsymbol{I}$, PNLMS reduces to NLMS. The PNLMS algorithm obtained in (5.12) is slightly different from the PNLMS algorithm in (Duttweiler, 2000) where the normalization factor in the update term is $\boldsymbol{x}^T(k)\boldsymbol{x}(k)$ rather than $\boldsymbol{x}^T(k)\boldsymbol{G}(k)\boldsymbol{x}(k)$. In practice this is not likely to cause any appreciable convergence differences.

For PNLMS the diagonal entries of the weighting matrix $\boldsymbol{G}(k)$ are calculated at every iteration using:

$$l_\infty(k) = \max\{\delta_p, |w_1(k)|, \ldots, |w_N(k)|\} \tag{5.13a}$$

$$\gamma_i(k) = \max\{\rho l_\infty(k), |w_i(k)|\}, \quad 1 \leq i \leq N \tag{5.13b}$$

$$g_i(k) = \frac{\gamma_i(k)}{\frac{1}{N}\sum_{j=1}^{N}\gamma_j(k)}, \quad 1 \leq i \leq N \tag{5.13c}$$

where δ_p and ρ are the PNLMS parameters that effect small-signal regularization. The parameter δ_p prevents the algorithm from misbehaving when the adaptive filter coefficients $\boldsymbol{w}(k)$ are very small as at initialization, and ρ prevents individual filter coefficients from freezing (i.e. being never adapted again) when their magnitude is much smaller than the largest coefficient l_∞. Typical values for the PNLMS parameters are $\delta_p = 0.01$ and $\rho = 5/N$. If $\rho \geq 1$, then we have $g_i(k) = 1$ (i.e. $\boldsymbol{G}(k) = \boldsymbol{I}$) and PNLMS reduces to NLMS.

μ-Law PNLMS

The μ-law PNLMS algorithm (Deng and Doroslovački, 2006) defines the diagonal weighting matrix $\boldsymbol{G}(k)$ in a different way, leading to improved convergence performance. The choice of $\boldsymbol{G}(k)$ is motivated by studying an approximately equivalent averaged system under the sufficiently small step-size assumption:

$$\boldsymbol{w}^a(k+1) = \boldsymbol{w}^a(k) + \mu\boldsymbol{G}(k)E\left\{\frac{\boldsymbol{x}(k)e(k)}{\epsilon + \boldsymbol{x}^T(k)\boldsymbol{G}(k)\boldsymbol{x}(k)}\right\} \tag{5.14}$$

For sufficiently large N, we can write:

$$\boldsymbol{w}^a(k+1) = \boldsymbol{w}^a(k) + \mu(k)\boldsymbol{G}(k)(\boldsymbol{p}(k) - \boldsymbol{R}\boldsymbol{w}^a(k)) \tag{5.15a}$$

where

$$\boldsymbol{R} = E\{\boldsymbol{x}(k)\boldsymbol{x}^T(k)\}, \boldsymbol{p} = E\{d(k)\boldsymbol{x}(k)\} \text{ and } \mu(k) = \mu/E\{\|\boldsymbol{x}(k)\|^2_{\boldsymbol{G}(k)}\} \tag{5.15b}$$

The coefficient error $\Delta\boldsymbol{w}^a(k) = \boldsymbol{w}_o - \boldsymbol{w}^a(k)$ at iteration k is given by:

$$\Delta\boldsymbol{w}^a(k) = \left(\prod_{i=0}^{k}(\boldsymbol{I} - \mu(i)\boldsymbol{G}(i)\boldsymbol{R})\right)\Delta\boldsymbol{w}^a(0) \tag{5.16}$$

Assuming a stationary white input signal $x(k)$ with variance σ_x^2 and zero initialization we have:

$$\Delta \boldsymbol{w}^a(k) = \left(\prod_{i=0}^{k} (\boldsymbol{I} - \sigma_x^2 \mu(i) \boldsymbol{G}(i)) \right) \boldsymbol{w}_o \tag{5.17}$$

It was shown in (Deng and Doroslovački, 2006) that setting the diagonal entries of $\boldsymbol{G}(k)$ to:

$$g_i \approx \frac{\ln \frac{|w_{o,i}|}{\varepsilon}}{\frac{1}{N} \sum_{j=1}^{N} \ln \frac{|w_{o,j}|}{\varepsilon}}, \quad i = 1, \ldots, N \tag{5.18}$$

gives the optimum convergence performance in that all coefficients converge to the ε-vicinity of the optimum solution $\boldsymbol{w}_o$ after the same number of iterations. Here $w_{o,i}$ denotes the ith entry of $\boldsymbol{w}_o$. Note that g_i does not change over time since it depends on the optimum solution. In practice, the optimum solution is not known and, therefore, the g_i must be estimated based on the current estimate of $\boldsymbol{w}_o$ (i.e. $\boldsymbol{w}(k)$). The μ-law PNLMS algorithm adopts this approach and uses the following time-varying weights:

$$g_i(k) = \frac{F(|w_i(k)|)}{\frac{1}{N} \sum_{j=1}^{N} F(|w_j(k)|)} \tag{5.19}$$

where:

$$F(x) = \ln\left(1 + \frac{x}{\varepsilon}\right) \tag{5.20}$$

is the μ-law function* for non-uniform quantization.

The weighting matrix $\boldsymbol{G}(k)$ for μ-law PNLMS is given by (cf. (5.13)):

$$F(|w_i(k)|) = \ln(1 + |w_i(k)|/\varepsilon) \tag{5.21a}$$

$$l_\infty(k) = \max\{\delta_p, F(|w_1(k)|), \ldots, F(|w_N(k)|)\} \tag{5.21b}$$

$$\gamma_i(k) = \max\{\rho l_\infty(k), F(|w_i(k)|)\}, \quad 1 \leq i \leq N \tag{5.21c}$$

$$g_i(k) = \frac{\gamma_i(k)}{\frac{1}{N} \sum_{j=1}^{N} \gamma_j(k)}, \quad 1 \leq i \leq N \tag{5.21d}$$

where δ_p and ρ are the PNLMS small-signal regularization parameters, and ε is a small positive number that is related to the steady-state MSE or additive noise variance. Typical values for the μ-law PNLMS parameters are $\delta_p = 0.01$, $\rho = 5/N$ and $\varepsilon = 0.001$.

*The μ-law function is defined as $F(x) = \ln(1 + \mu x)$. In order to avoid confusion with the step-size parameter μ, we do not substitute $\mu = 1/\varepsilon$ into (5.20).

Selective partial updates

The computational complexity of the PNLMS and μ-law PNLMS algorithms can be reduced by updating a subset of the filter coefficients at every iteration. We now show how to do this using the method of selective partial updates.

The selective-partial-update PNLMS algorithm solves the constrained optimization problem defined by:

$$\min_{\boldsymbol{I}_M(k)} \min_{\boldsymbol{w}_{\mathcal{M}}(k+1)} \|\boldsymbol{w}_{\mathcal{M}}(k+1) - \boldsymbol{w}_{\mathcal{M}}(k)\|^2_{\boldsymbol{G}^{-1}_{\mathcal{M}}(k)} \tag{5.22a}$$

subject to:

$$e_p(k) = \left(1 - \frac{\mu \|\boldsymbol{x}_{\mathcal{M}}(k)\|^2_{\boldsymbol{G}_{\mathcal{M}}(k)}}{\epsilon + \|\boldsymbol{x}_{\mathcal{M}}(k)\|^2_{\boldsymbol{G}_{\mathcal{M}}(k)}}\right) e(k) \tag{5.22b}$$

where $\boldsymbol{G}_{\mathcal{M}}(k)$ is the $M \times M$ diagonal matrix with weights selected by $\boldsymbol{I}_M(k)$.

For fixed $\boldsymbol{I}_M(k)$ define the cost function:

$$\begin{aligned} J(\boldsymbol{w}_{\mathcal{M}}(k+1), \lambda) &= \|\boldsymbol{w}_{\mathcal{M}}(k+1) - \boldsymbol{w}_{\mathcal{M}}(k)\|^2_{\boldsymbol{G}^{-1}_{\mathcal{M}}(k)} \\ &\quad + \lambda \left(\boldsymbol{x}^T_{\mathcal{M}}(k)(\boldsymbol{w}_{\mathcal{M}}(k+1) - \boldsymbol{w}_{\mathcal{M}}(k)) \right. \\ &\quad \left. - \frac{\mu \|\boldsymbol{x}_{\mathcal{M}}(k)\|^2_{\boldsymbol{G}_{\mathcal{M}}(k)}}{\epsilon + \|\boldsymbol{x}_{\mathcal{M}}(k)\|^2_{\boldsymbol{G}_{\mathcal{M}}(k)}} e(k) \right) \end{aligned} \tag{5.23}$$

where λ is a Lagrange multiplier. Solving $\partial J(\boldsymbol{w}_{\mathcal{M}}(k+1), \lambda)/\partial \boldsymbol{w}_{\mathcal{M}}(k+1) = \boldsymbol{0}$ and $\partial J(\boldsymbol{w}_{\mathcal{M}}(k+1), \lambda)/\partial \lambda = 0$ for $\boldsymbol{w}_{\mathcal{M}}(k+1)$ and λ results in:

$$\boldsymbol{w}_{\mathcal{M}}(k+1) = \boldsymbol{w}_{\mathcal{M}}(k) + \mu \frac{\boldsymbol{G}_{\mathcal{M}}(k)\boldsymbol{x}_{\mathcal{M}}(k)e(k)}{\epsilon + \boldsymbol{x}^T_{\mathcal{M}}(k)\boldsymbol{G}_{\mathcal{M}}(k)\boldsymbol{x}_{\mathcal{M}}(k)} \tag{5.24}$$

which can be written as:

$$\boldsymbol{w}(k+1) = \boldsymbol{w}(k) + \mu \frac{e(k)\boldsymbol{I}_M(k)\boldsymbol{G}(k)\boldsymbol{x}(k)}{\epsilon + \boldsymbol{x}^T(k)\boldsymbol{I}_M(k)\boldsymbol{G}(k)\boldsymbol{x}(k)} \tag{5.25}$$

Substituting the fixed-$\boldsymbol{I}_M(k)$ constrained optimization solution (5.24) into (5.22a) and ignoring ϵ we have:

$$\min_{\boldsymbol{I}_M(k)} \frac{1}{\boldsymbol{x}^T_{\mathcal{M}}(k)\boldsymbol{G}_{\mathcal{M}}(k)\boldsymbol{x}_{\mathcal{M}}(k)} \tag{5.26}$$

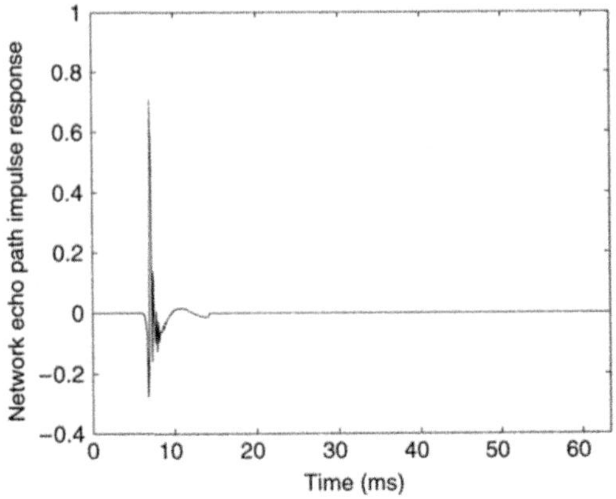

Figure 5.6 A sparse network echo impulse response specified in G.168.

which is equivalent to:

$$\max_{\boldsymbol{I}_M(k)} \boldsymbol{x}^T(k)\boldsymbol{I}_M(k)\boldsymbol{G}(k)\boldsymbol{x}(k) \tag{5.27}$$

Thus, the coefficient selection matrix for selective partial updates is given by:

$$\boldsymbol{I}_M(k) = \begin{bmatrix} i_1(k) & 0 & \cdots & 0 \\ 0 & i_2(k) & \ddots & \vdots \\ \vdots & \ddots & \ddots & 0 \\ 0 & \cdots & 0 & i_N(k) \end{bmatrix},$$

$$i_j(k) = \begin{cases} 1 & \text{if } g_j(k)x_j^2(k) \in \max_{1 \le l \le N}(g_l(k)x_l^2(k), M) \\ 0 & \text{otherwise} \end{cases} \tag{5.28}$$

To sum up, the selective-partial-update PNLMS and μ-law PNLMS algorithms are defined by the recursion (5.25) and coefficient selection matrix (5.28). The overheads introduced by coefficient selection can be alleviated by employing fast ranking algorithms (see Appendix A) and strobe-down methods (Duttweiler, 2000).

Simulation examples

Figure 5.6 shows the sparse echo network path used in the simulations. This is one of the network impulse responses specified in G.168 (ITU-T, 2002). The dispersion time of the echo is 8 ms and the flat delay is 6.25 ms. The time span of the echo response is 64 ms. The length of adaptive echo cancellers is set to $N = 512$ which covers the entire time span of the sparse network echo at an 8 kHz sampling rate. The full-update NLMS, PNLMS, μ-law PNLMS, and selective-partial-update versions of PNLMS and

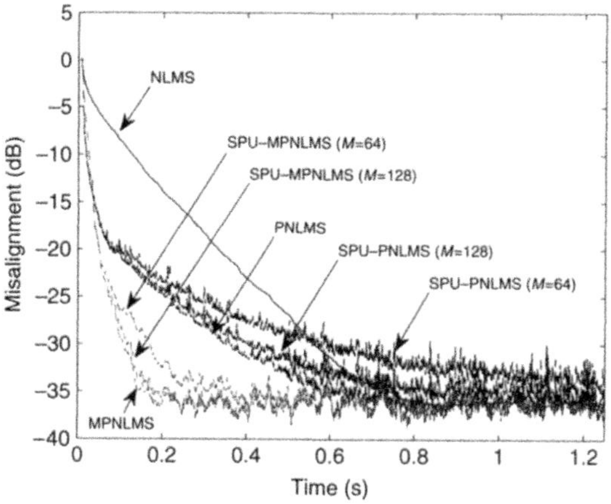

Figure 5.7 Misalignment of NLMS, PNLMS, μ-law PNLMS (MPNLMS) and selective-partial-update PNLMS and MPNLMS for $M = 64$ and $M = 128$.

μ-law PNLMS were simulated for a zero-mean white Gaussian input signal $x(k)$. The near-end noise is also zero-mean white Gaussian. The signal-to-noise ratio (SNR) for the echo signal is 30 dB.

The learning curves of the simulated adaptive filtering algorithms are shown in Figure 5.7. The simulation parameters for NLMS, PNMS and μ-law PNLMS are $\mu = 0.4$, $\rho = 5/512$, $\delta = 0.01$ and $\varepsilon = 0.001$. The selective-partial-update PNLMS and μ-law PNLMS were simulated using $M = 64$ and $M = 128$ corresponding to $1/8$ and $1/4$ of the filter coefficients being updated at every iteration, respectively. From Figure 5.7 we observe that the full-update μ-law PNLMS algorithm achieves the fastest convergence rate. The selective-partial-update μ-law PNLMS algorithm outperforms both NLMS and PNLMS for $M = 64$ and $M = 128$. We also remark that the convergence rate of the selective-partial-update μ-law PNLMS algorithm is expected to be almost identical to that of μ-law PNLMS for values of M larger than the length of the active network echo path segment (dispersion time), which is 64 in this case.

5.4 BLIND CHANNEL EQUALIZATION

In communication systems transmitted signals undergo distortion due to intersymbol interference (ISI) caused by time dispersive channels. ISI is one of the major obstacles to reliable high-speed data transmission over bandlimited channels (Qureshi, 1985). The role of a channel equalizer is to remove ISI from the received signal. In practice, the continuous-time channel output signal is sampled at twice the symbol rate T before being applied to the equalizer in order to avoid spectral aliasing and to simplify timing and carrier recovery. Oversampling of the channel output results in

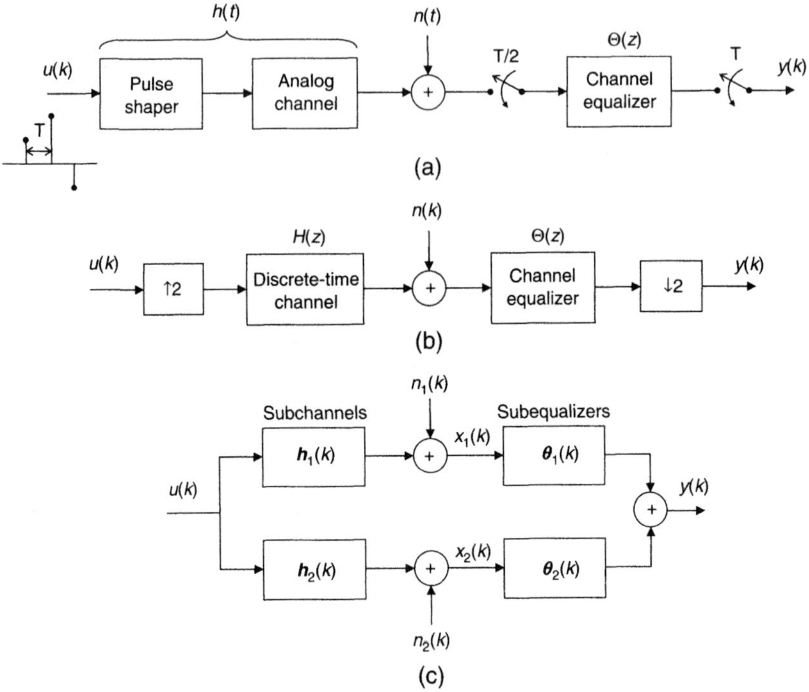

Figure 5.8 Fractionally spaced channel equalization: (a) baseband channel-equalizer model; (b) equivalent multirate model using discrete-time signals; (c) equivalent multichannel model.

fractionally spaced equalization (see Figure 5.8). The coefficients of the fractionally spaced equalizer are $T/2$-spaced. The equalizer output is downsampled to the symbol rate before being passed to the next processing block (possibly a decision device). For the same equalization time span, a fractionally spaced equalizer would require twice as many coefficients as a baud-rate equalizer that operates at the symbol rate. This implies doubling of the computational complexity of a fractionally spaced equalizer compared with a baud-rate equalizer. Despite this complexity penalty fractionally spaced equalization simplifies the timing and carrier recovery tremendously.

Figure 5.8 shows equivalent fractionally spaced channel equalization models. The multichannel model in Figure 5.8(c) assumes the subchannels have no common zeros. A fractionally spaced equalizer of finite length is capable of achieving perfect zero-forcing equalization for FIR channels provided that no common subchannel zeros exist (Johnson *et al.*, 1998). A common subchannel zero effectively causes a single FIR channel to be inserted between $u(k)$ and the subchannels $h_1(k)$ and $h_2(k)$. This means that zero-forcing equalization would necessitate inversion of the single FIR channel, resulting in an infinitely long FIR equalizer.

Sampling the continuous-time channel output signal at twice the symbol rate results in the multirate model shown in Figure 5.8(b). Here $H(z)$ is the transfer function of the discrete-time channel which includes the effects of both pulse shaper and analog channel:

$$H(z) = \sum_{i=0}^{2N_c-1} h_i(k)z^{-i} \tag{5.29}$$

The discrete-time channel is assumed be a possibly time-varying FIR system. The equalizer is defined as:

$$\Theta(z) = \sum_{i=0}^{2N-1} \theta_i(k)z^{-i} \tag{5.30}$$

Referring to the multichannel model in Figure 5.8(c), the subchannels:

$$\begin{aligned} \boldsymbol{h}_1(k) &= [h_{1,0}(k), h_{1,1}(k), \ldots, h_{1,N_c-1}(k)]^T \\ \boldsymbol{h}_2(k) &= [h_{2,0}(k), h_{2,1}(k), \ldots, h_{2,N_c-1}(k)]^T \end{aligned} \tag{5.31}$$

are related to the discrete-time channel $H(z)$ via:

$$H(z) = \sum_{i=0}^{N_c-1} h_{1,i}(k)z^{-2i} + z^{-1} \sum_{i=0}^{N_c-1} h_{2,i}(k)z^{-2i} \tag{5.32}$$

In other words, interleaving $\boldsymbol{h}_1(k)$ and $\boldsymbol{h}_2(k)$ gives $[h_0(k), h_1(k), \ldots, h_{2N_c-1}(k)]^T$. A similar relationship also exists between the subequalizers $\boldsymbol{\theta}_1(k), \boldsymbol{\theta}_2(k)$

$$\begin{aligned} \boldsymbol{\theta}_1(k) &= [\theta_{1,0}(k), \theta_{1,1}(k), \ldots, \theta_{1,N-1}(k)]^T \\ \boldsymbol{\theta}_2(k) &= [\theta_{2,0}(k), \theta_{2,1}(k), \ldots, \theta_{2,N-1}(k)]^T \end{aligned} \tag{5.33}$$

and $\Theta(z)$:

$$\Theta(z) = \sum_{i=0}^{N-1} \theta_{1,i}(k)z^{-2i} + z^{-1} \sum_{i=0}^{N-1} \theta_{2,i}(k)z^{-2i} \tag{5.34}$$

Defining the subchannel output regressor vectors as:

$$\boldsymbol{x}_1(k) = \begin{bmatrix} x_1(k) \\ x_1(k-1) \\ \vdots \\ x_1(k-N+1) \end{bmatrix}, \quad \boldsymbol{x}_2(k) = \begin{bmatrix} x_2(k) \\ x_2(k-1) \\ \vdots \\ x_2(k-N+1) \end{bmatrix} \tag{5.35}$$

the fractionally spaced equalizer output is given by:

$$\begin{aligned} y(k) &= \boldsymbol{x}_1^T(k)\boldsymbol{\theta}_1(k) + \boldsymbol{x}_2^T(k)\boldsymbol{\theta}_2(k) \\ &= \boldsymbol{x}^T(k)\boldsymbol{\theta}(k) \end{aligned} \tag{5.36}$$

where:

$$\boldsymbol{x}(k) = \begin{bmatrix} \boldsymbol{x}_1(k) \\ \boldsymbol{x}_2(k) \end{bmatrix}, \quad \boldsymbol{\theta}(k) = \begin{bmatrix} \boldsymbol{\theta}_1(k) \\ \boldsymbol{\theta}_2(k) \end{bmatrix} \tag{5.37}$$

The objective of equalization is to determine $\boldsymbol{\theta}(k)$ so that the equalizer output is a possibly delayed copy of the transmitted signal $u(k)$:

$$y(k) = u(k - D) \quad \forall k \tag{5.38}$$

where $D \geq 0$ is a constant equalization delay. Traditionally the equalizer utilizes training signals to learn $\boldsymbol{\theta}(k)$ that achieves or approximates (5.38). The use of training signals sacrifices bandwidth. Blind equalization aims to achieve channel equalization without utilizing any training, thereby saving bandwidth. In most blind equalization problems, the equalizer is allowed to be insensitive to phase rotations to facilitate separation of equalization from carrier recovery (Godard, 1980), resulting in simpler equalizer design. The constant modulus criterion is an example of phase-insensitive blind equalization criterion. For a binary input signal $u(k)$ taking on values ± 1, the constant modulus criterion is:

$$|y(k)|^2 = 1 \quad \forall k. \tag{5.39}$$

For other input constellations (5.39) can be generalized to:

$$|y(k)|^2 = R^2 \quad \forall k \tag{5.40}$$

where R is a constellation-dependent dispersion factor. For constant modulus constellations such as BPSK and M-PSK, R is simply equal to the magnitude of transmitted symbols (Treichler and Agee, 1983). For other constellations, R is obtained from higher-order moments of transmitted symbols (Godard, 1980).

The constant modulus algorithm (CMA) is a popular blind equalization algorithm. It aims to minimize the cost function:

$$J_{\text{CMA}}(\boldsymbol{\theta}(k)) = E\{(|y(k)|^2 - R^2)^2\} \tag{5.41}$$

Central to CMA is the concept of recovering the channel input modulus at the equalizer output. CMA is defined by the recursion:

$$\boldsymbol{\theta}(k+1) = \boldsymbol{\theta}(k) + \mu \boldsymbol{x}(k) y(k)(R^2 - |y(k)|^2), \quad k = 0, 1, \ldots \tag{5.42}$$

which must be initialized to a non-zero vector. Setting $\boldsymbol{\theta}(0) = \mathbf{0}$ leads to all zero updates, stalling the convergence process.

5.4.1 Normalized CMA

The normalized CMA (NCMA) is derived from the solution of a constrained optimization problem. For blind equalization employing the constant modulus criterion in (5.40) the constrained optimization problem takes the form (Hilal and Duhamel, 1992):

$$\min_{\boldsymbol{\theta}(k+1)} \|\boldsymbol{\theta}(k+1) - \boldsymbol{\theta}(k)\|_2^2 \tag{5.43a}$$

$$\text{subject to} \quad |\boldsymbol{x}^T(k)\boldsymbol{\theta}(k+1)| = R \tag{5.43b}$$

The solution of the above constrained optimization problem gives the full-update NCMA:

$$\boldsymbol{\theta}(k+1) = \boldsymbol{\theta}(k) + \frac{\mu}{\|\boldsymbol{x}(k)\|_2^2}(\text{sign}(y(k))R - y(k))\boldsymbol{x}(k) \tag{5.44}$$

where μ is a step-size parameter controlling the speed of convergence.

5.4.2 Selective-partial-update NCMA

The adaptive channel equalizer is the computationally most demanding building block of a digital receiver. This is mainly due to the filtering (convolution) and adaptation operations that must be performed at fast data rates. Partial coefficient updating provides a means of reducing the complexity requirements of the equalizer. In this section we present a selective-partial-update version of NCMA. The algorithm derivation is based on the solution of a modified version of the constrained optimization problem (Doğançay and Tanrıkulu, 2001c) in (5.43):

$$\min_{\boldsymbol{I}_M(k)} \min_{\boldsymbol{\theta}_{\mathcal{M}}(k+1)} \|\boldsymbol{\theta}_{\mathcal{M}}(k+1) - \boldsymbol{\theta}_{\mathcal{M}}(k)\|_2^2 \tag{5.45a}$$

$$\text{subject to} \quad |\boldsymbol{x}^T(k)\boldsymbol{\theta}(k+1)| = R \tag{5.45b}$$

For fixed $I_M(k)$ (5.45) becomes a constrained minimization problem over $\boldsymbol{\theta}_{\mathcal{M}}(k+1)$:

$$\min_{\boldsymbol{\theta}_{\mathcal{M}}(k+1)} \|\boldsymbol{\theta}_{\mathcal{M}}(k+1) - \boldsymbol{\theta}_{\mathcal{M}}(k)\|_2^2 \tag{5.46a}$$

$$\text{subject to} \quad |\boldsymbol{x}^T(k)\boldsymbol{\theta}(k+1)| = R \tag{5.46b}$$

which can be solved in a similar way to NCMA by using the method of Lagrange multipliers. The cost function to be minimized is:

$$J(\boldsymbol{\theta}_{\mathcal{M}}(k+1), \lambda) = \|\boldsymbol{\theta}_{\mathcal{M}}(k+1) - \boldsymbol{\theta}_{\mathcal{M}}(k)\|_2^2 + \lambda(R - |\boldsymbol{x}^T(k)\boldsymbol{\theta}(k+1)|) \tag{5.47}$$

where λ is a Lagrange multiplier. Setting $\partial J(\boldsymbol{\theta}_{\mathcal{M}}(k+1), \lambda)/\partial \boldsymbol{\theta}_{\mathcal{M}}(k+1) = \mathbf{0}$ and $\partial J(\boldsymbol{\theta}_{\mathcal{M}}(k+1), \lambda)/\partial \lambda = 0$, we have:

$$\boldsymbol{\theta}_{\mathcal{M}}(k+1) - \boldsymbol{\theta}_{\mathcal{M}}(k) - \frac{\lambda}{2}\text{sign}(\boldsymbol{x}^T(k)\boldsymbol{\theta}(k+1))\boldsymbol{x}_{\mathcal{M}}(k) = \mathbf{0} \tag{5.48a}$$

$$R - |\boldsymbol{x}^T(k)\boldsymbol{\theta}(k+1)| = 0 \tag{5.48b}$$

We can write the last equation as:

$$R - |\boldsymbol{x}^T(k)\boldsymbol{\theta}(k+1)| = 0 \tag{5.49a}$$

$$R - \text{sign}(\boldsymbol{x}^T(k)\boldsymbol{\theta}(k+1))\boldsymbol{x}^T(k)\boldsymbol{\theta}(k+1) = 0 \tag{5.49b}$$

$$R - \text{sign}(\boldsymbol{x}^T(k)\boldsymbol{\theta}(k+1))(\boldsymbol{x}_{\mathcal{M}}^T(k)\boldsymbol{\theta}_{\mathcal{M}}(k+1) + \bar{\boldsymbol{x}}_{\mathcal{M}}^T(k)\bar{\boldsymbol{\theta}}_{\mathcal{M}}(k+1)) = 0 \tag{5.49c}$$

where $\bar{\boldsymbol{x}}_{\mathcal{M}}^T(k)$ and $\bar{\boldsymbol{\theta}}_{\mathcal{M}}(k+1)$ denote the $(N-M) \times 1$ subvectors of $\boldsymbol{x}(k)$ and $\boldsymbol{\theta}(k)$ that are *not* selected by $I_M(k)$. Premultiplying both sides of (5.48a) with $\boldsymbol{x}_{\mathcal{M}}^T(k)$ we have:

$$\boldsymbol{x}_{\mathcal{M}}^T(k)\boldsymbol{\theta}_{\mathcal{M}}(k+1) = \boldsymbol{x}_{\mathcal{M}}^T(k)\boldsymbol{\theta}_{\mathcal{M}}(k) + \frac{\lambda}{2}\text{sign}(\boldsymbol{x}^T(k)\boldsymbol{\theta}(k+1))\|\boldsymbol{x}_{\mathcal{M}}(k)\|_2^2$$
$$= 0 \tag{5.50}$$

Substituting this into (5.49c) yields:

$$\frac{\lambda}{2} = \frac{R - \text{sign}(\boldsymbol{x}^T(k)\boldsymbol{\theta}(k+1))y(k)}{\|\boldsymbol{x}_{\mathcal{M}}(k)\|_2^2} \tag{5.51}$$

Note that $y(k) = \boldsymbol{x}_{\mathcal{M}}^T(k)\boldsymbol{\theta}_{\mathcal{M}}(k) + \bar{\boldsymbol{x}}_{\mathcal{M}}^T(k)\bar{\boldsymbol{\theta}}_{\mathcal{M}}(k+1)$. Plugging the last expression into (5.48a) and introducing a positive step-size μ as relaxation parameter, we obtain:

$$\boldsymbol{\theta}_{\mathcal{M}}(k+1) = \boldsymbol{\theta}_{\mathcal{M}}(k) + \frac{\mu}{\|\boldsymbol{x}_{\mathcal{M}}(k)\|_2^2}(R\text{sign}(y(k)) - y(k))\boldsymbol{x}_{\mathcal{M}}(k) \tag{5.52}$$

where we have replaced $\text{sign}(\boldsymbol{x}^T(k)\boldsymbol{\theta}(k+1))$ with $\text{sign}(y(k))$, assuming that the update term is sufficiently small so that it does not flip the sign of the resulting equalizer output for the current regressor vector.

Equation (5.52) solves the fixed-$\boldsymbol{I}_M(k)$ constrained minimization problem in (5.46). The selective-partial-update coefficient selection matrix is obtained by solving:

$$\min_{\boldsymbol{I}_M(k)} \left\| \frac{\mu}{\|\boldsymbol{x}_{\mathcal{M}}(k)\|_2^2}(R\text{sign}(y(k)) - y(k))\boldsymbol{x}_{\mathcal{M}}(k) \right\|_2^2 \tag{5.53}$$

which can be written as:

$$\max_{\boldsymbol{I}_M(k)} \boldsymbol{x}^T(k)\boldsymbol{I}_M(k)\boldsymbol{x}(k). \tag{5.54}$$

Finally, combining the fixed-$\boldsymbol{I}_M(k)$ recursion in (5.52) with the coefficient selection matrix defined above and introducing a regularization parameter results in the *selective-partial-update NCMA*:

$$\boldsymbol{\theta}(k+1) = \boldsymbol{\theta}(k) + \frac{\mu}{\epsilon + \boldsymbol{x}^T(k)\boldsymbol{I}_M(k)\boldsymbol{x}(k)} (R\text{sign}(y(k)) - y(k))\boldsymbol{I}_M(k)\boldsymbol{x}(k) \tag{5.55}$$

where:

$$\boldsymbol{I}_M(k) = \begin{bmatrix} i_1(k) & 0 & \cdots & 0 \\ 0 & i_2(k) & \ddots & \vdots \\ \vdots & \ddots & \ddots & 0 \\ 0 & \cdots & 0 & i_N(k) \end{bmatrix},$$

$$i_j(k) = \begin{cases} 1 & \text{if } |x_j(k)| \in \max_{1 \le l \le N}(|x_l(k)|, M) \\ 0 & \text{otherwise} \end{cases} \tag{5.56}$$

5.4.3 Simulation examples

In this section we present computer simulations for the full-update and selective-partial-update NCMA. The simulated channels are $T/2$-sampled terrestrial microwave channels obtained from the Rice University Signal Processing Information Base (SPIB) at http://spib.rice.edu/spib/microwave.html. In the simulations the full-update NCMA parameters are $N = 10$ (the total number of equalizer coefficients is $2N = 20$) and $\mu = 0.1$. The selective-partial-update NCMA is implemented using $M = 1$ (i.e. one out of 20 coefficients is updated at each iteration) and $\mu = 0.05$. Both algorithms are initialized to fractionally spaced 'centre-tap initialization':

$$\boldsymbol{\theta}_1(0) = \boldsymbol{\theta}_2(0) = [0, \ldots, 0, 1, 0, \ldots, 0]^T \tag{5.57}$$

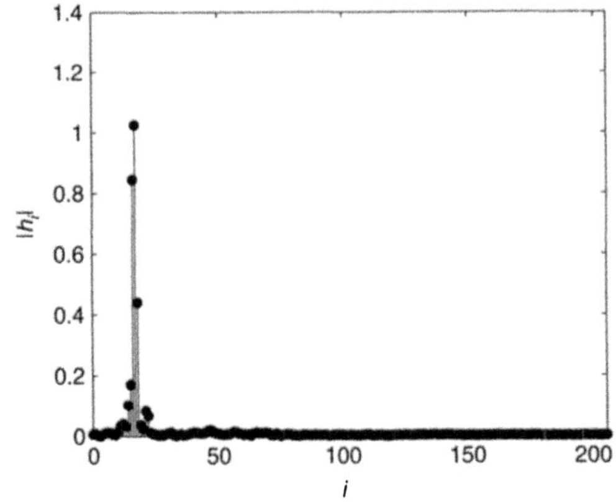

Figure 5.9 Magnitude of impulse response of SPIB channel 5.

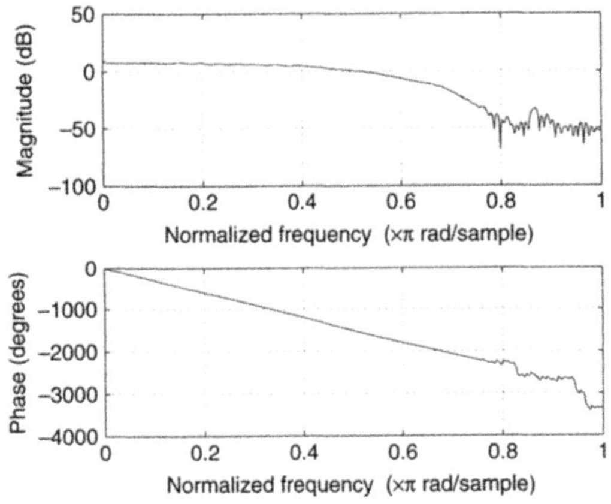

Figure 5.10 Frequency response of SPIB channel 5.

Since N is even, the number of zeros before and after 1 are not the same. We have 4 zeros before and 5 zeros after 1. The transmitted symbols $u(k)$ are drawn from the QPSK constellation. To measure equalization performance we use mean-square error defined as:

$$\text{MSE} = E\{(\text{sign}(y(k)) - y(k))^2\} \tag{5.58}$$

The simulated MSE curves use time averaging over a running window of length 100.

In the first simulation SPIB channel 5 is used. The impulse response magnitude and frequency response of SPIB channel 5 are shown in Figures 5.9 and 5.10, respectively. Figure 5.11 shows the time averaged MSE curves for the full-update and selective-

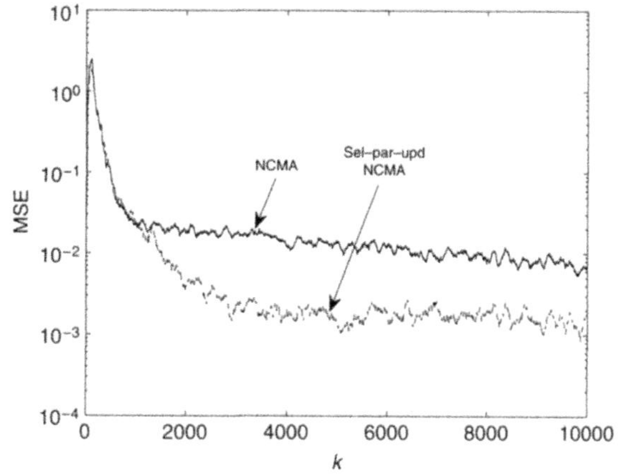

Figure 5.11 Time averaged MSE curves for the full-update and selective-partial-update NCMA with $M = 1$ (SPIB channel 5).

partial-update NCMA for 10000 iterations. We observe that despite updating only one coefficient out of 20, the selective-partial-update NCMA outperforms the full-update NCMA in this instance. Dispersion plots of equalizer outputs in the last 1000 iterations of convergence are shown in Figure 5.12.

The second simulation uses SPIB channel 10 with impulse response magnitude and frequency response shown in Figures 5.13 and 5.14, respectively. Figure 5.15 shows the time averaged MSE curves for the full-update and selective-partial-update NCMA for 10000 iterations. This time the selective-partial-update NCMA does not outperform the full-update NCMA. Dispersion plots of equalizer outputs in the last 1000 iterations of convergence are shown in Figure 5.16.

5.5 BLIND ADAPTIVE LINEAR MULTIUSER DETECTION

In direct-sequence code-division-multiple-access (DS-CDMA) communication systems the capacity and system performance can be greatly improved by employing joint detection of multiple users, which is commonly known as multiuser detection (MUD). Optimal techniques for MUD are computationally prohibitive (Verdu, 1986). Adaptive linear MUD techniques, on the other hand, provide a low-complexity sub-optimal solution with good performance. They also allow the receiver to keep track of changes in the system such as those caused by mobile users entering or exiting the system.

Several adaptive linear MUD algorithms have been proposed in the literature. The linear minimum mean-square-error (MMSE) detector developed in (Honig *et al.*, 1995) and (Madhow and Honig, 1994) can be used in either training or blind mode. In blind mode, it only requires the prior knowledge of the signatures and timing of

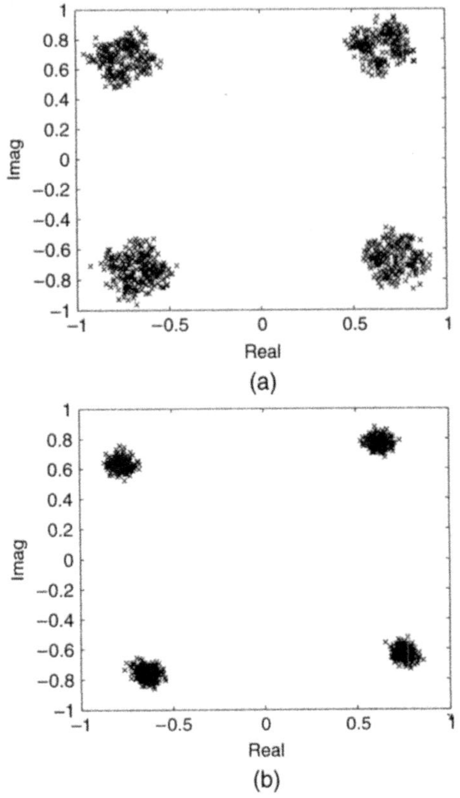

Figure 5.12 Dispersion of equalizer outputs in the last 1000 iterations of convergence for: (a) NCMA; and (b) selective-partial-update NCMA (SPIB channel 5).

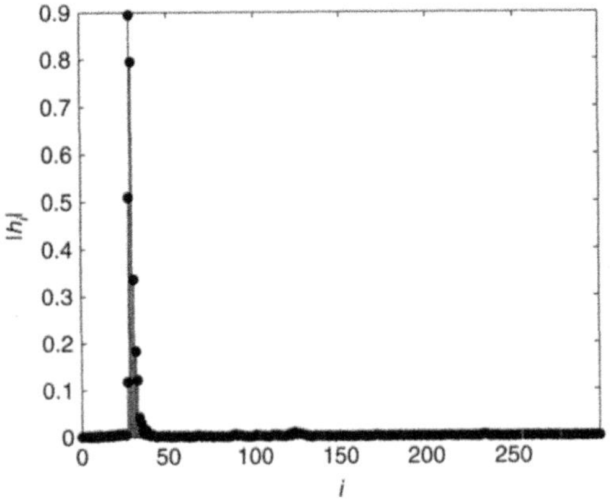

Figure 5.13 Magnitude of impulse response of SPIB channel 10.

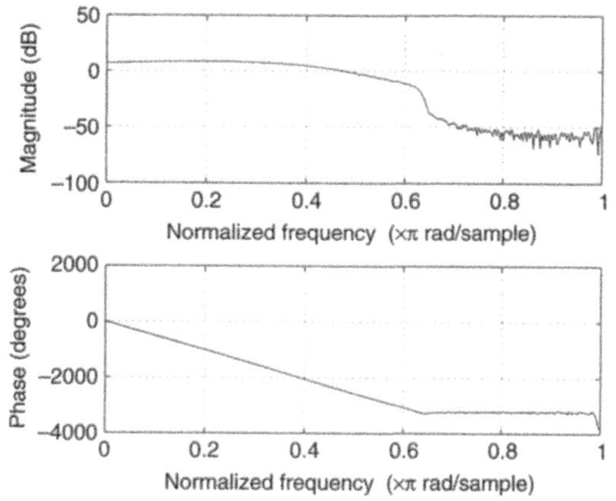

Figure 5.14 Frequency response of SPIB channel 10.

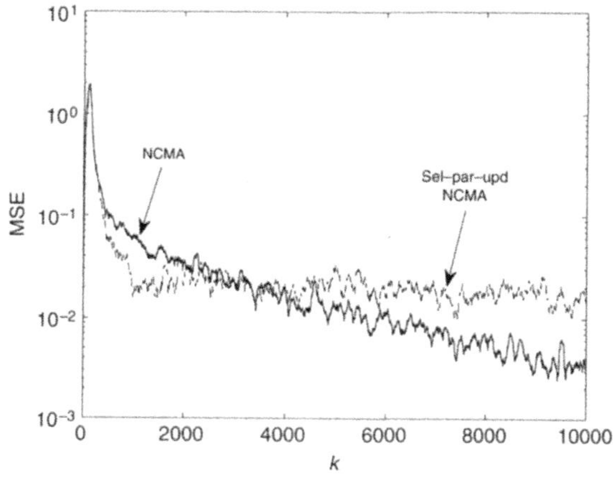

Figure 5.15 Time averaged MSE curves for the full-update and selective-partial-update NCMA with $M = 1$ (SPIB channel 10).

the users to be detected. This is particularly advantageous in the downlink where it is challenging for a mobile user to acquire knowledge of other users' signatures. The LMS algorithm has been utilized for simple and low-complexity implementation of blind linear MMSE detectors.

In this section we present a reduced complexity blind NLMS multiuser detection algorithm employing selective partial updates and sequential partial updates (Doğançay and Tanrıkulu, 2005). We show by simulation that, in the context of blind adaptive multiuser detection, the method of selective partial updates holds

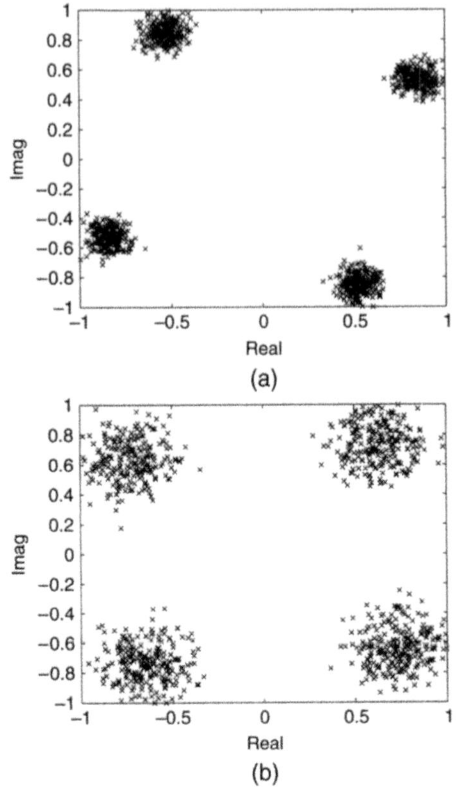

Figure 5.16 Dispersion of equalizer outputs in the last 1000 iterations of convergence for: (a) NCMA; and (b) selective-partial-update NCMA (SPIB channel 10).

the promise of improving the performance of the blind NLMS multiuser detection algorithm while reducing its computational complexity.

5.5.1 MUD in synchronous DS-CDMA

The synchronous DS-CDMA signal model for a K-user system using BPSK modulation is given by:

$$\boldsymbol{r}(i) = \sum_{k=1}^{K} A_k b_k(i) \boldsymbol{s}_k + \boldsymbol{n}(i) \tag{5.59}$$

where $\boldsymbol{r}(i)$ is the received signal vector at the ith symbol interval:

$$\boldsymbol{r}(i) = [r_0(i), r_1(i), \ldots, r_{N-1}(i)]^T$$

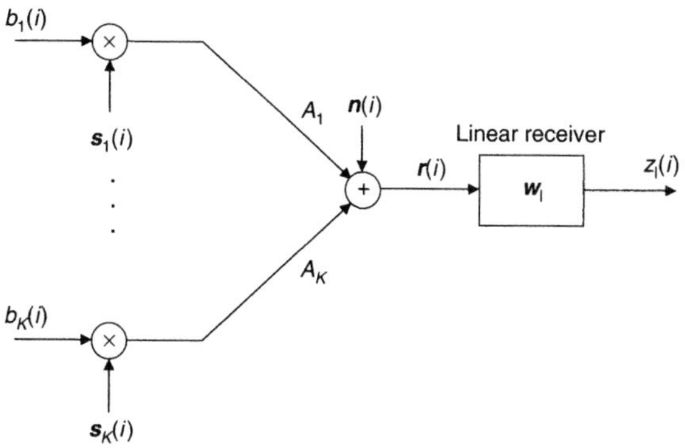

Figure 5.17 Linear receiver for multiuser detection.

N is the processing gain, A_k is the received complex signal amplitude for user k, $b_k(i) = \pm 1$ is the ith symbol of user k, $\boldsymbol{s}_k$ is the normalized signature of user k ($\|\boldsymbol{s}_k\|_2 = 1$)

$$\boldsymbol{s}_k = \frac{1}{\sqrt{N}}[c_{0,k}, c_{1,k}, \ldots, c_{N-1,k}]^T$$

and $\boldsymbol{n}(i) \sim \mathcal{N}_c(\boldsymbol{0}, \sigma^2 \boldsymbol{I}_N)$ is the additive complex Gaussian channel noise. The signature signal $\boldsymbol{s}_k$ can be an m-sequence or Gold sequence with period N. The latter has better cross-correlation properties (Gold, 1967).

As illustrated in Figure 5.17, a linear receiver aims to detect individual users from the received signal $\boldsymbol{r}(i)$ by means of linear filtering. A linear receiver for one of the users, say user l ($1 \leq l \leq K$), has the following form (Wang and Poor, 2004):

$$z_l(i) = \boldsymbol{w}_l^H \boldsymbol{r}(i)$$
$$\hat{b}_l(i) = \text{sign}(\Re\{A_l^* z_l(i)\})$$

where $\boldsymbol{w}_l$ is the linear receiver, $z_l(i)$ is the linear receiver output and $\hat{b}_l(i)$ is the detected symbol. Here $\Re\{\cdot\}$ denotes the real part. The output of the linear receiver can be written as:

$$z_l(i) = \underbrace{A_l(\boldsymbol{w}_l^H \boldsymbol{s}_l) b_l(i)}_{\text{desired signal}} + \underbrace{\sum_{\substack{1 \leq k \leq K \\ k \neq l}} A_k(\boldsymbol{w}_l^H \boldsymbol{s}_k) b_k(i)}_{\text{multiple access interference}} + \underbrace{\boldsymbol{w}_l^H \boldsymbol{n}(i)}_{\text{noise}} \tag{5.60}$$

An important performance measure of multiuser detection is the signal-to-interference-plus-noise ratio (SINR) at the receiver output:

$$\mathrm{SINR}(\boldsymbol{w}_l) = \frac{|A_l(\boldsymbol{w}_l^H \boldsymbol{s}_l)|^2}{\sum_{\substack{1 \le k \le K \\ k \ne l}} |A_k(\boldsymbol{w}_l^H \boldsymbol{s}_k)|^2 + \sigma^2 \|\boldsymbol{w}_l\|_2^2} \tag{5.61}$$

Setting $\boldsymbol{w}_l = \boldsymbol{s}_l$ results in a matched filter. The matched filter is optimal if the multiple access interference (MAI) is Gaussian. The near-far problem and the small number of users often rules out approximation of MAI as Gaussian.

It is desirable to perform multiuser detection in a blind mode (i.e., without the knowledge of other users' signatures) in order to save bandwidth especially in the downlink where the mobile users may not be aware of dynamic changes in the user traffic. A simple blind multiuser detection method can be formulated based on the solution of the constrained optimization problem:

$$\boldsymbol{m}_l = \arg \min_{\substack{\boldsymbol{w} \in \mathbb{C}^N \\ \boldsymbol{w}^H \boldsymbol{s}_l = 1}} E\{\|\boldsymbol{w}^H \boldsymbol{r}(i)\|_2^2\} \tag{5.62}$$

which is referred to as *minimum-output-energy* (MOE) detection (Honig *et al.*, 1995). The symbol $\mathbb{C}$ denotes the set of complex numbers. In (5.62) $\boldsymbol{m}_l$ is the blind linear receiver that demodulates user l and is given by:

$$\boldsymbol{m}_l = E^{-1}\{\boldsymbol{r}(i)\boldsymbol{r}^H(i)\}\boldsymbol{s}_l \tag{5.63}$$

which is termed the direct matrix inversion blind linear MMSE detector.

Decomposing $\boldsymbol{m}_l$ into orthogonal components:

$$\boldsymbol{m}_l = \boldsymbol{s}_l + \boldsymbol{P}_l \boldsymbol{x}_l, \quad \boldsymbol{s}_l \perp \boldsymbol{P}_l \boldsymbol{x}_l$$

with a projection matrix that is orthogonal to $\boldsymbol{s}_l$:

$$\boldsymbol{P}_l = \boldsymbol{I}_N - \boldsymbol{s}_l \boldsymbol{s}_l^H$$

such that $\boldsymbol{P}_l \boldsymbol{m}_l = \boldsymbol{P}_l \boldsymbol{x}_l$, (5.62) can be rewritten as an unconstrained optimization problem (Wang and Poor, 2004):

$$\boldsymbol{m}_l = \boldsymbol{s}_l + \boldsymbol{P}_l \arg \min_{\boldsymbol{x} \in \mathbb{C}^N} E\{\|(\boldsymbol{s}_l + \boldsymbol{P}_l \boldsymbol{x})^H \boldsymbol{r}(i)\|_2^2\} \tag{5.64}$$

The LMS algorithm solving (5.64) is given by (Wang and Poor, 2004):

$$\begin{aligned} \boldsymbol{x}_l(i+1) &= \boldsymbol{x}_l(i) - \mu((\boldsymbol{s}_l + \boldsymbol{P}_l\boldsymbol{x}_l(i))^H\boldsymbol{r}(i))^* \\ &\quad \boldsymbol{P}_l\boldsymbol{r}(i) & (5.65a) \\ z_l(i) &= (\boldsymbol{s}_l + \boldsymbol{P}_l\boldsymbol{x}_l(i))^H\boldsymbol{r}(i) & (5.65b) \\ \hat{b}_l(i) &= \operatorname{sign}(\Re\{A_l^* z_l(i)\}) & (5.65c) \end{aligned}$$

where μ is the LMS step-size parameter.

5.5.2 Blind multiuser NLMS algorithm

For the sake of simplicity, we will assume that A_k and $\boldsymbol{n}(i)$ are real-valued, and $A_k > 0$. The solution of the following constrained optimization problem then gives the normalized LMS (NLMS) algorithm for blind adaptive multiuser detection:

$$\min_{(\boldsymbol{s}_l+\boldsymbol{P}_l\boldsymbol{x}_l(i+1))^T\boldsymbol{r}(i)=0} \|\boldsymbol{x}_l(i+1) - \boldsymbol{x}_l(i)\|_2^2 \tag{5.66}$$

This optimization problem aims to modify $\boldsymbol{x}_l(i)$ in a minimal fashion so that the new vector $\boldsymbol{x}_l(i+1)$ minimizes the output energy for the current input. The constraint of the optimization problem $(\boldsymbol{s}_l + \boldsymbol{P}_l\boldsymbol{x}_l(i+1))^T\boldsymbol{r}(i) = 0$ is relaxed by the step-size parameter μ introduced to the NLMS recursion. In this sense we aim to bring $(\boldsymbol{s}_l + \boldsymbol{P}_l\boldsymbol{x}_l(i+1))^T\boldsymbol{r}(i)$ close to zero as in MOE detection while minimizing the change in $\boldsymbol{x}_l(i)$. The above optimization problem will also provide the basis for the selective-partial-update NLMS algorithm derived in the next section.

Using a Lagrange multiplier, the cost function to be minimized can be written as:

$$\begin{aligned} J(\boldsymbol{x}_l(i+1),\lambda) &= (\boldsymbol{x}_l(i+1) - \boldsymbol{x}_l(i))^T(\boldsymbol{x}_l(i+1) - \boldsymbol{x}_l(i)) \\ &\quad + \lambda(\boldsymbol{s}_l + \boldsymbol{P}_l\boldsymbol{x}_l(i+1))^T\boldsymbol{r}(i) & (5.67) \end{aligned}$$

Taking the gradient with respect to $\boldsymbol{x}_l(i+1)$ and λ and setting them equal to zero gives:

$$\begin{aligned} \frac{\partial J(\boldsymbol{x}_l(i+1),\lambda)}{\partial \boldsymbol{x}_l(i+1)} &= 2\boldsymbol{x}_l(i+1) - 2\boldsymbol{x}_l(i) + \lambda\boldsymbol{P}_l\boldsymbol{r}(i) = \mathbf{0} & (5.68a) \\ \frac{\partial J(\boldsymbol{x}_l(i+1),\lambda)}{\partial \lambda} &= (\boldsymbol{s}_l + \boldsymbol{P}_l\boldsymbol{x}_l(i+1))^T\boldsymbol{r}(i) = 0 & (5.68b) \end{aligned}$$

Solving (5.68a) for $\boldsymbol{x}_l(i+1)$, we obtain:

$$\boldsymbol{x}_l(i+1) = \boldsymbol{x}_l(i) - \frac{\lambda}{2}\boldsymbol{P}_l\boldsymbol{r}(i) \tag{5.69}$$

Substituting this into (5.68b) gives:

$$\left(s_l + P_l x_l(i) - \frac{\lambda}{2} P_l r(i)\right)^T r(i) = 0 \tag{5.70a}$$

$$(s_l + P_l x_l(i))^T r(i) - \frac{\lambda}{2} r^T(i) P_l r(i) = 0 \tag{5.70b}$$

$$\frac{\lambda}{2} = \frac{(s_l + P_l x_l(i))^T r(i)}{r^T(i) P_l r(i)}. \tag{5.70c}$$

Note that for real A_k and s_k, $P_l^T = P_l$ and $P_l^2 = P_l$. Using (5.69) and (5.70c), we obtain the NLMS algorithm for blind MUD:

$$x_l(i+1) = x_l(i) - \mu \frac{z_l(i) P_l r(i)}{r^T(i) P_l r(i)} \tag{5.71}$$

The only difference between the LMS and NLMS algorithms for blind MUD is that NLMS uses a normalization factor given by $r^T(i) P_l r(i) = \|P_l r(i)\|^2$.

5.5.3 Selective-partial-update NLMS for blind multiuser detection

The selective-partial-update version of the NLMS algorithm for blind MUD is given by the solution of the constrained optimization problem:

$$\min_{I_M(i)} \min_{(s_l + P_l x_l(i+1))^T r(i) = 0} \|x_{l,\mathcal{M}}(i+1) - x_{l,\mathcal{M}}(i)\|^2 \tag{5.72}$$

where $x_{l,\mathcal{M}}(i+1)$ is the $M \times 1$ subvector of $x_l(i+1)$ selected by $I_M(i)$. We first solve the above constrained optimization problem over $x_{l,\mathcal{M}}(i+1)$ for a given $I_M(i)$ by minimizing the cost function:

$$J(x_{l,\mathcal{M}}(i+1), \lambda) = \|x_{l,\mathcal{M}}(i+1) - x_{l,\mathcal{M}}(i)\|^2 + \lambda (s_l + P_l x_l(i+1))^T r(i) \tag{5.73}$$

Taking the gradient with respect to $x_{l,\mathcal{M}}(i+1)$ and λ and setting them to zero gives:

$$\frac{\partial J(x_{l,\mathcal{M}}(i+1), \lambda)}{\partial x_{l,\mathcal{M}}(i+1)} = 2x_{l,\mathcal{M}}(i+1) - 2x_{l,\mathcal{M}}(i) + \lambda P_{l,\mathcal{M}}^T r(i) = \mathbf{0} \tag{5.74a}$$

$$\frac{\partial J(x_{l,\mathcal{M}}(i+1), \lambda)}{\partial \lambda} = (s_l + P_l x_l(i+1))^T r(i) = 0 \tag{5.74b}$$

where $\boldsymbol{P}_{l,\mathcal{M}}$ is the $N \times M$ partition of $\boldsymbol{P}_l$ with columns selected by $\boldsymbol{I}_M(i)$. Solving (5.74a) for $\boldsymbol{x}_{l,\mathcal{M}}(i+1)$, we get:

$$\boldsymbol{x}_{l,\mathcal{M}}(i+1) = \boldsymbol{x}_{l,\mathcal{M}}(i) - \frac{\lambda}{2}\boldsymbol{P}_{l,\mathcal{M}}^T \boldsymbol{r}(i) \tag{5.75}$$

Substituting this into (5.74b) gives:

$$\left(\boldsymbol{s}_l + \boldsymbol{P}_l\boldsymbol{x}_l(i) - \frac{\lambda}{2}\boldsymbol{P}_{l,\mathcal{M}}\boldsymbol{P}_{l,\mathcal{M}}^T\boldsymbol{r}(i)\right)^T \boldsymbol{r}(i) = 0 \tag{5.76a}$$

$$(\boldsymbol{s}_l + \boldsymbol{P}_l\boldsymbol{x}_l(i))^T\boldsymbol{r}(i) - \frac{\lambda}{2}\boldsymbol{r}^T(i)\boldsymbol{P}_{l,\mathcal{M}}\boldsymbol{P}_{l,\mathcal{M}}^T\boldsymbol{r}(i) = 0 \tag{5.76b}$$

$$\frac{\lambda}{2} = \frac{(\boldsymbol{s}_l + \boldsymbol{P}_l\boldsymbol{x}_l(i))^T\boldsymbol{r}(i)}{\boldsymbol{r}^T(i)\boldsymbol{P}_{l,\mathcal{M}}\boldsymbol{P}_{l,\mathcal{M}}^T\boldsymbol{r}(i)} \tag{5.76c}$$

Using (5.75) and (5.76c), we obtain the following adaptation algorithm:

$$\boldsymbol{x}_{l,\mathcal{M}}(i+1) = \boldsymbol{x}_{l,\mathcal{M}}(i) - \mu\frac{(\boldsymbol{s}_l + \boldsymbol{P}_l\boldsymbol{x}_l(i))^T\boldsymbol{r}(i)}{\|\boldsymbol{P}_{l,\mathcal{M}}^T\boldsymbol{r}(i)\|^2}\boldsymbol{P}_{l,\mathcal{M}}^T\boldsymbol{r}(i) \tag{5.77}$$

The coefficient selection matrix $\boldsymbol{I}_M(i)$ is obtained by solving:

$$\min_{\boldsymbol{I}_M(i)} \left\| \mu\frac{(\boldsymbol{s}_l + \boldsymbol{P}_l\boldsymbol{x}_l(i))^T\boldsymbol{r}(i)}{\|\boldsymbol{P}_{l,\mathcal{M}}^T\boldsymbol{r}(i)\|^2}\boldsymbol{P}_{l,\mathcal{M}}^T\boldsymbol{r}(i)\right\|_2^2 \tag{5.78}$$

which can be written as:

$$\max_{\boldsymbol{I}_M(i)} \boldsymbol{r}^T(i)\boldsymbol{P}_l\boldsymbol{I}_M(i)\boldsymbol{P}_l\boldsymbol{r}(i) \tag{5.79}$$

Thus, the selective-partial-update NLMS algorithm for blind MUD is given by:

$$\boldsymbol{x}_l(i+1) = \boldsymbol{x}_l(i) - \mu\frac{(\boldsymbol{s}_l + \boldsymbol{P}_l\boldsymbol{x}_l(i))^T\boldsymbol{r}(i)}{\boldsymbol{r}^T(i)\boldsymbol{P}_l\boldsymbol{I}_M(i)\boldsymbol{P}_l\boldsymbol{r}(i)}\boldsymbol{I}_M(i)\boldsymbol{P}_l\boldsymbol{r}(i) \tag{5.80}$$

where:

$$\boldsymbol{I}_M(i) = \begin{bmatrix} j_1(i) & 0 & \cdots & 0 \\ 0 & j_2(i) & \ddots & \vdots \\ \vdots & \ddots & \ddots & 0 \\ 0 & \cdots & 0 & j_N(i) \end{bmatrix},$$

$$j_n(i) = \begin{cases} 1 & \text{if } |q_{l,n}(i)| \in \max_{1 \le m \le N}(|q_{l,m}(i)|, M) \\ 0 & \text{otherwise} \end{cases} \tag{5.81}$$

and:

$$\boldsymbol{q}_l(i) = \begin{bmatrix} q_{l,1}(i) \\ \vdots \\ q_{l,N}(i) \end{bmatrix} = \boldsymbol{P}_l \boldsymbol{r}(i) \tag{5.82}$$

The computational complexity of the full-update blind MUD NLMS algorithm in (5.71) is $3N$ multiplications for computing the output $z_l(i)$, and $4N$ multiplications and one division for the update term per iteration. The selective-partial-update NLMS algorithm in (5.80) has the same complexity as NLMS for computing $z_l(i)$, but requires $2(N + M)$ multiplications, one division and $\min(2N + 2\min(M, N - M)\log_2 N, 2M + 2(N - M)\log_2 M)$ comparisons for the update term per iteration. We assume the heapsort algorithm is used to find the M maxima of $|q_{l,j}(i)|$, $j = 1, \ldots, N$.

5.5.4 Simulation examples

In the simulation examples we consider a synchronous DS-CDMA system with $K = 10$ users. Each user is assigned a unique signature of length $N = 63$ (i.e. the processing gain or spreading factor is 63). The signature sequences are short codes obtained from m-sequences with period $N = 63$. We assume that user 1 is the user to be detected (i.e. $l = 1$). The other nine users are therefore treated as multiple access interference. Out of these nine users, six have 10 dB MAI and three have 20 dB MAI (i.e. $A_k^2/A_1^2 = 10$ for $k = 2, \ldots, 7$, and $A_k^2/A_1^2 = 100$ for $k = 8, 9, 10$). This represents a serious near-far problem. The desired signal to ambient noise ratio is 20 dB.

Figure 5.18 shows the SINR performance of the full-update, selective-partial-update and sequential-partial-update NLMS algorithms. For selective and sequential partial updates we have $M = 1$, which means only one out of 63 coefficients is updated at each iteration. The sequential-partial-update NLMS algorithm selects the coefficients to be updated in a round-robin fashion:

$$\boldsymbol{x}_l(i+1) = \boldsymbol{x}_l(i) - \mu \boldsymbol{I}_M(i) \frac{z_l(i) \boldsymbol{P}_l \boldsymbol{r}(i)}{\boldsymbol{r}^T(i) \boldsymbol{P}_l \boldsymbol{r}(i)} \tag{5.83}$$

where $\boldsymbol{I}_M(i)$ is given by ((4.14)). All algorithms were initialized to the signature sequence of user 1. Therefore, the SINR curves in Figure 5.18 show the performance improvement that can be achieved by replacing the matched filter with an NLMS

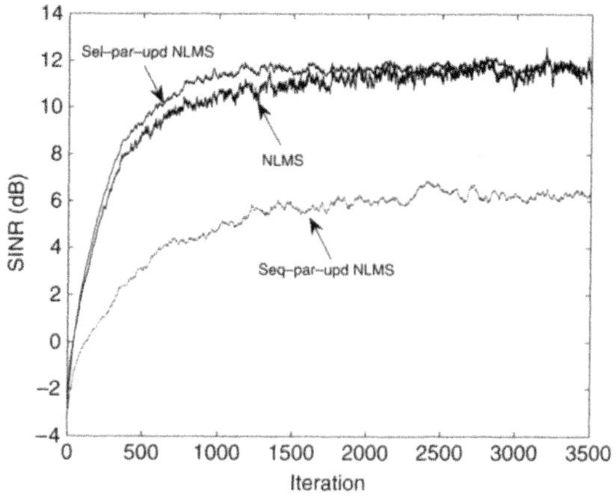

Figure 5.18 Convergence of SINR for full-update, selective-partial-update and sequential-partial-update NLMS. Partial-update NLMS algorithms update one out of 63 coefficients at each iteration ($M = 1$, $N = 63$).

algorithm. We observe that the selective-partial-update NLMS algorithm has the best MUD performance slightly outperforming the full-update NLMS algorithm despite having $M = 1$. The sequential-partial-update NLMS algorithm, on the other hand, performs poorly since its performance heavily depends on M. In general, the smaller M the slower the convergence rate of the sequential-partial-update NLMS algorithm.

Appendix | A

Overview of fast sorting algorithms

A.1 INTRODUCTION

In this appendix we summarize the operation and computational complexity of sorting algorithms for use in data-dependent partial coefficient updates. Data-dependent partial-update algorithms require ranking of regressor data or a well-defined function of regressor data. In the context of complexity reduction, which is the chief objective of partial coefficient updates, the computational complexity of ranking is an important consideration. Efficient fast sorting algorithms are of particular interest in the design of partial-update adaptive filters. In this appendix we provide an overview of computationally efficient running min/max and sorting algorithms, as well as a traditional sorting algorithm that does not exploit the shift structure of data. Running sorting algorithms are suitable for adaptive filtering applications where the regressor vector or its relevant function which is used for ranking has a shift structure. On the other hand, traditional sorting algorithms are applicable to ranking problems where the data does not have a shift structure as in transform-domain adaptive filters.

A.2 RUNNING MIN/MAX AND SORTING ALGORITHMS

A.2.1 Divide-and-conquer approaches

Let the minimum or maximum of a regressor vector with shift structure $\boldsymbol{x}(k)$ be denoted by:

$$y(k) = \mathcal{T}(x(k), x(k-1), \ldots, x(k-N+1)) \tag{A.1}$$

where $\mathcal{T}$ is the min or max operator. Suppose $N = 2^m$ where m is an integer. Then applying the divide-and-conquer method (Pitas, 1989) to (A.1) we have:

$$y(k) = \mathcal{T}(\mathcal{T}(x(k), \ldots, x(k - N/2 + 1)),$$
$$\mathcal{T}(x(k - N/2), \ldots, x(k - N + 1))) \tag{A.2}$$

which replaces finding the min/max of $\boldsymbol{x}(k)$ with three simpler min/max operations. Another application of the divide-and-conquer method to $y(k - N/2) = \mathcal{T}(x(k - N/2), \ldots, x(k - 3N/2 + 1))$ gives:

$$y(k - N/2) = \mathcal{T}(\mathcal{T}(x(k - N/2), \ldots, x(k - N + 1)),$$
$$\mathcal{T}(x(k - N), \ldots, x(k - 3N/2 + 1))) \tag{A.3}$$

We observe that $\mathcal{T}(x(k - N/2), \ldots, x(k - N + 1))$ is common to $y(k)$ in (A.2) and $y(k - N/2)$ in (A.3). Continuing like this until we obtain min/max partitions of size two, it can be shown that the min/max problem in (A.1) is equivalent to:

$$y(k) = \mathcal{T}(x(k), x(k - 1), y^{[\log_2 N - 1]}(k - 2), y^{[\log_2 N - 2]}(k - 4), \ldots,$$
$$y^{[2]}(k - N/4), y^{[1]}(k - N/2)) \tag{A.4}$$

where:

$$y^{[l]}(k) = \mathcal{T}(x(k), \ldots, x(k - (N/2^l) + 1)) \tag{A.5}$$

Note that $y^{[\log_2 N - 1]}(k - 2), y^{[\log_2 N - 2]}(k - 4), \ldots, y^{[2]}(k - N/4), y^{[1]}(k - N/2)$ in (A.4) would have already been computed previously thanks to the shifting property of $\boldsymbol{x}(k)$ and therefore do not require additional comparisons. Thus, from (A.4) we see that the complexity of the divide-and-conquer approach to running min/max computation is:

$$C(k) = \log_2 N \tag{A.6}$$

For $N \neq 2^m$ the previous algorithm can still be used after some modification. Let us first express N as a binary number:

$$N = \sum_{i=0}^{m-1} b_i 2^i, \quad b_i \in \{0, 1\}, \quad m = \lceil \log_2 N \rceil \tag{A.7}$$

where $\lceil x \rceil$ rounds up x if it is not integer (e.g. $\lceil 8/3 \rceil = 3$). Depending on whether $b_0 = 0$ or $b_0 = 1$ (A.1) can be divided into two min/max problems as follows:

$$y(k) = \begin{cases} \mathcal{T}(y^{[1]}(k), y^{[1]}(k - N/2)) & \text{if } b_0 = 0 \\ \mathcal{T}(y^{[1]}(k), x(k - \lfloor N/2 \rfloor), y^{[1]}(k - \lfloor N/2 \rfloor - 1)) & \text{if } b_0 = 1 \end{cases} \tag{A.8}$$

where $\lfloor x \rfloor$ denotes rounding towards zero (e.g. $\lfloor 8/3 \rfloor = 2$). We observe that the min/max problems arising from partitioning of the original problem are also common to previous min/max computations and thus they need not be computed again. Repeating the partitioning of the original regressor vector until we obtain subsequences of size two or three, we obtain:

$$y^{[j]}(k) = \begin{cases} \mathcal{T}(y^{[j+1]}(k), y^{[j+1]}(k - N/2^{j+1})) & \text{if } b_j = 0 \\ \mathcal{T}(y^{[j+1]}(k), x(k - \lfloor N/2^{j+1} \rfloor), \\ y^{[j+1]}(k - \lfloor N/2^{j+1} \rfloor - 1)) & \text{if } b_j = 1 \end{cases} \tag{A.9}$$

where $j \in \{0, 1, \ldots, m - 3\}$, and:

$$y^{[m-2]}(k) = \begin{cases} \mathcal{T}(x(k), x(k-1)) & \text{if } b_{m-2} = 0 \\ \mathcal{T}(x(k), x(k-1), x(k-2)) & \text{if } b_{m-2} = 1 \end{cases} \tag{A.10}$$

Thus, the min/max problem (A.1) takes $m - 1$ steps where each step requires either one or two additional comparisons depending on the value of b_j. In the worst case, (i.e. $b_j = 1, j = 0, 1, \ldots, m - 1$) the total number of comparisons will be:

$$C(k) = 2(m - 1) \tag{A.11}$$

The running sorting of $\boldsymbol{x}(k)$ is defined by:

$$\boldsymbol{y}(k) = \mathcal{S}(x(k), x(k-1), \ldots, x(k - N + 1)) \tag{A.12}$$

where $\mathcal{S}(\cdot)$ denotes the (ascending) sorting operator:

$$\boldsymbol{y}(k) = \begin{bmatrix} y_{(1)}(k) \\ y_{(2)}(k) \\ \vdots \\ y_{(N)}(k) \end{bmatrix}, \quad y_{(j)}(k) \leq y_{(j+1)}(k), \quad 1 \leq j \leq N - 1 \tag{A.13}$$

Assuming $N = 2^m$, where m is an integer, the application of the divide-and-conquer method to the sorting problem in (A.12) results in:

$$\boldsymbol{y}(k) = \mathcal{M}(\boldsymbol{y}^{[1]}(k), \boldsymbol{y}^{[1]}(k - N/2)) \tag{A.14}$$

where $\mathcal{M}(\cdot)$ denotes merging of two sorted sequences and:

$$y^{[l]}(k) = \mathcal{S}(x(k), \ldots, x(k - (N/2^l) + 1)) \tag{A.15}$$

The complexity of merging of two sorted sequences of length N is $2N - 1$. Repeating partitioning until the size of sequences is two, we see that the computation of $\boldsymbol{y}(k)$ in (A.14) takes $\log_2 N$ steps. At step $j, 0 \leq j \leq \log_2 N - 1$, merging of two sequences of length $N/2^{j+1}$ is required. This takes $N/2^j - 1$ comparisons. Taking into account merging, the total computational complexity of running sorting is:

$$C(k) = 2(N - 1) - \log_2 N \tag{A.16}$$

A.2.2 Maxline algorithm

Suppose $y(k-1) = \mathcal{T}(\boldsymbol{x}(k-1))$ is known and we wish to compute $y(k) = \mathcal{T}(\boldsymbol{x}(k))$. Noting that $\boldsymbol{x}(k)$ is obtained from $\boldsymbol{x}(k-1)$ by dropping $x(k-N)$ and inserting $x(k)$, for the running maximum computation (i.e., $\mathcal{T}(\cdot)$ is the max operator) we have the following relations (Pitas, 1989):

$$y(k) = \begin{cases} x(k) & \text{if } x(k) \geq y(k-1) \\ y(k-1) & \text{if } x(k) < y(k-1) \text{ and } x(k-N) < y(k-N) \\ \mathcal{T}(x(k), \ldots, x(k-N+1)) & \text{if } x(k) < y(k-1) \text{ and } x(k-N) = y(k-1) \end{cases} \tag{A.17}$$

This algorithm is known as the *maxline algorithm*. The number of comparisons required is 1, 2 and $N + 1$ for the cases in (A.17), respectively. The worst-case complexity of the maxline algorithm is $N+1$ comparisons. It is shown in Pitas (1989) that the mean number of comparisons is approximately 3 for uniform i.i.d. input data. The algorithm (A.17) can be easily modified to obtain a running min algorithm.

A.2.3 The Gil–Werman algorithm

Consider the running max finding problem in (A.1) where N is assumed to be odd with no loss of generality. The Gil-Werman algorithm (Gil and Werman, 1993) proceeds as follows. Given the sequence $x(k+N-1), x(k+N-2), \ldots, x(k+1), x(k), x(k-1), \ldots, x(k-N+1)$ which has $2N - 1$ elements, compute:

$$\begin{aligned} R_{N-1} &= x(k-1) \\ R_{N-2} &= \mathcal{T}(x(k-1), x(k-2)) \\ &\vdots \\ R_2 &= \mathcal{T}(x(k-1), x(k-2), \ldots, x(k-N+2)) \\ R_1 &= \mathcal{T}(x(k-1), x(k-2), \ldots, x(k-N+2), x(k-N+1)) \end{aligned} \tag{A.18}$$

and:

$$
\begin{aligned}
S_0 &= x(k) \\
S_1 &= \mathcal{T}(x(k+1), x(k)) \\
S_2 &= \mathcal{T}(x(k+2), x(k+1), x(k)) \\
&\vdots \\
S_{N-2} &= \mathcal{T}(x(k+N-2), \ldots, x(k)) \\
S_{N-1} &= \mathcal{T}(x(k+N-1), x(k+N-2), \ldots, x(k))
\end{aligned}
\tag{A.19}
$$

Then the maxima of $\boldsymbol{x}(k), \boldsymbol{x}(k+1), \ldots, \boldsymbol{x}(k+N-1)$ are given by:

$$
\begin{aligned}
y(k) &= \mathcal{T}(R_1, S_0) \\
y(k+1) &= \mathcal{T}(R_2, S_1) \\
&\vdots \\
y(k+N-2) &= \mathcal{T}(R_{N-1}, S_{N-2}) \\
y(k+N-1) &= S_{N-1}
\end{aligned}
\tag{A.20}
$$

The number of comparisons required to compute R_i in (A.18), S_i in (A.19) and $y(k), \ldots, y(k+N-1)$ in (A.20) is $N-2$, $N-1$ and $N-1$, respectively. The total computational complexity is, therefore, $3N-4$ comparisons. The average complexity per output is:

$$\bar{C}(k) = (3N-4)/N = 3 - 4/N \tag{A.21}$$

which is independent of the statistics of $x(k)$. The individual outputs have non-uniform complexities:

$$
\begin{aligned}
y(k) &= \mathcal{T}(R_1, S_0) && N-1 \text{ comparisons} \\
y(k+1) &= \mathcal{T}(R_2, S_1) && 2 \text{ comparisons} \\
&\vdots \\
y(k+N-2) &= \mathcal{T}(R_{N-1}, S_{N-2}) && 2 \text{ comparisons} \\
y(k+N-1) &= S_{N-1} && 1 \text{ comparison}
\end{aligned}
$$

Thus, even though the average complexity is constant, the maximum complexity is the same as the worst-case complexity of the maxline algorithm. A modification of the Gil-Werman algorithm was proposed in (Gevorkian *et al.*, 1997) to lower its average complexity while maintaining the same worst-case average complexity as the Gil-Werman algorithm.

A.2.4 Sortline algorithm

The sortline algorithm (Pitas, 1989) is the running sorting version of the maxline algorithm. Suppose $\boldsymbol{y}(k-1) = S(\boldsymbol{x}(k-1))$ is known. Then $\boldsymbol{y}(k)$ is obtained from:

$$\boldsymbol{y}(k) = (\boldsymbol{y}(k-1) \ominus x(k-N)) \oplus x(k) \tag{A.22}$$

where $\ominus$ and $\oplus$ denote deletion and insertion in a sorted sequence. In the sortline algorithm, these operations are implemented as binary delete and binary insert using binary search techniques. The binary search in a sequence of N numbers takes a maximum of $\lceil \log_2 N \rceil$ comparisons*. The worst-case complexity of the sortline algorithm is:

$$C(k) = \lceil \log_2 N \rceil + \lceil \log_2 (N-1) \rceil + 2 \approx 2\lceil \log_2 N \rceil + 2 \tag{A.23}$$

where $\lceil \log_2 N \rceil$ is the complexity of deletion, $\lceil \log_2 (N-1) \rceil$ is the complexity of insertion, and 2 comparisons are needed to compare the new data $x(k)$ with the maximum and minimum of $\boldsymbol{y}(k-1) \ominus x(k-N)$ before insertion is applied.

A.3 HEAPSORT ALGORITHM

Given a sequence of N numbers, the heapsort algorithm (Knuth, 1973; Sedgewick, 1998) firstly creates a *heap* structure whereby numbers are placed in a binary tree with each node larger than equal to its children. The root of the heap has the maximum number. The heap can also be constructed by having each node less than or equal to its children, in which case the root will have the minimum number of the sequence. An example heap is shown in Figure A.1. When a number is extracted from the heap, it is necessary to modify the heap so that it maintains its structure (i.e. each node is larger than equal to its children). This modification is referred to as *heapifying*.

Using the bottom-up heap construction, the complexity of creating a heap is less than $2N$ comparisons. Once the heap is built, sorting is done by applying *delete the maximum* operations starting from the root of the heap. Each *delete the maximum* operation extracts the root and then restores the heap structure with a maximum complexity of $2 \log_2 N$ comparisons. Sorting the entire sequence requires:

$$C = 2N + 2N \log_2 N \tag{A.24}$$

comparisons in the worst case. Finding the M maxima is easily done by terminating sorting after M or $N - M$ *delete the maximum* operations, whichever is smaller. If

*To be more precise the complexity of binary search is at most $\lfloor \log_2 N \rfloor + 1$ comparisons (Sedgewick, 1998), which differs from $\lceil \log_2 N \rceil$ only if N is even. Even so in light of approximation made in (A.23) this difference will be ignored.

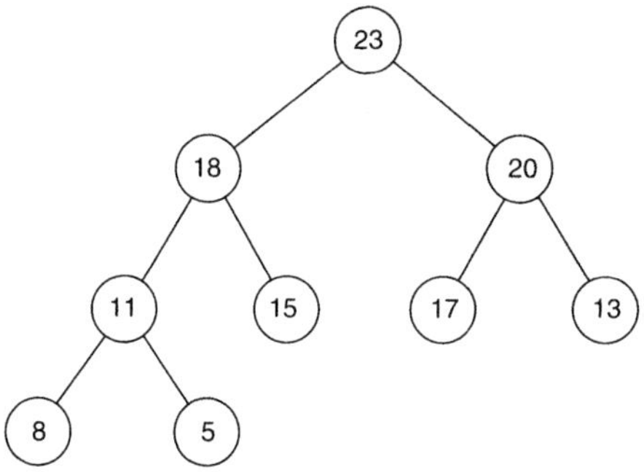

Figure A.1 A heap for a sequence of $N = 9$ numbers. The root of the heap has the maximum number. Each node is larger than its children. The nodes of the heap are filled from left to right.

$N - M < M$, a minimum-oriented heap is constructed using the min operator with the root of the heap having the smallest number, and *delete the maximum* operation is replaced with *delete the minimum* operation. In other words, finding the M maxima is replaced with finding the $N - M$ minima. The worst-case complexity of finding the M maxima is given by:

$$C = 2N + 2\min(M, N - M)\log_2 N \tag{A.25}$$

An alternative approach to finding the M maxima is to build a minimum-oriented heap of size M and then to perform $N - M$ *replace the minimum* operations (i.e., *insert* followed by *delete the minimum* operations) over the $N - M$ remaining numbers (Sedgewick, 1998). The complexity of this approach is:

$$C = 2M + 2(N - M)\log_2 M \tag{A.26}$$

which is smaller than (A.25) when N is large and M is small.

The worst-case complexity of heapsort is much smaller than other competing sorting techniques such as quicksort with a worst-case complexity of $O(N^2)$, even though its average complexity may be larger for long sequences. The heapsort algorithm requires no additional storage (memory), which is attractive from the practical point of view. In hardware implementations the complexity is often governed by the worst-case scenario. Since heapsort has a worst-case complexity slightly larger than its average complexity, it is the sorting algorithm of choice in data-dependent partial-coefficient-update applications where the data to be sorted does not possess a shift structure, thereby ruling out the sortline algorithm.

References

Aboulnasr, T. and Mayya, K. Complexity reduction of the NLMS algorithm via selective coefficient update. *IEEE Trans. on Signal Processing*, 47(5):1421–1424, May 1999.

Anderson, B.D.O., Bitmead, R.R., Johnson Jr., C.R., Kokotovic, P.V., Kosut, R.L., Mareels, I.M.Y., Praly, L., and Riedle, B.D. *Stability of Adaptive Systems: Passivity and Averaging*. MIT Press, Cambridge, MA, 1986.

Benesty, J., Gänsler, T., Morgan, D.R., Sondhi, M.M., and Gay, S.L. *Advances in Network and Acoustic Echo Cancellation*. Springer-Verlag, Berlin, Germany, 2001.

Benveniste, A., Metivier, M., and Priouret, P. *Adaptive Algorithms and Stochastic Approximations*. Springer-Verlag, New York, 1990.

Bondyopadhyay, P.K. Application of running Hartley transform in adaptive digital filtering. *Proc. IEEE*, 76(10):1370–1372, October 1988.

Breining, C., Dreiseitel, P., Hänsler, E., Mader, A., Nitsch, B., Puder, H., Schertler, T., Schmidt, G., and Tilp, J. Acoustic echo control: an application of very-high-order adaptive filters. *IEEE Signal Processing Magazine*, 16(4):42–69, July 1999.

Deng, H. and Doroslovački, M. Proportionate adaptive algorithms for network echo cancellation. *IEEE Trans. on Signal Processing*, 54(5):1794–1803, May 2006.

Diniz, P.S.R. *Adaptive Filtering: Algorithms and Practical Implementation*. Kluwer, Boston, 2nd edition, 2002.

Doğançay, K. Generalized subband decomposition adaptive filters for sparse systems. *Signal Processing*, 83(5):1093–1103, May 2003a.

Doğançay, K. Complexity considerations for transform-domain adaptive filters. *Signal Processing*, 83(6):1177–1192, June 2003b.

Doğançay K. and Naylor P.A. Recent advances in partial update and sparse adaptive filters. In *Proc. 13th European Signal Processing Conference, EUSIPCO 2005*, Antalya, Turkey, September 2005.

Doğançay K. and Tanrıkulu O. Selective-partial-update NLMS and affine projection algorithms for acoustic echo cancellation. In *Proc. of IEEE Int. Conf. on Acoustics, Speech, and Signal Processing, ICASSP 2000*, volume 1, pages 448–451, Istanbul, Turkey, June 2000.

Doğançay, K. and Tanrıkulu, O. Adaptive filtering algorithms with selective partial updates. *IEEE Trans. on Circuits and Systems II*, 48(8):762–769, August 2001a.

Doğançay K. and Tanrıkulu O. Generalised subband decomposition with selective partial updates and its application to acoustic echo cancellation. In *Proc. Int. Workshop on Acoustic Echo and Noise Control, IWAENC 2001*, pages 104–107, Darmstadt, Germany, September 2001b.

Doğançay K. and Tanrıkulu O. Normalised constant modulus algorithm with selective partial updates. In *Proc. of IEEE Int. Conf. on Acoustics, Speech, and Signal Processing, ICASSP 2001*, volume IV, pages 2181–2184, Salt Lake City, Utah, May 2001c.

Doğançay K. and Tanrıkulu O. Generalized subband decomposition LMS algorithm employing selective partial updates. In *Proc. IEEE Int. Conference on Acoustics, Speech, and Signal Processing, ICASSP 2002*, volume II, pages 1377–1380, Orlando, Florida, May 2002.

Doğançay K. and Tanrıkulu O. Reduced complexity NLMS algorithm for blind adaptive multiuser detection. In *Proc. 13th European Signal Processing Conference, EUSIPCO 2005*, Antalya, Turkey, September 2005.

Douglas, S.C. A family of normalized LMS algorithms. *IEEE Signal Processing Letters*, 1(3):49–51, March 1994.

Douglas S.C. Analysis and implementation of the max-NLMS adaptive filter. In *Conf. Record of 29th Asilomar Conf. on Signals, Systems, and Computers*, volume 1, pages 659–663, Pacific Grove, CA, USA, October 1995.

Douglas, S.C. Adaptive filters employing partial updates. *IEEE Trans. on Circuits and Systems II*, 44(3):209–216, March 1997.

Duttweiler, D.L. Proportionate normalized least-mean-squares adaptation in echo cancelers. *IEEE Trans. on Speech and Audio Processing*, 8(5):508–518, September 2000.

Eyre, J. and Bier, J. The evolution of DSP processors. *IEEE Signal Processing Magazine*, 17(2):43–51, March 2000.

Farhang-Boroujeny, B. Order of *N* complexity transform domain adaptive filters. *IEEE Trans. on Circuits and Systems II*, 42(7):478–480, July 1995.

Ferguson, T.S. *A Course in Large Sample Theory*. Chapman and Hall, London, 1996.

Gay S.L. An efficient, fast converging adaptive filter for network echo cancellation. In *Conf. Record of Asilomar Conf. on Signals, Systems and Computers, 1998*, volume 1, pages 394–398, Pacific Grove, CA, November 1998.

Gay S.L. and Tavathia S. The fast affine projection algorithm. In *Proc. of IEEE Int. Conf. on Acoustics, Speech, and Signal Processing, ICASSP '95*, volume 5, pages 3023–3026, Detroit, MI, USA, May 1995.

Gevorkian, D.Z., Astola, J.T., and Atourian, S.M. Improving Gil-Werman algorithm for running min and max filters. *IEEE Trans. on Pattern Analysis and Machine Intelligence*, 19(5):526–529, May 1997.

Gil, J. and Werman, M. Computing 2-D min, median, and max filters. *IEEE Trans. on Pattern Analysis and Machine Intelligence*, 15(5):504–507, May 1993.

Godard, D.N. Self-recovering equalization and carrier tracking in two-dimensional data communication systems. *IEEE Trans. Commun.*, 28:1867–1875, November 1980.

Godavarti, M. and Hero III, A.O. Partial update LMS algorithms. *IEEE Trans. on Signal Processing*, 53(7):2382–2399, July 2005.

Gold, R. Optimal binary sequences for spread spectrum multiplexing. *IEEE Trans. on Information Theory*, 13:619–621, October 1967.

Gollamudi, S., Nagaraj, S., Kapoor, S., and Huang, Y.F. Set-membership filtering and a set-membership normalized LMS algorithm with an adaptive step size. *IEEE Signal Processing Letters*, 5(4):111–114, April 1998.

Haykin, S. *Adaptive Filter Theory*. Prentice Hall, New Jersey, 3rd edition, 1996.

Hilal K. and Duhamel P. A convergence study of the constant modulus algorithm leading to a normalized-CMA and block-normalized-CMA. In *Proc. EUSIPCO-92*, pages 135–138, Brussels, Belgium, August 1992.

Honig, M.L., Madhow, U., and Verdu, S. Blind adaptive multiuser detection. *IEEE Trans. on Information Theory*, 41(4):944–960, July 1995.

Hwang, K.-Y. and Song, W.-J. An affine projection adaptive filtering algorithm with selective regressors. *IEEE Trans. on Circuits and Systems–II: Express Briefs*, 54(1): 43–46, January 2007.

ITU-T, *Digital Network Echo Cancellers, Recommendation ITU-T G.168*. International Telecommunication Union, Series G: Transmission Systems and Media, Digital Systems and Networks, 2002.

Johnson Jr., C.R., Schniter, P., Endres, T.J., Behm, J.D., Brown, D.R., and Casas, R.A. Blind equalization using the constant modulus criterion: a review. *Proc. IEEE*, 86(10):1927–1950, October 1998.

Knuth, D.E. *The Art of Computer Programming: Sorting and Searching, volume 3*. Addison-Wesley, Reading, MA, 2nd edition, 1973.

Kratzer S.G. and Morgan D.R. The partial-rank algorithm for adaptive beamforming. In *Real-Time Signal Processing VIII, SPIE Proceedings*, volume 564, pages 9–16, Bellingham, WA, January 1985.

Le Cadre, J.-P. and Jauffret, C. On the convergence of iterative methods for bearings-only tracking. *IEEE Trans. on Aerospace and Electronic Systems*, 35(3):801–818, July 1999.

Lee, E.A. and Messerschmitt, D.G. *Digital Communication*. Kluwer Academic Publishers, Boston, MA, 2nd edition, 1994.

Liu, Ray K.J., Wu, A.-Y., Raghupathy, A., and Chen, J. Algorithm-based low-power and high-performance multimedia signal processing. *Proc. IEEE*, 86(6): 1155–1202, June 1998.

Ljung, L. *System Identification: Theory for the User*. Prentice Hall, Upper Saddle River, NJ, 2nd edition, 1999.

Madhow, U. and Honig, M. MMSE interference suppression for direct-sequence spread-spectrum CDMA. *IEEE Trans. on Commun.*, 42(12):3178–3188, December 1994.

Mareels, I. and Polderman, J.W. *Adaptive Systems: An Introduction*. Birkhauser, Cambridge, MA, 1996.

Messerschmitt D., Hedberg D., Cole C., Haoui A. and Winship P. Digital voice echo canceller with a TMS32020. Application report: SPRA129, Texas Instruments, 1989.

Nagumo, J. and Noda, A. A learning method for system identification. *IEEE Trans. on Automatic Control*, 12(3):282–287, June 1967.

Narayan, S.S., Peterson, A.M., and Narasimha, M.J. Transform domain LMS algorithm. *IEEE Trans. on Acoustics, Speech, and Signal Processing*, 31(3):609–615, June 1983.

Oppenheim, A.V., Schafer, R.W., and Buck, J.R. *Discrete-Time Signal Processing*. Prentice Hall, Upper Saddler River, N.J., 2nd edition, 1999.

Petraglia, M.R. and Mitra, S.K. Adaptive FIR filter structure based on the generalized subband decomposition of FIR filters. *IEEE Trans. on Circuits and Systems II*, 40(6):354–362, June 1993.

Pitas, I. Fast algorithms for running ordering and max/min calculation. *IEEE Trans. on Circuits and Systems*, 36(6):795–804, June 1989.

Press, W.H., Flannery, B.P., Teukolsky, S.A., and Vetterling, W.T. *Numerical Recipes in C: The Art of Scientific Computing*. Cambridge University Press, Cambridge, NY, 2nd edition, 1993.

Qureshi, S.U.H. Adaptive equalization. *Proc. IEEE*, 1349–1387, September 1985.

Sayed, A.H. *Fundamentals of Adaptive Filtering*. IEEE/Wiley, NJ, 2003.

Schertler T. Selective block update of NLMS type algorithms. In *Proc. of IEEE Int. Conf. on Acoustics, Speech, and Signal Processing, ICASSP '98*, volume 3, pages 1717–1720, Seattle, USA, May 1998.

Sedgewick, R. *Algorithms in C, Parts 1–4*. Addison-Wesley, Boston, MA, 3rd edition, 1998.

Shynk, J.J. Frequency-domain and multirate adaptive filtering. *IEEE Signal Processing Magazine*, 9(1):14–37, January 1992.

Slock, D.T.M. On the convergence behavior of the LMS and the normalized LMS algorithms. *IEEE Trans. on Signal Processing*, 41(9):2811–2825, September 1993.

Solo, V. and Kong, X. *Adaptive Signal Processing Algorithms: Stability and Performance*. Prentice Hall, Englewood Cliffs, NJ, 1995.

Storn R. Echo cancellation techniques for multimedia applications—a survey. Technical Report TR-96-046, International Computer Science Institute, November 1996. ftp.icsi.berkeley.edu.

Tanrıkulu O. and Doğançay K. Selective-partial-update proportionate normalized least-mean-squares algorithm for network echo cancellation. In *Proc. IEEE Int. Conference on Acoustics, Speech, and Signal Processing, ICASSP 2002*, volume II, pages 1889–1892, Orlando, Florida, May 2002.

Treichler, J.R. and Agee, B.G. A new approach to multipath correction of constant modulus signals. *IEEE Trans. on Acoustics, Speech, Signal Processing*, ASSP-31: 459–472, April 1983.

Verdu, S. Minimum probability of error for asynchronous gaussian multiple-access channels. *IEEE Trans. on Information Theory*, 32(1):85–96, January 1986.

Wang, X. and Poor, H.V. *Wireless Communication Systems: Advanced Techniques for Signal Reception*. Prentice Hall, Upper Saddle River, NJ, 2004.

Werner, S., de Campos, M.L.R., and Diniz, P.S.R. Partial-update NLMS algorithms with data-selective updating. *IEEE Trans. on Signal Processing*, 52(4):938–949, April 2004.

Werner, S. and Diniz, P.S.R. Set-membership affine projection algorithm. *IEEE Signal Processing Letters*, 8(8):231–235, August 2001.

Widrow, B. and Lehr, M. 30 years of adaptive neural networks: perceptron, madaline, and backpropagation. *Proc. IEEE*, 78(9):1415–1442, September 1990.

Widrow, B. and Stearns, S.D. *Adaptive Signal Processing*. Prentice Hall, Englewood Cliffs, NJ, 1985.

Wong, K.Y. and Polak, E. Identification of linear discrete time systems using the instrumental variable method. *IEEE Trans. on Automatic Control*, 12(6):707–718, December 1967.

Index

Printed and bound by CPI Group (UK) Ltd, Croydon, CR0 4YY
08/05/2025
01864879-0003